T0329085

Genomic and Precision Medicine

Genomic and Precision Medicine

Foundations, Translation, and Implementation

Third Edition

Edited by

Geoffrey S. Ginsburg
Duke University, Durham, NC, United States

Huntington F. Willard
Marine Biological Laboratory, Woods Hole, MA, United States;
University of Chicago, Chicago, IL, United States

AMSTERDAM • BOSTON • HEIDELBERG • LONDON
NEW YORK • OXFORD • PARIS • SAN DIEGO
SAN FRANCISCO • SINGAPORE • SYDNEY • TOKYO
Academic Press is an imprint of Elsevier

Academic Press is an imprint of Elsevier
125 London Wall, London EC2Y 5AS, United Kingdom
525 B Street, Suite 1800, San Diego, CA 92101-4495, United States
50 Hampshire Street, 5th Floor, Cambridge, MA 02139, United States
The Boulevard, Langford Lane, Kidlington, Oxford OX5 1GB, United Kingdom

British Library Cataloguing-in-Publication Data
A catalogue record for this book is available from the British Library

Library of Congress Cataloging-in-Publication Data
A catalog record for this book is available from the Library of Congress

ISBN: 978-0-12-800681-8

For Information on all Academic Press publications
visit our website at https://www.elsevier.com

  Working together
to grow libraries in
developing countries

www.elsevier.com • www.bookaid.org

Publisher: Mica Haley
Acquisition Editor: Peter Linsley
Editorial Project Manager: Lisa Eppich
Production Project Manager: Edward Taylor
Designer: Matthew Limbert

Typeset by MPS Limited, Chennai, India

Contents

18. Molecular Genetic Testing and the Future of Clinical Genomics

S.H. Katsanis and N. Katsanis

19. Bringing Genomics to Medicine: Ethical, Policy, and Social Considerations

Laura Lyman Rodriguez and Elyse Galloway

List of Contributors

Joanne Armstrong Aetna, Sugar Land, TX, United States

Katrina Armstrong Massachusetts General Hospital, Harvard Medical School, Boston, MA, United States

Samuel J. Aronson Partners HealthCare Personalized Medicine, Cambridge, MA, United States

Adam C. Berger US Food and Drug Administration, Silver Spring, MD, United States

Zhaohui Chen Cedars Sinai Medical Center, Los Angeles, CA, United States

Andrew Dervan University of Washington, Seattle, WA, United States

Katherine Donigan US Food and Drug Administration, Silver Spring, MD, United States

Edward S. Dove University of Edinburgh, Edinburgh, United Kingdom; Global Alliance for Genomics and Health, Toronto, Canada

Eric Faulkner Evidera, Bethesda, MD, United States; University of North Carolina at Chapel Hill, Chapel Hill, NC, United States

Benjamin French University of Pennsylvania, Philadelphia, PA, United States

Elyse Galloway National Human Genome Research Institute, National Institutes of Health, Bethesda, MD, United States

Geoffrey S. Ginsburg Duke University, Durham, NC, United States

Elizabeth A. Grice University of Pennsylvania, Philadelphia, PA, United States

Ken J. Hampel University of Vermont Medical Center, Burlington, VT, United States

N. Katsanis Duke University Medical Center, Durham, NC, United States

S.H. Katsanis Duke University, Durham, NC, United States

Stephen E. Kimmel University of Pennsylvania, Philadelphia, PA, United States

Bartha M. Knoppers Global Alliance for Genomics and Health, Toronto, Canada; McGill University, Montreal, Canada

Graeme T. Laurie University of Edinburgh, Edinburgh, United Kingdom; Global Alliance for Genomics and Health, Toronto, Canada

Liis Leitsalu University of Tartu, Tartu, Estonia

Debra G.B. Leonard University of Vermont Medical Center, Burlington, VT, United States; University of Vermont College of Medicine, Burlington, VT, United States

Elizabeth Mansfield US Food and Drug Administration, Silver Spring, MD, United States

Daniel R. Masys University of Washington, Seattle, WA, United States

Robert McDonough Aetna, Sugar Land, TX, United States

Jacquelyn S. Meisel University of Pennsylvania, Philadelphia, PA, United States

Andres Metspalu University of Tartu, Tartu, Estonia

Michael F. Murray Genomic Medicine Institute, Geisinger Health System, Forty Fort, PA, United States

Lori A. Orlando Duke University, Durham, NC, United States

Philip R.O. Payne Washington University in St. Louis, St. Louis, MO, United States

Alanna Kulchak Rahm Genomic Medicine Institute, Geisinger Health System, Forty Fort, PA, United States

Arti K. Rai Duke Law Center for Innovation Policy, Durham, NC, United States

Timothy E. Reddy Duke University School of Medicine, Durham, NC, United States

Shelby D. Reed Duke University, Durham, NC, United States

Laura Lyman Rodriguez National Human Genome Research Institute, National Institutes of Health, Bethesda, MD, United States

R. Ryanne Wu Duke University, Durham, NC, United States

Jay Shendure University of Washington, Seattle, WA, United States

Nikoletta Sidiropoulos University of Vermont Medical Center, Burlington, VT, United States; University of Vermont College of Medicine, Burlington, VT, United States

Steven Tjoe US Food and Drug Administration, Silver Spring, MD, United States

Jennifer E. Van Eyk Cedars Sinai Medical Center, Los Angeles, CA, United States

David L. Veenstra University of Washington, Seattle, WA, United States

Deepak Voora Duke University, Durham, NC, United States

Robyn Ward University of Queensland, Brisbane, QLD, Australia

Huntington F. Willard Marine Biological Laboratory, Woods Hole, MA, United States; University of Chicago, Chicago, IL, United States

Marc S. Williams Genomic Medicine Institute, Geisinger Health System, Danville, PA, United States

Chapter 1

The Human Genome: Foundation for Genomic and Precision Medicine

Huntington F. Willard[1,2]

[1]Marine Biological Laboratory, Woods Hole, MA, United States, [2]University of Chicago, Chicago, IL, United States

Chapter Outline

INTRODUCTION

That genetic variation can influence health and disease has been a central, if not broadly practiced, principle of medicine for over a hundred years. What has limited full application of this principle until recently has been the special nature and presumed rarity of clinical circumstances or conditions to which genetic variation was relevant. Now, however, with the availability of a reference sequence of the human genome and a growing number of personal genome sequences from both asymptomatic and symptomatic individuals, with emerging appreciation of the extent of genome variation among different individuals and different populations worldwide, and with a growing understanding of the role of common as well as rare variation in disease, we are increasingly able to begin to exploit the impact

Genomic and Precision Medicine. DOI: http://dx.doi.org/10.1016/B978-0-12-800681-8.00001-3

of that variation on human health on a broad scale, in the context of genomic and precision medicine [1].

Variation in the human genome has long been the cornerstone of the field of human genetics (Box 1.1), and its study led to the establishment of the medical specialty of medical genetics. The general nature and frequency of gene variants in the human genome became apparent with the classic work over 50 years ago on the incidence of polymorphic protein variants in populations of healthy individuals, work that is the conceptual forerunner to the much larger and detailed efforts that mark modern human genetics and genomics. Such data underlie the conclusion that virtually every individual has his or her own unique constitution of gene products, the implications of which provide a foundation for what today we call personalized or precision medicine as a modern application of what the British physician Archibald Garrod called "chemical individuality" in the very early years of the last century [2].

In this chapter, the organization, variation, and expression of the human genome are presented as a foundation for the chapters to follow on human genomics, on genome technology and informatics, on approaches in translational genomics and, finally, on the principles of genomic and precision medicine as applied to health and disease. Given limitations on the number of references, this introductory chapter favors reviews and papers describing recent advances; for more comprehensive references, readers are pointed to the relevant chapters elsewhere in this volume.

THE HUMAN GENOME

The typical human genome consists of approximately 3 billion (3×10^9) base pairs of DNA, divided among the 24 types of nuclear chromosomes (22 autosomes, plus the sex chromosomes, X and Y) and the much smaller mitochondrial chromosome (Table 1.1).

Individual chromosomes can best be visualized and studied at metaphase in dividing cells, and karyotyping of patient chromosomes has been a valuable and routine clinical laboratory procedure for a half century, albeit at levels of resolution that fall well short of most pathologic DNA variants (Fig. 1.1). The ultimate resolution, of course, comes from direct sequence analysis, and an increasing number of new technologies have facilitated comparisons of individual genomes with the reference human genome sequence, enabling clinical sequencing of patient samples to search for novel variants or mutations that might be of clinical importance.

Genes in the Human Genome

While the human genome contains an estimated 20,000 protein-coding genes, the coding segments of those genes—the exons—comprise less than 2% of

BOX 1.1 Genetics and Genomics in Precision Medicine

Throughout this and the many other chapters in these volumes, the terms "genetics" and "genomics" are used repeatedly, both as nouns and in their adjectival forms. While these terms seem similar, they in fact describe quite distinct (though frequently overlapping) approaches in biology and in medicine. Having said that, there are inconsistencies in the way the terms are used, even by those who work in the field. To some, genetics is a subfield of genomics; to others, genomics is a subfield of genetics!

Here, we provide operational definitions to distinguish the various terms and the subfields of medicine to which they contribute.

The field of *genetics* is the scientific study of heredity and of the genes that provide the physical, biological, and conceptual bases for heredity and inheritance. To say that something—a trait, a disease, a code, or an information—is "genetic" refers to its basis in genes and in DNA.

Heredity refers to the familial phenomenon whereby traits (including clinical traits) are transmitted from generation to generation, due to the transmission of genes from parent to child. A disease that is said to be inherited or hereditary is certainly genetic; however, not all genetic diseases are hereditary (witness cancer, which is always a genetic disease, but is only occasionally an inherited disease).

Genomics is the scientific study of a genome or genomes. A *genome* is the complete DNA sequence, referring to the entire genetic information of a gamete, an individual, a population, or a species. As such, it is a subfield of genetics when describing an approach taken to study genes. The word "genome" originated as an analogy with the earlier term "chromosome," referring to the physical entities (visible under the microscope) that carry genes from one cell to its daughter cells or from one generation to the next. "Genomics" gave birth to a series of other "-omics" that refer to the comprehensive study of the full complement of genome products—for example, proteins (hence, *proteomics*), transcripts (*transcriptomics*), or metabolites (*metabolomics*). The essential feature of the "-omes" is that they refer to the complete collection of genes or their derivative proteins, transcripts, or metabolites, not just to the study of individual entities. The distinguishing characteristics of genomics and the other "omics" are their comprehensiveness and scale, their integration with and dependence on technology development, an emphasis on rapid data release and availability, and an awareness of the policy and ethical implications of such work in research, in the practice of medicine, and increasingly in the social arena [3].

By analogy with genetics and genomics, *epigenetics* and *epigenomics* refer to the study of factors that affect gene (or, more globally, genome) function, but without an accompanying change in genes or the genome. The *epigenome* is the comprehensive set of epigenetic changes in a given individual, tissue, tumor, or population. It is the paired combination of the genome and the epigenome that appear to best characterize and determine one's phenotype.

(Continued)

BOX 1.1 Genetics and Genomics in Precision Medicine (Continued)

Medical Genetics is the application of genetics to medicine with a particular emphasis on inherited disease. Medical genetics is a broad and varied field, encompassing many different subfields, including clinical genetics, biochemical genetics, cytogenetics, molecular genetics, the genetics of common diseases, and genetic counseling. Medical Genetics and Genomics is one of 24 medical specialties recognized by the American Board of Medical Specialties, the medical organization overseeing physician certification in the United States.

Genetic Medicine is a term used to refer to the application of genetic principles to the practice of medicine and thus overlaps medical genetics. However, genetic medicine is somewhat broader, as it is not limited to the specialty of Medical Genetics and Genomics, but is relevant to health professionals in many, if not all, specialties and subspecialties. Both medical genetics and genetic medicine approach clinical care largely through consideration of individual genes and their effects on patients and their families.

By contrast, *Genomic Medicine* refers to the use of large-scale genomic information and to consideration of the full extent of an individual's genome and other "omes" in the practice of medicine and medical decision making. The principles and approaches of genomic medicine are relevant well beyond the traditional purview of medical genetics and include, as examples, gene expression profiling to characterize tumors or to define prognosis in cancer, genotyping variants in the set of genes involved in drug metabolism or action to determine an individual's correct therapeutic dosage, scanning the entire genome for millions of variants that influence one's susceptibility to disease, or analyzing multiple protein or RNA biomarkers to detect exposure to potential pathogens, to monitor therapy and to provide predictive information in presymptomatic individuals.

Finally, *Precision Medicine* refers to the rapidly advancing field of health care that is informed by each person's unique clinical, genetic, genomic, and environmental information [1]. The goals of precision medicine are to take advantage of a molecular understanding of disease—and a vast array of data from sources that range from genomes to Electronic Medical Records to mobile health devices—to optimize preventive health care strategies and drug therapies while people are still well or at the earliest stages of disease. Because these factors are different for every person, the nature of disease, its onset, its course, and how it might respond to drug or other interventions are as individual as the people who have them and the communities in which they live. In order for precision medicine to be used by health care providers and their patients, these findings, analyzed with significant input from the data sciences, must be translated into precision diagnostic tests and targeted therapies, using implementation tools that are compatible with the modern health care environment [4]. Since the overarching goal is to optimize medical care and outcomes for each individual, treatments, medication types and dosages, and/or prevention strategies may differ from person to person—resulting in customization and precise targeting of patient care.

The principles underlying *Genomic and Precision Medicine* and their applications to the practice of clinical medicine are presented throughout the chapters that comprise this and other volumes.

TABLE 1.1 Characteristics of the Reference Human Genome[a]

Length of the human genome (base pairs)	3,096,649,726
Number of known protein-coding genes	20,441
Average gene density (number of genes/Mb)[b]	6.6[c]
Number of ncRNA genes	22,219
Number of known short sequence variants[d]	156,148,362
Number of known structural variants[e]	4,485,861

[a]From Ensembl, database GRCh38, version 85.38 (accessed August 2016).
[b]Mb, mega base pairs.
[c]Protein-coding genes only.
[d]For example, SNPs, substitutions, in/dels.
[e]For example, CNVs, inversions.

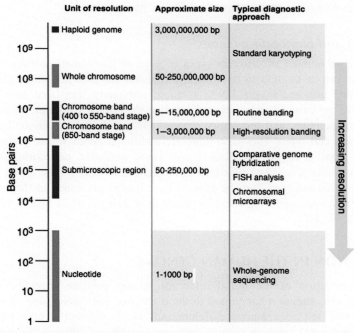

FIGURE 1.1   Spectrum of resolution in chromosome and genome analysis. The typical resolution and range of effectiveness are given for various diagnostic approaches used routinely in clinical and research practice. FISH, fluorescence in situ hybridization. Source: *From Nussbaum RL, McInnes RR, Willard HF. Genetics in medicine. 8th ed. Philadelphia, PA: W.B. Saunders, Co.; 2016. p. 546 [2], with permission.*

the genome; most of the genome consists of DNA that lies between genes, far from genes or in vast areas spanning several million base pairs (Mb) that appear to contain no genes at all. Despite much progress in gene identification and genome annotation, it is nearly certain that there are some genes, including clinically relevant genes, that are currently undetected or that display characteristics that we do not currently recognize as being associated with genes. Nonetheless, the statement that the vast majority of the genome consists of spans of DNA that are nongenic, of no obvious function, and of uncertain clinical relevance remains true.

In addition to being relatively sparse in the genome, genes are distributed quite nonrandomly along the different human chromosomes. Some chromosomes are relatively gene-rich, while others are quite gene-poor, ranging from approximately 3 genes/Mb of DNA to more than 20 genes/Mb (excluding the Y chromosome and the tiny mitochondrial chromosome). And even within a chromosome, genes tend to cluster in certain regions and not in others, a point of clear clinical significance when evaluating genome integrity, dosage, or arrangement in different patient samples.

Coding and Noncoding Genes

There are a number of different types of gene in the human genome. Most genes known or thought to be clinically relevant are protein-coding and are transcribed into messenger RNAs that are ultimately translated into their respective proteins; their products comprise the list of enzymes, structural proteins, receptors, and regulatory proteins that are found in various human tissues and cell types. However, there are additional genes whose functional product appears to be the RNA itself. These so-called noncoding RNAs (ncRNAs) have a range of functions in the cell, and many do not as yet have any identified function. But the genes whose transcripts make up the collection of ncRNAs could represent as many as a half of all identified human genes (Table 1.1).

VARIATION IN THE HUMAN GENOME

With completion of the initial reference human genome sequence some 15 years ago, attention has turned to the discovery and cataloguing of variation in that sequence among different individuals (including both healthy individuals and those with various diseases) and among different populations [5−8]. Any given individual carries 4−5 million sequence variants that are known to exist in multiple forms (i.e., are polymorphic) in our species. In addition, there are countless very rare variants, many of which probably exist in only a single or a few individuals. In fact, given the number of individuals in our species, essentially each and every base pair in the human genome is expected to vary in someone somewhere around the globe. It is for this reason that the original human genome sequence is considered only a "reference" sequence, derived as a consensus of the limited number of genomes whose sequencing was part of the Human Genome Project, but actually identical to no individual's genome.

Types of Variation

Early estimates were that any two randomly selected individuals have sequences that are 99.9% identical or, put another way, that an individual genome would be heterozygous at approximately 3—5 million positions, with different bases at the maternally and paternally inherited copies of that particular sequence position. The majority of these differences involve simply a single unit in the DNA code and are referred to as single-nucleotide polymorphisms (SNPs) (Table 1.1). The remaining variation consists of insertions or deletions (in/dels) of (usually) short sequence stretches, variation in the number of copies of repeated elements or inversions in the order of sequences at a particular locus in the genome (Fig. 1.2). Any and all of these types of variation can influence disease and thus must be accounted for in

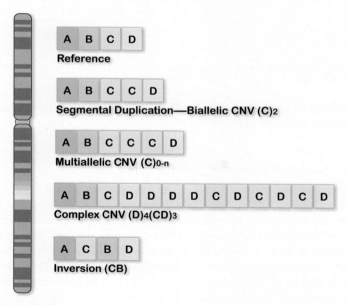

Chromosome

FIGURE 1.2 Schematic representation of different types of structural polymorphism in the human genome, leading to deletions, duplications, inversions, and CNV changes relative to the reference arrangement. Source: *From Estivill X, Armengol L. Copy number variants and common disorders: filling the gaps and exploring complexity in genome-wide association studies. PLoS Genet 2007;3:e190 [9], with permission.*

TABLE 1.2 Common Variation in the Human Genome

Type of Variation	Size Range (approx.)[a]	Effect(s) in Biology and Medicine
Single-nucleotide polymorphisms	1 bp	Nonsynonymous → functional change in encoded protein?
		Others → potential regulatory variants?
		Most → no effect? ("neutral")
Copy number variants (CNVs)	10 kb to 1 Mb	Gene dosage variation → functional consequences? Most → no effect or uncertain effect
Insertion/deletion polymorphisms (in/dels)	1 bp to 1 Mb	In coding sequence: frameshift mutation? → functional change
		Most → uncertain effects
Inversions	Few bp to 100 kb	? break in gene sequence
		? long-range effect on gene expression
		? indirect effects on reproductive fitness
		Most → no effect? ("neutral")
Segmental duplications	10 kb to >1 Mb	Hotspots for recombination → polymorphism (CNVs)

[a]bp, base pair; kb, kilobase pair; Mb, megabase pair.

any attempt to understand the contribution of genetics to human health and to precision medicine (Table 1.2).

Copy Number Variation

Over the past decade, a number of important studies have focused on the prevalence of structural variants in the genome, which, in any given genome, collectively account for far more variation in genome sequence (expressed in terms of the amount of genomic DNA affected) than do SNPs [10]. The most common type of structural variation involves changes in the local copy number of sequences (including genes) in the genome. This variation is based on blocks of different sequences (hence, not defining a particular family of sequences) that are present in multiple copies, often with extraordinarily high sequence conservation, in many different locations around the genome. Rearrangements between such duplicated segments are a source of significant variation between individuals in the number of copies of these DNA sequences and these are generally referred to as copy number variants (CNVs) (Fig. 1.2). When the duplicated regions contain genes, genomic rearrangements can result in the deletion of the region (and the genes) between the copies and thus give rise to disease. It is of considerable ongoing interest to evaluate the role of CNVs and other structural variants in the etiology of a range of clinical conditions.

De Novo Mutations

While much emphasis is placed on inherited genome variation, all such variation had to originate as a de novo or new change occurring in germ cells. At that point, such a variant would be quite rare in the population (occurring just once), and its ultimate frequency in the population over time depends on chance and on the principles of Mendelian inheritance and population genetics. While there have been many efforts to estimate the human mutation rate, the ability to sequence genomes directly provides a robust method for measuring such rates genome-wide, by, for example, comparing the sequence of an offspring's genome (or a portion of that genome) with that of his or her parents [11].

Such studies have shown that any individual carries an estimated 30–70 new mutations per genome that were not present in the genomes of his or her parents. This rate, however, varies from gene to gene around the genome and from individual to individual and is dependent on the age of the parents [12,13]. Overall, the rate, combined with considerations of population growth and dynamics, predicts that there must be an enormous number of relatively new (and thus rare) mutations in the current worldwide population of 7 billion individuals [14].

Conceptually similar studies have explored de novo mutations in CNVs, where the generation of a new length variant depends on recombination, rather than on errors in DNA synthesis to generate a new base pair. Indeed, the measured rate of formation of new CNVs is orders of magnitude higher than that of base substitutions [15].

Variation in Individual Genomes

The most extensive current inventory of the amount and type of variation to be expected in any given genome comes from the direct analysis of individual diploid human genomes. Any given human genomes typically carry about 5 million SNPs, many of which are previously unknown. This suggests that the number of SNPs described for our species is still incomplete, although presumably the fraction of such novel SNPs will decrease as more and more genomes from more and more populations are sampled.

Typically, each genome carries thousands of nonsynonymous SNPs, that is, variants that encode a different amino acid in thousands of protein-coding genes around the genome. These measurements underscore the potential impact of gene and genome variation on human biology and on medicine.

The most comprehensive study to date compared several thousand genome sequences from the 1000 Genomes Project [16]. This study concluded that each genome carries 100 or more likely loss-of-function mutations, about 10,000 nonsynonymous changes, and some 500,000 variants that overlap known gene regulatory regions.

These and other findings indicate—perhaps surprisingly—that thousands of genes in the human genome are highly tolerant to many mutations that appear likely to result in a loss of function [5,7,8,17,18]. Within the clinical setting, this awareness has important implications for the interpretation of data from sequencing of patient material, particularly when predicting the impact of mutations in genes of currently unknown function.

Notwithstanding the remarkable amount of information on genomes and genome variation over the past decade, it is clear that we are still in a mode of discovery; no doubt many millions of additional SNPs and other variants remain to be uncovered, as does the degree to which any of them might impact an individual's phenotype in the context of wellness and health care. The broad question of "what is normal?"—an essential concept in human biology and in clinical medicine—remains very much an open question when it comes to the human genome.

Variation in Populations

Taking advantage of major technological developments (including whole-exome and whole-genome sequencing) that have greatly increased the throughput of genotyping on a genome-wide scale, a number of large-scale projects have gathered genotypic information on millions of SNPs and structural variants in many thousands of individuals from hundreds of populations worldwide (e.g., Refs. [7,10,16,19]). Two major conclusions emerge: First, some 85−90% of the common variation found in our species is shared among different population groups; a relative minority of common variants are specific to or highly enriched/depleted in genomes from a particular population. And second, most variable sites in the genome are rare, not common, and are private to specific populations rather than ancient and shared among populations [14].

These findings reflect an explosion of population growth from an ancestral population of likely fewer than 10,000 individuals, with Eurasians diverging from an ancestral African population an estimated 38−64 thousand years ago. Their diaspora has been shaped by various patterns of migration around the globe, both in ancient times and continuing to within just the last few generations. It is now possible to trace or reconstruct the history and the genetic/geographic origins of many population groups around the world. These findings are of innate interest to specific groups, but also have profound implications for health care delivery to different groups of individuals worldwide.

EXPRESSION OF THE HUMAN GENOME

A key question in exploring the function of the human genome is to understand how proper expression of our 20,000—40,000 genes is determined, how it can be influenced by either genetic variation or by environmental exposures or inputs, and by what mechanisms such alterations in gene expression can lead to pathology evident in the practice of clinical medicine. The control of gene activity—in development, in different tissues, during the cell cycle, and during the lifetime of an individual both in sickness and in health—is determined by a complex interplay of genetic and epigenetic features [2].

By "genetic" features, we here refer to those found in the genome sequence (see Box 1.1), which plays a role, of course, in determining the identity of each gene, its particular form (alleles), its level of expression (requiring a consideration of various regulatory elements), and its particular genomic landscape (three-dimensional configuration, base composition, chromatin composition). By "epigenetic" features, here we mean packaging of the DNA into chromatin, in which it is complexed with a variety of histones as well as innumerable nonhistone proteins that influence the accessibility and activity of genes and other genomic sequences. The structure of chromatin—unlike the genome sequence itself—is highly dynamic and underlies the control of gene expression that shapes in a profound way both cellular and organismal function (see chapter: The Functional Genome: Epigenetics and Epigenomics).

It has been appreciated for decades that there is high variability in gene expression levels among individuals. Much of this is due to differences among the genes themselves, a result consistent with local sequence variation influencing the expression of such genes. It is likely that the discovered regulatory variants will correlate with these patterns of epigenetic modification [20].

The Genome in Three-Dimensional Space

In contrast to the impression one gets when viewing the genome as a linear string of sequence, the genome adopts a highly ordered and dynamic arrangement within the three-dimensional space of a nucleus (Fig. 1.3). This three-dimensional architecture is highly predictive of the transcriptome map and reflects megabase-sized domains (loops) evident from examining coordinated patterns of gene expression at the chromosome level, as well as dynamic interactions between different intra- and interchromosomal points of contact within the nucleus [24—27].

An initial glimpse into the folding principles imbedded in the genome comes from work that builds on technical advances to map and sequence points of contact around the genome in the context of three-dimensional space [23,27,28] (Fig. 1.3). The biophysical, epigenomic, and/or sequence properties that facilitate or specify the orderly and dynamic packaging of

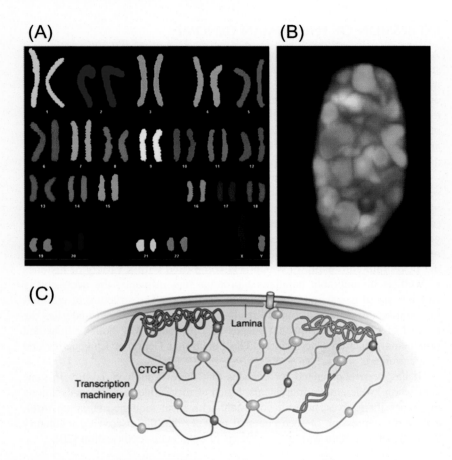

FIGURE 1.3 Organization of the human genome in metaphase and interphase. (A) Individual metaphase chromosomes from a normal male. Chromosomes are colored differently after hybridization to chromosome-specific DNA. (B) Arrangement of chromosomes in an interphase nucleus. Each chromosome maintains its own territory, with minimal mixing. (C) Long-range intra- and interchromosomal interactions, reflecting chromosome and genome architecture within the nucleus. Parts of only two chromosomes are shown, tethered to the nuclear lamina or tethered together by transcriptional machinery or regulatory factors such as the transcription factor CTCF. Source: *From Levy S, Sutton G, Ng P, Feuk L, Halpern A, et al. The diploid genome sequence of an individual human. PLoS Biol 2007 [21];5:e254; Bolzer A, Kreth G, Solovei I, Koehler D, Saracoglu K, et al. Three-dimensional maps of all chromosomes in human male fibroblast nuclei and prometaphase rosettes. PLoS Biol 2005;3:e157 [22]; Van Steensel B, Dekker J. Genomics tools for unraveling chromosome architecture. Nat Biotechnol 2010;28:1089−95 [23], with permission.*

each chromosome during each cell cycle without reducing the genome to a tangled mess within the nucleus remain unknown and are the subject of much investigation in genome biology, with relevance to clinical medicine when a genome is rearranged and genes are then found in different locations with different three-dimensional properties.

Overall, such data support a view of gene expression as the sum of several different (but interrelated) effects, including gene sequence, regulatory sequences and their epigenetic packaging, organization of the genome into domains, programmed interactions between different parts of the genome, and dynamic three-dimensional packaging in the nucleus. All must coordinate in an efficient and hierarchical fashion, and disruption of any one—due either to genetic change or to disease-related processes—might be expected to alter the overall cellular phenotype.

GENES, GENOMES, AND DISEASE

In the context of genomic and precision medicine, an overriding question is to what extent variation in the sequence and/or expression of one's genome influences the likelihood of disease onset, determines or signals the natural history of disease, and/or provides clues relevant to the management of wellness or disease. As just discussed, variation in one's constitutional genome can have a number of different direct or indirect effects on gene expression, thus contributing to the likelihood of disease.

Comprehensive catalogues of genomic and other "omic" data, from sequence to functional elements encoded in the genome, to interacting networks of RNAs and proteins, and to metabolites, carbohydrates and small molecules in a variety of cell and tissue types, are emerging (Table 1.3) [29−31]. The integrative nature of physiology and medicine, aided in major part by advances in the data sciences, lends itself well to "omic" approaches that seek to gather comprehensive datasets that can be queried informatically to gain insights into patterns that promise to reveal distinctive insights about health or disease (see chapter: From Data to Knowledge: An Introduction to Biomedical Informatics).

Clinical Sequencing in Search of Causal Variants

The development of new massively parallel sequencing technologies (see chapter: The State of Whole-Genome Sequencing) has extended earlier efforts at genome-wide genotyping of common SNPs to uncover rare or novel variants responsible for (or at least statistically associated with) a disease or phenotype of interest (e.g., Refs. [5,7,8]). Three conceptually related sequencing-based strategies have emerged to relate or implicate particular variants in a range of clinical conditions or phenotypes: targeted resequencing of specific genes (or exons), whole-exome sequencing, or whole-genome

TABLE 1.3 Personalized "Omic" Signatures of Health or Disease

Dataset	"Omic" Approach	Technology Platform or Approach
Human genome sequence	Genomics	SNPs; other genome variants; complete personal genome sequences
Gene expression profiles	Transcriptomics	Sequencing of RNA products (RNA-seq)
Protein abundance	Proteomics	Protein arrays of specific protein products
Metabolites	Metabolomics	Analysis of hundreds to thousands of metabolites
Chromatin	Epigenomics	Array- or sequence-based assessment of chromatin modification
Gene networks, interactions	Systems biology	Large-scale interactions among genes or proteins
Carbohydrates, glycomedicine	Glycomics	Comprehensive assessment of carbohydrates and protein modification
Microbiomes	Metagenomics	Analysis of viral, fungal, and bacterial communities in human specimens
Genomic and clinical data	Informatics	Integrated databases of "omic" data and electronic health records

sequencing of samples either from individuals or groups of patients with distinctive phenotypes or from individuals within particular populations.

Targeted Resequencing

In the targeted sequencing approach, original efforts focused on one or several genes that were believed to be strong candidates for the phenotype under study. However, as reduced costs have made it more cost-effective to sequence whole exomes, targeted resequencing, with its inherent bias toward candidate genes that "seem plausible," has fallen out of favor in most studies.

Whole-Exome Sequencing

The approach of isolating and sequencing entire exomes from patient samples has increasingly become the favored approach (see chapter: The State of Whole-Genome Sequencing), both for research and for clinical care. At least several hundred patient samples have been examined in searches for potentially causative variants, either targeted to particular phenotypes or broadened in population-based cohorts to any clinical finding or laboratory value that covaries with a genome variant [7,8].

Exome sequencing does, however, present challenges, especially when needing to distinguish between legitimate disease associations and population stratification. A potentially overwhelming number of rare or novel variants need to be evaluated for potential functional importance before one can conclude with confidence that a particular variant is causal. This task appears tractable for Mendelian disorders within families, but is significantly more challenging for complex diseases and traits at the moment. Nonetheless, with several years of experience, clinical whole-exome sequencing is now being offered as a diagnostic service at an increasing number of academic medical centers, and there are ongoing efforts to determine how best to integrate sequencing into clinical care on a routine basis [4,32,33].

Whole-Genome Sequencing

These challenges are even more pressing for whole-genome sequencing. Issues of disease penetrance, possible allele–allele interactions around the genome, coding and noncoding variants, and protective versus predisposing alleles all have to be taken into consideration [5,34]. At this point, as the lines between research and clinical care "blur" [35], it seems fair to say that the major challenges confronting whole-genome sequencing over time will fall in the categories of informatics and clinical utility, not technical capacity or even cost (see chapter: The State of Whole-Genome Sequencing).

Notwithstanding these challenges, whole-genome sequencing has been used to positive effect for at least some conditions, ranging from Mendelian disorders to complex disease to cancer.

Microbiomes and Microbiota

The human genome is not the only genome relevant to the practice of medicine. Both in states of health and disease, our own genome is vastly outnumbered by the genomes of a host of microorganisms, many living peacefully and continuously on various body surfaces as commensal or symbiotic microbiomes, others wreaking havoc as adventitious viral, bacterial, or fungal pathogens (see chapter: The Human Microbiome).

The genomes of thousands of microorganisms have been determined and are being utilized to provide rapid diagnostic tests in clinical settings, to predict antibiotic or antifungal efficacy, to identify the source of airborne, water, or soil contaminants, to monitor hospital or community environments, and to better understand the contribution of communities of microbes—the microbiota—and various environmental exposures to diverse human phenotypes.

The human colon contains hundreds of microbial species comprising some 10^{13} to 10^{14} microorganisms. Each adult's gut provides a unique environment whose origins and impact on human disease are being explored. The microbiotic gene set is significantly different from that of the human genome and thus has the capacity to alter the metabolic profile of different individuals or different populations, with clinically meaningful effects on drug metabolism, toxicity, and efficacy, from early development to the end of life. The application of a variety of "omics" approaches to microbiomes is now impacting clinical diagnostics, for example to identify unknown viral infections or to diagnose antibiotic-resistance infections.

The field of metagenomics explores this heterogeneous ecosystem by comprehensive sequence analysis of the collected genomes from biological specimens (such as stool, urine, sputum, water sources, and air), followed by both taxonomic and bioinformatic analysis to deconvolute the many genomes contained in such specimens and to define the different organisms, their genes, and genome variants. The metagenomic approach is particularly informative for characterizing organisms that cannot be cultured in standard microbiology laboratories. The dynamic balance among the many species that make up the microbiome points to cooperative networks of communities of microbiota, evolving over time and in response to different environments [36].

A number of diseases have been associated with large-scale imbalances in the gut microbiome, such as Crohn's disease, ulcerative colitis, antibiotic-resistant diarrhea, and obesity. In parallel, extensive population studies have revealed phenotypic microbiome covariates (such as stool consistency, body mass index, age and sex, diet, and medication use) that are associated with the composition of diverse communities of microbiota in stool samples [37,38]. These initial studies of several thousand individuals provide a foundation for further understanding microbe $\times$ host interactions in the context of variable and dynamic environments [39].

Undoubtedly, the states of health and disease are determined in part by the balance of genomes both within us and external to us. The full complement of genomic information from both of these sources of genomes will both provide insights into defining the states of health and disease and contribute to a basis for precision in both the prevention of disease and its treatment.

FROM GENOMES TO PRECISION MEDICINE

Of all the promises of the current scientific, social, and medical revolution stemming from advances in our understanding of the human genome and its variation, genomic and precision medicine may be the most eagerly awaited. The prospect of examining an individual's entire genome (or at least a significant fraction of it) in order to make individualized risk predictions and treatment decisions is an attractive, albeit challenging, one [3,4,40,41].

Having access to the reference human genome sequence has been transformative for the fields of human genetics and genome biology [6], but by itself is an insufficient prerequisite for genomic medicine. As discussed in subsequent chapters, equally important are the various complementary technologies to reliably capture and utilize information on individual genomes, their epigenetic modification, and their derivatives the transcriptome, proteome, microbiome, and metabolome for health and disease status (Table 1.3). Each of these technologies provides information that, in combination with phenotypic data and evaluation of environmental triggers, will contribute to assessment of individual risks and guide clinical management and decision making.

This chapter has focused on the organization, variation, and expression of the human genome as a foundation for the chapters to follow on principles of human genomics, on new and clinically relevant technologies in genomics and informatics, on approaches in translational genomics and, finally, on applications of genomic and precision medicine to medical practice, from primary care to specialty care of health and disease. While the challenges both for the health care profession and for society remain significant, the development of comprehensive, cost-efficient, and high-throughput technologies, combined with powerful tools in biomedical and clinical informatics for analyzing and storing vast amounts of data, will enable the application and widespread availability growing field of genomic and precision medicine.

ABBREVIATIONS

bp base pair
CNV copy number variant
Gb gigabase, one billion base pairs of DNA
in/del polymorphism involving insertion or deletion of one or more base pairs
kb kilobase pair, one thousand base pairs
Mb megabase, one million base pairs of DNA
ncRNA noncoding RNA
SNP single-nucleotide polymorphism

REFERENCES

[1] Collins FS, Varmus H. A new initiative on precision medicine. N Engl J Med 2015;372:793–5.
[2] Nussbaum RL, McInnes RR, Willard HF. Genetics in medicine. 8th ed. Philadelphia, PA: W.B. Saunders, Co; 2016:546 pp.
[3] Green ED, Guyer MS, National Human Genome Research Institute. Charting a course for genomic medicine from base pairs to bedside. Nature 2011;470:204–13.
[4] Aronson SJ, Rehm HL. Building the foundation for genomics in precision medicine. Nature 2015;526:336–42.

[5] Gudbjartsson DF, Helgason H, Gudjonsson SA, Zink F, Oddson A, et al. Large-scale whole-genome sequencing of the Icelandic population. Nat Genet 2015;47:435−44.

[6] Lander ES. Initial impact of the sequencing of the human genome. Nature 2011;470:187−97.

[7] The UK10K Consortium. The UK10K project identifies rare variants in health and disease. Nature 2015;526:82−90.

[8] Dewey FE, Murray MF, Overton JD, Habegger L, Leader JB, et al. Distribution and clinical impact of functional variants in 50,726 whole exome sequences from the DiscovEHR Study. Science 2016; in press.

[9] Estivill X, Armengol L. Copy number variants and common disorders: filling the gaps and exploring complexity in genome-wide association studies. PLoS Genet 2007;3:e190.

[10] Sudmant PH, Mallick S, Nelson BJ, Hormozdiari F, Krumm N, et al. Global diversity, population stratification, and selection of human copy number variation. Science 2015;349, aab3761.

[11] Shendure J, Akey JM. The origins, determinants, and consequences of human mutations. Science 2015;349:1478−83.

[12] Rahbari R, Wuster A, Lindsay SJ, Hardwick RJ, Alexandrov LB, et al. Timing, rates, and spectra of human germline mutation. Nat Genet 2016;48:126−33.

[13] Conrad DF, Keebler JEM, DePristo MA, Lindsay SJ, Zhang Y, et al. Variation in genome-wide mutation rates within and between human families. Nat Genet 2011;43:712−14.

[14] Olson MV. Human genetic individuality. Annu Rev Genomics Hum Genet 2012;13:1−27.

[15] Itsara A, Wu H, Smith JD, Nickerson DA, Romieu I, et al. De novo rates and selection of large copy number variation. Genome Res 2011;20:1469−81.

[16] The 1000 Genomes Project Consortium. A global reference for human genetic variation. Nature 2015;526:68−74.

[17] Lek M, Karczweski KJ, Minikel EV, Samocha KE, Banks E, et al. Analysis of protein-coding genetic variation in 60,706 humans. Nature 2016;536:285−91.

[18] Narasimhan VM, Hunt KA, Mason D, Baker CL, Karczewski KJ, et al. Health and population effects of rare gene knockouts in adult humans with related parents. Science 2016;352:474−7.

[19] Sudmant PH, Rausch T, Gardner EJ, Handsaker RE, Abyzov A, et al. An integrated map of structural variation in 2,504 human genomes. Nature 2015;526:75−80.

[20] Waszak SM, Delaneau O, Gschwind AR, Kilpinen H, Raghav SK, et al. Population variation and genetic control of modular chromatin architecture in humans. Cell 2015;162:1−12.

[21] Levy S, Sutton G, Ng P, Feuk L, Halpern A, et al. The diploid genome sequence of an individual human. PLoS Biol 2007;5:e254.

[22] Bolzer A, Kreth G, Solovei I, Koehler D, Saracoglu K, et al. Three-dimensional maps of all chromosomes in human male fibroblast nuclei and prometaphase rosettes. PLoS Biol 2005;3:e157.

[23] Van Steensel B, Dekker J. Genomics tools for unraveling chromosome architecture. Nat Biotechnol 2010;28:1089−95.

[24] Nichols MH, Corces VG. A CTCF code for 3D genome architecture. Cell 2015;162:703−5.

[25] Whalen S, Truty RM, Pollard KS. Enhancer-promoter interactions are encoded by complex genomic signatures on looping chromatin. Nat Genet 2016;48:488−96.

[26] Dekker J, Mirny L. The 3D genome as moderator of chromosomal communication. Cell 2016;164:1110−21.

[27] Rao SSP, Huntley MH, Durand NC, Stamenova EK, Bochkov ID, et al. A 3D map of the human genome at kilobase resolution reveals principles of chromatin looping. Cell 2015;159:1665—80.

[28] Kind J, Pagie L, de Vries SS, Nahidiazar L, Dey SS, et al. Genome-wide maps of nuclear lamina interactions in single human cells. Cell 2015;163:134—47.

[29] Huttlin EL, Ting L, Bruckner RJ, Gebreab F, Gygi MP, et al. The BioPlex network: a systematic exploration of the human interactome. Cell 2015;162:425—40.

[30] Rolland T, Taşan M, Charloteaux B, Pevzner SJ, Zhong Q, et al. A proteome-scale map of the human interactome network. Cell 2014;159:1212—26.

[31] Vidal M, Cusick ME, Barabasi A-L. Interactome networks and human disease. Cell 2011;144:986—98.

[32] Rehm HL, Berg JS, Brooks LD, Bustamante CD, Evans JP, et al. ClinGen—the Clinical Genome Resource. N Engl J Med 2015;372:2235—42.

[33] Yang Y, Muzny DM, Xia F, Niu Z, Person R, et al. Molecular findings among patients referred for clinical whole-exome sequencing. JAMA 2014;312:1870—9.

[34] Gloyn AL, McCarthy MI. Variation across the allele frequency spectrum. Nat Genet 2010;42:648—50.

[35] Angrist M, Jamal L. Living laboratory: whole-genome sequencing as a learning healthcare enterprise. Clin Genet 2014;87:311—18.

[36] Rakoff-Nahoum S, Foster KR, Comstock LE. The evolution of cooperation within gut microbiota. Nature 2016;533:255—9.

[37] Zhernakova A, Kurilshikov A, Bonder MJ, Tigchelaar EF, Schirmer M, et al. Population-based metagenomics analysis reveals markers for gut microbiome composition and diversity. Science 2016;352:565—70.

[38] Falony G, Joossens M, Vieira-Silva S, Wang J, Darzi Y, et al. Population-level analysis of gut microbiome variation. Science 2016;352:560—4.

[39] Gilbert JA, Quinn RA, Debelius J, Xu ZZ, Morton J, et al. Microbiome-wide association studies link dynamic microbial consortia to disease. Nature 2016;535:94—103.

[40] Gonzaga-Jauregui C, Lupski JR, Gibbs RA. Human genome sequencing in health and disease. Annu Rev Med 2012;63:35—61.

[41] Manolio TA, Abramowicz M, Al-Mulla F, Anderson W, Balling R, et al. Global implementation of genomic medicine: we are not alone. Sci Transl Med 2015;7 290ps13.

Chapter 2

The Functional Genome: Epigenetics and Epigenomics

Timothy E. Reddy
Duke University School of Medicine, Durham, NC, United States

Chapter Outline

INTRODUCTION

As outlined in the previous chapter, the primary sequence of the human genome is the order in which the nucleotides adenine (A), cytosine (C), guanine (G), and thymine (T) occur along an individual's chromosomes. The genome sequence of an individual is almost entirely the same in every cell. How a common genome sequence generates the panoply of distinct cell types and environmental responses is therefore a major area of research. The answer, in part, relies on the addition of chemical modifications to the genome sequence and to the proteins that package the genome inside of cells. Collectively, those modifications are known as the epigenome.

Genomic and Precision Medicine. DOI: http://dx.doi.org/10.1016/B978-0-12-800681-8.00002-5

While the genome is largely constant between different cells of a person's body, the epigenome can vary drastically between cell types and can change over time and in response to the environment. There is also evidence that changes in the epigenome can be transmitted from parent to offspring, creating the potential for heritable transmission without modifying the genome sequence itself. This chapter will describe the basic components of the epigenome; how the epigenome is established and maintained; the role of the epigenome in health and disease; and current and future directions in epigenetic and epigenomic research.[1]

THE COMPOSITION OF THE EPIGENOME

The epigenome has two primary components: chemical modifications of the individual nucleotides that make up the genome and chemical modifications to the histone proteins that package the genome in cells. Collectively, those modifications are referred as "epigenetic marks." Nucleotide modifications in the human genome are primarily limited to the methylation and hydroxy-methylation of cytosine, the latter of which was first demonstrated to occur in humans in 2009 [1]. In sharp contrast, the histone modifications present in human cell are far more diverse. While many epigenetic marks have been associated with changes in how the genome is used in a cell, it is important to note that direct causality between epigenetic marks and genome function is typically not well established. Overcoming technical hurdles to interrogate potential causal relationships remains an active area of investigation.

Chemically Modified Nucleotides

Various covalently modified nucleotides occur throughout the tree of life. In humans, those modifications are thought to be limited to the addition of a methyl group to the 5' carbon in cytosine (5-methylcytosine, 5mC) and the subsequent addition of a hydroxy group to that methyl group to form 5-hydroxymethylcytosine (5hmC). The process by which these modifications occur is summarized in Fig. 2.1.

5-Methylcytosine

Human cytosine methylation occurs almost exclusively when a cytosine is followed by a guanine in the genome. The pair is commonly referred to as a CpG dinucleotide, where the "p" represents the phosphodiester bond between adjacent nucleotides. 5mC was first discovered as an indicator of tightly

---

1. The terms "epigenetics" and "epigenomics" are often used interchangeably. In this chapter, we will use "epigenetics" to refer to the study of chemical modifications at individual loci in the genome, and "epigenomics" to refer to the study of how those modifications are established and altered across the entire human genome. The scale at which epigenetics becomes epigenomics is not well defined, hence the inconsistent and often interchangeable use of the terms.

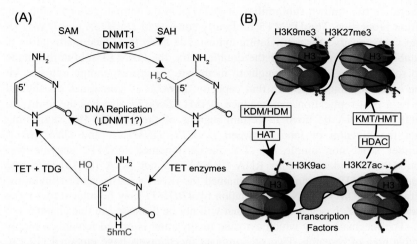

FIGURE 2.1   Overview of the enzymatic steps and enzymes involved in establishing and modifying the epigenome. Each arrow indicates an enzymatic step and is labeled with enzyme(s) known or thought to catalyze that step. (A) Schematic of the methylation and demethylation of cytosine in the human genome. (B) Schematic of the modification of histone methylation and acetylation. The *DNMT1* and *DNMT3* enzymes catalyze the transfer of a methyl group from SAM to the 5' carbon of cysteine, resulting in 5mC and *S*-adenosyl-L-homocysteine (SAH). The mechanisms by which 5mC is converted back to cytosine are less clear. One mechanism involves rounds of DNA replication that, perhaps with reduced *DNMT1* levels, fail to reproduce the methylation state of the parent cell. A second possible mechanism may involve the conversion of 5mC to 5hmC and other modified forms of cysteine by the ten-eleven translocation methylcytosine dioxygenase 1 (*TET1*). 5hmC may then be directly converted to cysteine via subsequent unknown enzymatic steps. Alternatively, 5hmC may evade recognition by the pathway that recapitulates 5mC after mitosis, leading to a DNA-synthesis dependent loss of 5mC.

compacted regions of the genome known as heterochromatin [2]. Genes in heterochromatin are generally not expressed, and 5mC was therefore considered an epigenetic mark that silences gene expression. More recent studies have expanded on those initial findings to reveal that the 5mC may have different effects in different regions of the genome. Outside of genes, 5mC is typically associated with the exclusion of regulatory proteins known as transcription factors that control the expression of nearby genes. That finding is consistent with the role of 5mC in silencing gene expression. Conversely, within genes, there is growing evidence that 5mC is positively correlated with expression levels [3]. While it is not yet known if gene body methylation is a cause or an effect of increased gene expression, it is clear that the relationship between 5mC and gene expression is complex and dependent on genomic location.

The Establishment and Maintenance of 5mC

Cytosines are methylated by the DNA methyltransferase (DNMT) family of enzymes [4]. The DNMTs transfer a methyl group from *S*-adenosyl-L-methionine (SAM) to the 5' carbon of cytosine residues in the genome [5].

There are two known subfamilies of DNMTs in the human genome with different properties. The DNMT3 subfamily consists of *DNMT3A* and *DNMT3B* and is responsible for methylating cytosine de novo [6]. Once methylation is established on one strand of the double helix, the DNMT1 subfamily, made up solely of *DNMT1*, is thought to methylate the corresponding cytosine on the opposite strand [7,8]. In that capacity, *DNMT1* is sometimes considered to be a post-replication maintenance DNMT. *DNMT1* also has the ability to methylate DNA de novo in certain regions of the genome [9], suggesting that the division of labor between *DNMT1*, *DNMT3A*, and *DNMT3B* is neither strict nor simple [4]. The formerly-named DNMT2 enzyme was found to methylate transfer RNA (tRNA) and not DNA [10,11]. For that reason, DNMT2 has since been renamed the tRNA aspartic acid methyltransferase 1 (TRDMT1). RNA methylation by *TRDMT1* may be related to a more extensive RNA-based epigenome that will not be discussed in this chapter.

The mechanism by which cytosine methylation is erased remains a highly active area of research. Two predominant mechanisms have emerged [12,13]. The passive model of cytosine demethylation relies on a failure to methylate newly synthesized DNA during mitosis, perhaps involving the inhibition of *DNMT1*. If the cytosine methylation is systematically not established at a given genomic location after DNA replication, then the number of chromosomes with cytosine methylation at that location in the two daughter cells will be reduced by half. With subsequent rounds of mitosis and new DNA synthesis, the vast majority of daughter cells will eventually have nonmethylated cytosines at that location [13]. The appeal of the passive model in humans is that it does not depend on the presence of an active cytosine demethylation enzyme or pathway that has long eluded discovery. The requirement for mitosis, however, restricts passive demethylation to dividing cells and does not explain, for example, the rapid loss of methylation that occurs on the paternal genome after a sperm cell fertilizes an egg but before the first rounds of mitotic cell division [14,15].

More recently, increasing evidence also supports a second, active mechanism of demethylation that does not depend on cell division. The base-excision-repair hypothesis states that methylated cytosines are excised from the genome and replaced by nonmodified cytosines. Meanwhile, the enzymatic demethylation hypothesis states that the methyl group on 5mC is directly removed from the cytosine by a demethylase enzyme. Demethylation by base-excision repair has been demonstrated in plants, but has not been shown to occur in the human genome [16]. Enzymatic removal of the methyl group from 5mC by a single enzyme is thought to be energetically unlikely. Instead, the possibility of a demethylation pathway that relies on successive modifications of the 5mC by several enzymes has recently gained attention. The ten-eleven translocation (TET) enzymes convert 5mC into 5hmC, which may be the first step in a chain of further modifications that ultimately result in unmodified cytosine [16–18]. Human enzymes that convert 5hmC or 5hmC derivatives to cytosine, however, have not yet been found.

Monoallelic Cytosine Methylation

Over the majority of the genome, DNA methylation occurs on both copies of the same chromosome or not at all. If a cytosine is methylated on the maternally inherited chromosome, then the homologous cytosine on the paternally inherited allele is also methylated. There are two major exceptions to that rule, both of which are essential for human health: X inactivation and imprinting.

The defining genetic difference between typical females and males is the presence of two X chromosomes in the female genome, and one X and one Y in the male genome. The number of functional copies of the X is held more or less equivalent between the sexes, however, through a process known as X inactivation [19,20]. In female cells, a combination of epigenetic mechanisms silence most (but not all) of the gene expression from one of the two X chromosomes. That process begins with the expression of the long noncoding RNA gene XIST from one of the two X chromosomes. XIST expression triggers extensive cytosine methylation, heterochromatin formation, and gene silencing on the same X chromosome [21−23]. The initial choice of which X to inactivate is typically but not always random and is made independently by a small number of cells early in embryogenesis [20,24,25]. After cells make the initial choice of which X to inactivate, that choice is transmitted to daughter cells, leading to mosaic patterns of which X is inactivated in the adult. A classic visual example is the patches of color in the coats of calico cats. Those patches are created by mosaic inactivation of coat color genes located on the feline X chromosome.

Monoallelic DNA methylation and gene silencing also occur via a process known as imprinting [26]. Unlike for X inactivation, imprinting occurs in a parent-of-origin dependent manner. In some loci, the maternal chromosome is inactivated and the paternal chromosome is active; in other loci, the reverse is true. A classic example is the H19 locus which, as for X inactivation, requires the expression of a long noncoding RNA from the imprinted allele. In the H19 locus, the long noncoding RNA H19 is expressed from the maternally inherited chromosome, leading to silencing of gene expression and extensive cytosine methylation in the surrounding region on the same chromosome [27−29]. How exactly the H19 RNA directs imprinting is not fully understood and may involve mechanisms distinct from those that cause X inactivation [26]. There are currently over 100 loci in the human genome with evidence for imprinting, and the rate at which imprinted loci were discovered rapidly increased with the ability to detect monoallelic gene expression using genome-wide approaches [30,31]. The rate of discovery has recently slowed, however, and there is growing evidence that nearly all imprinted genes in humans have been identified [32,33].

5-Hydroxymethylcytosine

The discovery of 5hmC in the human brain was the first evidence that a second nucleotide modification was present in the human epigenome. In contrast to 5mC which is found in every cell type, it is currently thought that 5hmC is predominantly limited to the brain and to early stages of development [1,34,35]. Those findings are very recent, however, and there is much to be learned about the distribution of 5hmC in the rest of the human body. As described above, one potential biological function of 5hmC is as part of a demethylation pathway. Another is that binding of regulatory factors to 5hmC may influence gene regulation [36]. Both of these hypotheses are currently areas of active investigation.

HISTONE MODIFICATIONS

The human genome is wrapped around complexes of histone proteins in a structure known as the nucleosome. Covalent modifications to the histone proteins—the second part of the epigenome—have been associated with how tightly the nucleosomes are packed together and whether the proteins that control gene expression can bind the genome [37,38]. Regions in which the nucleosomes are tightly packed are known as heterochromatin. As discussed earlier, heterochromatin is also highly enriched for 5mC and depleted for expressed genes [39,40]. Genomic regions that are not heterochromatic are called euchromatin. In the euchromatic regions of the genome, nucleosomes are less densely packed, allowing the binding of RNA polymerase and regulatory proteins to control gene expression. Differences between heterochromatin or euchromatin may be an important component of cell and tissue differentiation [41].

The Histone Code

The histones that form the core of the nucleosome are an octameric complex that consists of two copies each of histone H2A, H2B, H3, and H4. The histones are encoded in numerous clusters throughout the genome, and there are several slightly variant versions of each histone protein in the genome [42]. Within each histone protein, there are numerous sites where the amino acids can be modified by the covalent addition of small molecule groups. The most well-understood histone modifications involve the addition of between one and three methyl groups or an acetyl group to specific lysine residues on the N-terminal tail of histone H3. Those modifications will be discussed in some depth and are diagrammed in Fig. 2.2. However, many other modifications have been discovered across all four core histone proteins. Those modifications include arginine methylation, serine phosphorylation, sumolyation, ubiquitination, etc. [38]. The modifications are typically written using a shorthand notation that specifies (1) the histone, (2) the amino acid, and (3) the covalent modification. For example, the acetylation of histone H3 lysine 27 is expressed as H3K27ac. Similarly, H3K4me3 means that lysine 4 on histone H3 has three methyl groups.

A growing body of evidence suggests that different histone modifications each associate with different states of the genome, a hypothesis known as the histone code [43,44]. For example, heterochromatin is not only enriched for

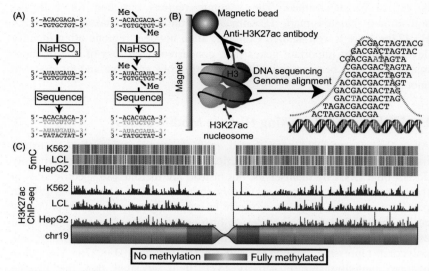

FIGURE 2.2 **Methods for observing the epigenome. (A) DNA methylation can be mea-sured with bisulfite sequencing, in which unmethylated cytosine is converted to uracil by sodium bisulfite (NaHSO₃), whereas methylated cytosine is protected from the reaction. Sequencing of the converted DNA can then reveal the NaHSO₃-induced mutations that mark the unmethylated cytosines. (B) Chromatin Immunoprecipitation (ChIP) can be used to isolate histones with specific epigenetic modifications such as H3K27ac. In ChIP, an anti-body specific to the modification is conjugated to a magnetic bead, enabling immunoprecip-itation of the histones with that modification. The DNA bound to that histone can then be observed with high-throughput DNA sequencing. Mapping the resulting sequences to the genome can then be used to determine the genomic location of the modified histone. That combination of ChIP with high-throughput sequencing is known as ChIP-seq. (C) Both bisulfite sequencing and ChIP-seq can be used to measure the epigenome genome-wide. As an example, the distribution of cytosine methylation (top) and H3K27ac (bottom) across human chromosome 19 as determined by the ENCODE project is shown.** In heterochromatic regions of the genome, depicted at top, nucleosomes are tightly packed and typically marked with H3K9me3 and H3K27me3. During the transition to euchromatin, depicted at the bottom, a combina-tion of histone demethylases and histone acetyltransferases act to replace the methyl modification with acetyl modifications. One hypothesis is that the acetyl groups neutralize the positive charge on histone tails, leading to less tightly bound histones and a chromatin state that is more permissive to the binding of transcription factors. In the transition from euchromatin to heterochromatin, the histone modifications are reversed by histone deacetylases and histone methyltransferases.

5mC but is also enriched for H3K27me3 and H3K9me3 [45,46]. Meanwhile, in euchromatin, different sets of histone marks have been associated with different genomic activities. For example, the start sites of actively expressed genes are enriched in H3K4me3 and H3K9ac [47–49]. Different modifica-tions of the same amino acid are sometimes associated with opposing activi-ties. For example, the mutually exclusive modifications H3K27ac and H3K27me3 are enriched at active and poised distal regulatory elements, respectively [50,51]. Finally, marks on different amino acids that signal con-tradictory activities may be indicative of regions for which the terminal state is yet to be determined. The genomic regions enriched for both the activation-associated mark H3K4me3 and the repression-associated mark H3K27me3 are a well-studied example. Such bivalent regions often identify developmentally important genes that, concurrent with loss of one of those marks, will later become actively expressed in one cell type and repressed in another [51−55].

Establishing and Removing Histone Modifications

As in the case of DNA methylation, multiple enzymes contribute to establishing and modifying histone modifications [38]. Those enzymes typically do not bind DNA by themselves, but instead are recruited to specific sites in the genome as part of larger regulatory complexes. The E1A-binding protein p300, for example, plays a major role in regulating genes from distal enhancers. The *EP300* gene includes a histone acetyltransferase domain that establishes H3K27ac when recruited to enhancers [56]. Meanwhile, the polycomb repressive complexes establish and maintain H3K27me3 via their histone methyltransferase activity [57]. It is possible that the histone modifications established by these enzymes are not necessary for altering genome activity and are instead simply a consequence of their recruitment to the genome for other purposes. However, growing evidence suggests that some histone modifications do indeed mechanistically contribute to the regulation of nearby gene expression [58,59].

EPIGENETIC MECHANISMS OF DISEASE

The epigenome has a major role in gene regulation during differentiation and development [60−62]. It is therefore not surprising that numerous diseases also involve disruption of the enzymes that establish and maintain the epigenome [63−65]. Examples of some of the most well studied epigenetic diseases are described below.

Diseases Involving DNA Methylation

Global changes in DNA methylation are a hallmark of cancer, especially at advanced stages [66−69]. The extent to which that methylation contributes causally to cancer remains to be fully understood. Growing evidence indicates that, in some instances, changes in cytosine methylation established prior to tumor development can serve as early markers of tumor formation [70,71]. Whether causative or not, the association between changes in DNA methylation and cancer has prognostic value in some cases [72,73] and has been used for early detection [74−76]. Chemotherapeutic drugs have also been developed that target DNA methylation. The most well known of those drugs are based on the cytosine analog 5-azacytidine that globally reduces genomic 5mC [77]. While the primary mechanism by which 5-azacytidine blocks cancer cell replication is through inhibiting DNA synthesis, losses in cytosine methylation may contribute to additional beneficial effects [78,79]. The inhibition of DNA methylation by 5-azacytidine has also been harnessed by researchers to reactivate the expression of genes that have been silenced during differentiation or other processes [80,81].

Mutations in the enzymes that establish and maintain 5mC and in the proteins that bind 5mC have also been implicated in severe but rare diseases (Table 2.1). Mutations in each of the DNMT1 and DNMT3 genes have been connected to developmental neurological, sensory, and immune defects [82–85]. Meanwhile, one of the most well-studied epigenetic diseases, Rett syndrome, results not from mutations in the DNMTs but instead in a protein that binds 5mC in the genome. Children born with Rett syndrome have severely affected neurological development that manifests as impaired communication and motor skills with onset within the first year of life [86]. In nearly all cases, Rett syndrome results from the spontaneous mutation of the methyl-CpG binding protein *MECP2* [87]. While much of the underlying mechanism has yet to be resolved, a leading hypothesis is that binding of *MECP2* to methylated regions of the genome contributes to the regulation of nearby genes in a way that is essential for neuronal development. MECP2 is located on the X chromosome. Likely due to prenatal lethality in males who only have one X, nearly all Rett syndrome occurs in females.

TABLE 2.1 Diseases Caused by Mutations in Enzymes That Alter 5mC and 5hmC

Gene	Enzymatic Activity	Disease Associated with Mutation	Reference (PubMed ID)
Genes Involved or Potentially Involved in DNA Methylation			
DNMT1	DNA methyltransferase	Cerebellar ataxia, deafness, and narcolepsy, autosomal dominant	22328086
		Neuropathy, hereditary sensory, type IE	21532572, 23365052
DNMT3A	DNA methyltransferase	Tatton-Brown-Rahman syndrome	24614070
DNMT3B	DNA methyltransferase	Immunodeficiency-centromeric instability-facial anomalies syndrome 1	10647011 and others
TET2	Methylcytosine dioxygenase	Myelodysplastic syndrome, somatic	21057493, 19474426, and others

Defects in imprinting can also contribute to disease either through a loss of imprinting leading to aberrantly increased gene expression or through silencing of both alleles leading to a loss of gene expression. Beckwith−Wiedemann syndrome (BWS), a rare developmental disorder, is an example of imprinting loss leading to increased gene expression and developmental disorder [88,89]. Children with BWS have abnormally large birth size and a predisposition to cancer. The primary cause of BWS involves loss of imprinting in the maternally imprinted H19 locus described above [90,91]. In some cases, loss of imprinting results from the inheritance of two active copies of H19 from the father. In other cases, genetic mutations in the region lead to a loss of imprinting [89]. Regardless of the mechanism, with imprinting lost, genes in the region become expressed from both chromosomes as opposed to from a single paternal copy. The result is elevated expression of a growth factor and a cell-cycle gene in the region, which likely explains the physical manifestation of BWS [92−94].

Examples of imprinting-related loss of gene expression leading to developmental abnormalities include Prader−Willi syndrome (PWS) and Angelman syndrome (AS) [95]. In both cases, the underlying cause involves a combination of genetic mutation on one copy of a gene and imprinting of the other copy. While both PWS and AS are caused by mutations in the same locus on chromosome 15, the genes involved and the phenotypes are distinct. In PWS, loss-of-function mutations on the paternal chromosome combined with imprinting on the maternal chromosome results in no functional copies of several genes in the child. Conversely, in AS, genetic loss-of-function mutations occur on the maternally inherited copy of UBE3A while the paternal copy is epigenetically silenced. Together, the two examples show that, because of a lack of redundancy in the genes expressed from imprinted loci, mutations in the single expressed copy of imprinted genes can have severe effects on phenotype.

Diseases Involving Histone Modifications

There are dozens of histone-modifying enzymes known to exist in the human genome, but relatively few human phenotypes have been linked to mutations in those enzymes (Table 2.2). Speculatively, that could be due to either redundancy in the histone-modifying proteins or, conversely, due to the severity of consequences associated with those mutations. However, as genome sequencing has become more common, an increasing number of rare congenital disorders have been mapped to specific histone-modifying enzymes. Three examples are provided here. Weaver syndrome, characterized by a combination of skeletal and cognitive abnormalities, is an example in which the causal mutation is in an enzyme *EZH2* that is responsible for the methylation of H3K27 [96]. Mutations that abrogate the function or expression of the histone deacetylase *HDAC4* have been linked to

TABLE 2.2 Diseases Caused by Mutations in Enzymes That Alter Histone Acetylation or Methylation

Gene	Enzymatic Activity	Disease Associated with Mutation	Reference (PubMed ID)
Genes Involved in Histone Acetylation			
CBP	Histone acetyltransferase	Rubinstein—Taybi syndrome	7630403
P300	Histone acetyltransferase	Rubinstein—Taybi syndrome 2	17299436, 7630403, 19353645
		Colorectal cancer, somatic	10700188, 21390126, and others
KAT6B	Histone acetyltransferase	Genitopatellar syndrome	22077973, 22265017
		SBBYSS syndrome	23436491
HDAC4	Histone deacetylase	Brachydactyly-mental retardation syndrome	
HDAC6	Histone deacetylase	Chondrodysplasia	20181727
HDAC8	Histone deacetylase	Cornelia de Lange syndrome 5	22885700
		Wilson—Turner syndrome	22889856
Genes Involved in Histone Methylation			
EZH2	Histone methyltransferase	Weaver syndrome	22177091
KMT2A	Histone methyltransferase	Wiedemann-Steiner syndrome	22795537
KMT2D	Histone methyltransferase	Kabuki syndrome 1	20711175
EHMT1	Histone methyltransferase	Kleefstra syndrome	15805155, 19264732, 16826528
KDM5C	Histone demethylase	Mental retardation, X-linked, syndromic, Claes—Jensen type	15586325, 21575681
KDM6A	Histone demethylase	Kabuki syndrome 2	22197486

brachydactyly-mental retardation syndrome, a syndrome that includes developmental delays and skeletal and craniofacial abnormalities [97]. In cases of X-linked Claes—Jensen type mental retardation, mutations in the lysine demethylase KDM5C are thought to be causative [98]. Together, these results show that defects in histone-modifying enzymes can and do cause disease, but those diseases appear to have extremely low prevalence in the human population.

Several other disorders are associated not with the histone-modifying enzymes directly, but instead with mutations in other genes that lead to altered histone modifications. A classic example is the neurodegenerative and autosomal dominant Huntington's disease (HD) [99]. The genetic cause of HD is the expansion of trinucleotide CAG repeats in the DNA sequence of the huntingtin (*HTT*) gene. Those repeats lead to poly-glutamines in the HTT protein, ultimately leading to progressive neurodegenerative consequences. Patients with HD are now well documented to have epigenetic abnormalities, including a loss of histone acetylation [100,101]. The mechanism by which mutant HTT contributes to that hypoacetylation is still being determined. Current evidence supports a model in which mutant HTT alters genomic recruitment of histone-modifying genes [102]. That mechanism may generalize to other neurodegenerative diseases [102,103].

Small molecules that inhibit histone deacetylases, known as HDAC inhibitors, have been developed under the hypothesis that correcting the epigenome will improve outcomes in so-affected patients. In early successes, HDAC inhibitors have been approved as a second-line treatment for certain types of lymphomas, and several additional cancer trials are currently underway [104,105]. There is also evidence that HDAC inhibitors improve HD-associated symptoms in animal models, but human trials have not been performed [106—109]. Basic research studies have also used HDAC inhibitors to investigating the cause and effect relationships between the epigenome and gene regulation [110,111]. Those studies have met challenges due to the broad genome-wide effects of those compounds combined with potential effects resulting from inhibition of the deacetylation of nonhistone proteins. It is likely that more targeted approaches currently under development will overcome some of those challenges, leading to a more nuanced understanding of the cause and effect of epigenetic histone modifications.

EPIGENETIC RESPONSES TO THE ENVIRONMENT

Even though the epigenome is largely established during cell differentiation early in development, interactions between the environment and the epigenome are important at all stages in life [112]. The epigenome both contributes to determining how cells in the body respond to the environment and, vice-versa, the environment can lead to changes in the epigenome. Environmental effects on the epigenome can accumulate over a lifetime and potentially contribute to changes in disease risk later in life. To provide deeper insight into the interactions between the epigenome and the environment, several examples will be discussed in detail.

Epigenetic Effects on Environmental Responses

It is well established that changes in histone modifications are associated with changes in gene regulation. While establishing cause and effect relationships remains challenging, recent studies of steroid hormone responses show that preestablished epigenetic marks contribute to determining the gene expression responses to those hormones. One of the classic mechanisms by which some steroid hormones act involves binding to receptors inside of cells and causing those receptors to bind the genome and regulate gene expression [113−115]. The specific places that those receptors bind and the genes that they regulate differ between cell types. Several hypotheses about the determinants of such tissue-specific responses have been posited [116]. Among the possible explanations, recent studies have shown that epigenetic states associated with gene activation are strongly predictive of where in the genome hormone receptors bind in different types of cells [117−122]. Meanwhile, in the relatively small number of genomic regions where hormone receptors bind despite an absence of those epigenetic marks, histone acetylation is typically gained after hormone treatment [123−126]. Together, these results demonstrate the likely possibility that the epigenome both controls and is controlled by the binding of regulatory proteins that govern responses to the environment.

Epigenetic Changes Over a Lifetime

Studies of human populations and mouse models clearly show that maternal environment during gestation can have epigenetic impacts on the developing fetus. One such example is the Dutch famine during the winter of 1944−5, also known as the Dutch hunger winter. German blockades of Dutch ports combined with frozen canals due to extremely low temperatures led to a severe shortage of food in German-held areas of the Netherlands. Millions of Dutch were exposed to severe famine lasting from November of 1944 until liberation in May of 1944.

Studies of the effects that the famine on survivors revealed striking health consequences on babies who were in utero at the time. For example, those babies had substantially elevated risk for metabolic and cardiovascular disease later in life [127,128]. Because the famine affected people of all ages and social classes, attribution of the effects can be made to the famine independent of socioeconomic explanations. The Dutch hunger winter studies therefore provide some of the strongest evidence that the maternal environment during pregnancy can have substantial and long-lasting effects on the health of the offspring. Several studies have expanded upon those initial findings, and it is now known that a wide variety of exposures both during gestation and immediately after birth can have life-long impacts on the child's long-term health [129−131].

A key next challenge, then, is to determine the mechanisms by which fetal environment impacts life-long health. Strong evidence supports the hypothesis that the epigenome is a major contributor to those effects. Follow-up studies of the Dutch hunger winter found changes in DNA methylation at both imprinted and nonimprinted loci 60 years after the famine [132,133]. A similar effect has been demonstrated in mice, in which maternal grooming during the first weeks after birth contributes to epigenetic reprogramming of the stress response pathways and worse health later in life. The effects of maternal grooming can be partially reversed by histone deacetylase inhibitors, a finding that strengthens the argument that the fetal environment acting on the epigenome contributes to the long-term effects [134].

Exposure to environmental agents later in life can also lead to epigenetic changes that may contribute to health [135−138]. Because the epigenome is responsive to a wide variety of environmental exposures, it is unsurprising that studies have found substantial age-related epigenetic changes [139−141]. Taken together, it is now clear that epigenetic reprogramming occurs in early development, continues throughout life, and is likely involved in a wide variety of traits and diseases.

HERITABILITY OF THE EPIGENOME

The epigenome can be transmitted both via genetic and nongenetic mechanisms. Genetic transmission can trivially involve a mutation that disrupts a CpG dinucleotide, thus preventing methylation of that cytosine. Additional studies have shown that non-CpG mutations can also alter methylation in a region [142]. In those cases, the change in DNA sequence may alter the recruitment of regulatory factors that, in turn, may contribute to altered epigenetic state. Finally, it is important to note that cytosine methylation increases the rate of cytosine to thymidine mutations and, via that mechanism, the epigenome can influence the genome sequence over long timescales [143].

Examples of epigenetic inheritance that do *not* have a genetic basis are more elusive. One clear example is imprinting, where the epigenetic state is transmitted to child in a parent-dependent manner [144]. There are also growing examples in which environmental exposures in one generation lead to changes in disease risk in subsequent generations [145]. One example already discussed is the Dutch hunger winter study, in which the second generation was also shown to have increased adverse health outcomes [146]. Several studies have elaborated on that finding to show that epigenetic state established via other environmental signals can be passed on to offspring both via the maternal and the paternal lineage [144,147−150]. An intuitive potential mechanism is that incomplete erasure of DNA methylation during early development allows some methylation to be directly passed to the off-spring [15,151]. It is also possible that parental DNA methylation state is transmitted via more complex pathways involving protein intermediates. With rapidly increasing ability to measure genetic epigenetic changes genome-wide with few cells, the required studies to dissect those mechanisms may soon be possible.

GENOME-WIDE OBSERVATION OF THE EPIGENOME

The ability to measure the epigenome has been revolutionized over the past decade. Whereas it has long been possible to measure the epigenome at specific targeted locations in the human genome, technological advances in DNA sequencing [152,153] have made it possible to measure the epigenome at every location in the genome in a single experiment [154,155]. Examples of two of the most commonly used strategies and the data that they produce are shown in Fig. 2.2. That increase in measurement ability has been a major driver for new understanding into the basic principles governing the epigenome. The result is that we now have highly detailed maps of the epigenetic differences between individuals [156−159]; between different cells and tissues in the human body [160,161]; of the epigenetic responses to the environment [131]; and of epigenetic alterations that may be contributing to a variety of diseases [162−164]. The ability to harness such genome-wide approaches to study the epigenome is quickly becoming an essential skill for epigenetic and epigenomic research.

Integrative Epigenomics and Other Future Areas of Study

The ability to comprehensively measure the state of the epigenome has led to a new era of discovery. Major efforts such as the Roadmap Epigenome Project [159] and the ENCODE Project [165] have coordinated multinational teams of researchers to map a wide diversity of epigenetic marks across multiple cell types. As a result of those and related efforts, vast public repositories of epigenetic information are now freely available. Researchers have begun to rely on those data in various ways such as predicting how specific genes are regulated and investigating how gene regulation contributes to disease [166,167]. Initial studies to integrate across different epigenetic assays have confirmed that certain combinations of epigenetic marks are often acquired at the same genomic locations, and that many of those marks are associated with the regulation of nearby genes [168]. Those projects have revealed numerous new associations between the genome, the epigenome, gene regulation, and human health. Ongoing studies are actively investigating possible mechanisms explaining those newly identified associations. Meanwhile, as sequencing of whole human genomes is becoming routine, an increasing number of rare diseases may be mapped to mutations in epigenome-modifying enzymes that are currently not implicated in disease. Combined with expanding application of high-throughput epigenomic assays to various health and basic research questions, there is likely to soon be major advances in understanding how the epigenome contributes to health and diseases.

REFERENCES

[1] Kriaucionis S, Heintz N. The nuclear DNA base 5-hydroxymethylcytosine is present in Purkinje neurons and the brain. Science 2009;324(5929):929−30.

[2] Miller OJ, Schnedl W, Allen J, Erlanger BF. 5-Methylcytosine localised in mammalian constitutive heterochromatin. Nature 1974;251(5476):636−7.

[3] Ball MP, Li JB, Gao Y, Lee JH, LeProust EM, Park IH, et al. Targeted and genome-scale strategies reveal gene-body methylation signatures in human cells. Nat Biotechnol 2009;27(4):361−8.

[4] Bestor TH. The DNA methyltransferases of mammals. Hum Mol Genet 2000;9(16):2395−402.

[5] Adams RL, McKay EL, Craig LM, Burdon RH. Mouse DNA methylase: methylation of native DNA. Biochim Biophys Acta 1979;561(2):345−57.

[6] Okano M, Bell DW, Haber DA, Li E. DNA methyltransferases Dnmt3a and Dnmt3b are essential for de novo methylation and mammalian development. Cell 1999;99(3):247−57.

[7] Glickman JF, Flynn J, Reich NO. Purification and characterization of recombinant baculovirus-expressed mouse DNA methyltransferase. Biochem Biophys Res Commun 1997;230(2):280−4.

[8] Song J, Rechkoblit O, Bestor TH, Patel DJ. Structure of DNMT1-DNA complex reveals a role for autoinhibition in maintenance DNA methylation. Science 2011;331 (6020):1036−40.

[9] Yoder JA, Soman NS, Verdine GL, Bestor TH. DNA (cytosine-5)-methyltransferases in mouse cells and tissues. Studies with a mechanism-based probe. J Mol Biol 1997;270(3):385−95.

[10] Goll MG, Kirpekar F, Maggert KA, Yoder JA, Hsieh CL, Zhang X, et al. Methylation of tRNAAsp by the DNA methyltransferase homolog Dnmt2. Science 2006;311 (5759):395−8.

[11] Durdevic Z, Mobin MB, Hanna K, Lyko F, Schaefer M. The RNA methyltransferase Dnmt2 is required for efficient Dicer-2-dependent siRNA pathway activity in *Drosophila*. Cell Rep 2013;4(5):931−7.

[12] Chen ZX, Riggs AD. DNA methylation and demethylation in mammals. J Biol Chem 2011;286(21):18347−53.

[13] Franchini DM, Schmitz KM, Petersen-Mahrt SK. 5-Methylcytosine DNA demethylation: more than losing a methyl group. Annu Rev Genet 2012;46:419−41.

[14] Mayer W, Niveleau A, Walter J, Fundele R, Haaf T. Demethylation of the zygotic paternal genome. Nature 2000;403(6769):501−2.

[15] Morgan HD, Santos F, Green K, Dean W, Reik W. Epigenetic reprogramming in mammals. Hum Mol Genet 2005;R47−58 14 Spec No 1.

[16] Zhu JK. Active DNA demethylation mediated by DNA glycosylases. Annu Rev Genet 2009;43:143−66.

[17] He YF, Li BZ, Li Z, Liu P, Wang Y, Tang Q, et al. Tet-mediated formation of 5-carboxylcytosine and its excision by TDG in mammalian DNA. Science 2011;333 (6047):1303−7.

[18] Ito S, Shen L, Dai Q, Wu SC, Collins LB, Swenberg JA, et al. Tet proteins can convert 5-methylcytosine to 5-formylcytosine and 5-carboxylcytosine. Science 2011;333 (6047):1300−3.

[19] Barr ML, Bertram EG. A morphological distinction between neurones of the male and female, and the behaviour of the nucleolar satellite during accelerated nucleoprotein synthesis. Nature 1949;163(4148):676.

[20] Lyon MF. Gene action in the X-chromosome of the mouse (*Mus musculus* L.). Nature 1961;190:372−3.

[21] Brown CJ, Ballabio A, Rupert JL, Lafreniere RG, Grompe M, Tonlorenzi R, et al. A gene from the region of the human X inactivation centre is expressed exclusively from the inactive X chromosome. Nature 1991;349(6304):38−44.

[22] Clemson CM, McNeil JA, Willard HF, Lawrence JB. XIST RNA paints the inactive X chromosome at interphase: evidence for a novel RNA involved in nuclear/chromosome structure. J Cell Biol 1996;132(3):259−75.

[23] Penny GD, Kay GF, Sheardown SA, Rastan S, Brockdorff N. Requirement for Xist in X chromosome inactivation. Nature 1996;379(6561):131−7.

[24] Amos-Landgraf JM, Cottle A, Plenge RM, Friez M, Schwartz CE, Longshore J, et al. X chromosome-inactivation patterns of 1,005 phenotypically unaffected females. Am J Hum Genet 2006;79(3):493−9.

[25] Augui S, Nora EP, Heard E. Regulation of X-chromosome inactivation by the X-inactivation centre. Nat Rev Genet 2011;12(6):429−42.

[26] Ferguson-Smith AC. Genomic imprinting: the emergence of an epigenetic paradigm. Nat Rev Genet 2011;12(8):565−75.

[27] Bartolomei MS, Zemel S, Tilghman SM. Parental imprinting of the mouse H19 gene. Nature 1991;351(6322):153−5.

[28] Forne T, Oswald J, Dean W, Saam JR, Bailleul B, Dandolo L, et al. Loss of the maternal H19 gene induces changes in Igf2 methylation in both cis and trans. Proc Natl Acad Sci USA 1997;94(19):10243−8.

[29] Davis TL, Yang GJ, McCarrey JR, Bartolomei MS. The H19 methylation imprint is erased and re-established differentially on the parental alleles during male germ cell development. Hum Mol Genet 2000;9(19):2885−94.

[30] Morison IM, Ramsay JP, Spencer HG. A census of mammalian imprinting. Trends Genet 2005;21(8):457−65.

[31] Pollard KS, Serre D, Wang X, Tao H, Grundberg E, Hudson TJ, et al. A genome-wide approach to identifying novel-imprinted genes. Hum Genet 2008;122(6):625−34.

[32] Wang X, Soloway PD, Clark AG. A survey for novel imprinted genes in the mouse placenta by mRNA-seq. Genetics 2011;189(1):109−22.

[33] DeVeale B, van der Kooy D, Babak T. Critical evaluation of imprinted gene expression by RNA-Seq: a new perspective. PLoS Genet 2012;8(3):e1002600.

[34] Tahiliani M, Koh KP, Shen Y, Pastor WA, Bandukwala H, Brudno Y, et al. Conversion of 5-methylcytosine to 5-hydroxymethylcytosine in mammalian DNA by MLL partner TET1. Science 2009;324(5929):930−5.

[35] Wen L, Tang F. Genomic distribution and possible functions of DNA hydroxymethylation in the brain. Genomics 2014;104(5):341−6.

[36] Jin SG, Kadam S, Pfeifer GP. Examination of the specificity of DNA methylation profiling techniques towards 5-methylcytosine and 5-hydroxymethylcytosine. Nucleic Acids Res 2010;38(11):e125.

[37] Kouzarides T. Chromatin modifications and their function. Cell 2007;128(4):693−705.

[38] Bannister AJ, Kouzarides T. Regulation of chromatin by histone modifications. Cell Res 2011;21(3):381−95.

[39] Grewal SI, Moazed D. Heterochromatin and epigenetic control of gene expression. Science 2003;301(5634):798−802.

[40] Straub T. Heterochromatin dynamics. PLoS Biol 2003;1(1):E14.

[41] Gaspar-Maia A, Alajem A, Meshorer E, Ramalho-Santos M. Open chromatin in pluripotency and reprogramming. Nat Rev Mol Cell Biol 2011;12(1):36−47.

[42] Andrews AJ, Luger K. Nucleosome structure(s) and stability: variations on a theme. Annu Rev Biophys 2011;40:99−117.

[43] Strahl BD, Allis CD. The language of covalent histone modifications. Nature 2000;403 (6765):41−5.

[44] Berger SL. Histone modifications in transcriptional regulation. Curr Opin Genet Dev 2002;12(2):142−8.

[45] Trojer P, Reinberg D. Facultative heterochromatin: is there a distinctive molecular signature? Mol Cell 2007;28(1):1−13.

[46] Hansen KH, Bracken AP, Pasini D, Dietrich N, Gehani SS, Monrad A, et al. A model for transmission of the H3K27me3 epigenetic mark. Nat Cell Biol 2008;10(11):1291−300.

[47] Liang G, Lin JC, Wei V, Yoo C, Cheng JC, Nguyen CT, et al. Distinct localization of histone H3 acetylation and H3-K4 methylation to the transcription start sites in the human genome. Proc Natl Acad Sci USA 2004;101(19):7357−62.

[48] Bernstein BE, Kamal M, Lindblad-Toh K, Bekiranov S, Bailey DK, Huebert DJ, et al. Genomic maps and comparative analysis of histone modifications in human and mouse. Cell 2005;120(2):169−81.

[49] Hon GC, Hawkins RD, Ren B. Predictive chromatin signatures in the mammalian genome. Hum Mol Genet 2009;18(R2):R195−201.

[50] Creyghton MP, Cheng AW, Welstead GG, Kooistra T, Carey BW, Steine EJ, et al. Histone H3K27ac separates active from poised enhancers and predicts developmental state. Proc Natl Acad Sci USA 2010;107(50):21931−6.

[51] Rada-Iglesias A, Bajpai R, Swigut T, Brugmann SA, Flynn RA, Wysocka J. A unique chromatin signature uncovers early developmental enhancers in humans. Nature 2011;470 (7333):279−83.

[52] Azuara V, Perry P, Sauer S, Spivakov M, Jorgensen HF, John RM, et al. Chromatin signatures of pluripotent cell lines. Nat Cell Biol 2006;8(5):532−8.

[53] Bernstein BE, Mikkelsen TS, Xie X, Kamal M, Huebert DJ, Cuff J, et al. A bivalent chromatin structure marks key developmental genes in embryonic stem cells. Cell 2006;125 (2):315−26.

[54] Mikkelsen TS, Ku M, Jaffe DB, Issac B, Lieberman E, Giannoukos G, et al. Genome-wide maps of chromatin state in pluripotent and lineage-committed cells. Nature 2007;448 (7153):553−60.

[55] Mohn F, Weber M, Rebhan M, Roloff TC, Richter J, Stadler MB, et al. Lineage-specific polycomb targets and de novo DNA methylation define restriction and potential of neuronal progenitors. Mol Cell 2008;30(6):755−66.

[56] Vo N, Goodman RH. CREB-binding protein and p300 in transcriptional regulation. J Biol Chem 2001;276(17):13505−8.

[57] Schwartz YB, Pirrotta V. A new world of polycombs: unexpected partnerships and emerging functions. Nat Rev Genet 2013;14(12):853−64.

[58] Witzgall R, O'Leary E, Leaf A, Onaldi D, Bonventre JV. The Kruppel-associated box-A (KRAB-A) domain of zinc finger proteins mediates transcriptional repression. Proc Natl Acad Sci USA 1994;91(10):4514−18.

[59] Mendenhall EM, Williamson KE, Reyon D, Zou JY, Ram O, Joung JK, et al. Locus-specific editing of histone modifications at endogenous enhancers. Nat Biotechnol 2013;31(12):1133−6.

[60] Reik W, Dean W, Walter J. Epigenetic reprogramming in mammalian development. Science 2001;293(5532):1089−93.

[61] Agger K, Cloos PA, Christensen J, Pasini D, Rose S, Rappsilber J, et al. UTX and JMJD3 are histone H3K27 demethylases involved in HOX gene regulation and development. Nature 2007;449(7163):731−4.

[62] Cantone I, Fisher AG. Epigenetic programming and reprogramming during development. Nat Struct Mol Biol 2013;20(3):282−9.

[63] Hendrich B, Bickmore W. Human diseases with underlying defects in chromatin structure and modification. Hum Mol Genet 2001;10(20):2233−42.

[64] Portela A, Esteller M. Epigenetic modifications and human disease. Nat Biotechnol 2010;28(10):1057−68.

[65] Brookes E, Shi Y. Diverse epigenetic mechanisms of human disease. Annu Rev Genet 2014;48:237−68.

[66] Laird PW, Jaenisch R. The role of DNA methylation in cancer genetic and epigenetics. Annu Rev Genet 1996;30:441−64.

[67] Baylin SB, Herman JG. DNA hypermethylation in tumorigenesis: epigenetics joins genetics. Trends Genet 2000;16(4):168−74.

[68] Jones PA, Baylin SB. The epigenomics of cancer. Cell 2007;128(4):683−92.

[69] Baylin SB, Jones PA. A decade of exploring the cancer epigenome-biological and translational implications. Nat Rev Cancer 2011;11(10):726−34.

[70] Palmisano WA, Divine KK, Saccomanno G, Gilliland FD, Baylin SB, Herman JG, et al. Predicting lung cancer by detecting aberrant promoter methylation in sputum. Cancer Res 2000;60(21):5954−8.

[71] Anjum S, Fourkala EO, Zikan M, Wong A, Gentry-Maharaj A, Jones A, et al. A BRCA1-mutation associated DNA methylation signature in blood cells predicts sporadic breast cancer incidence and survival. Genome Med 2014;6(6):47.

[72] Wei SH, Balch C, Paik HH, Kim YS, Baldwin RL, Liyanarachchi S, et al. Prognostic DNA methylation biomarkers in ovarian cancer. Clin Cancer Res 2006;12(9):2788−94.

[73] Brock MV, Hooker CM, Ota-Machida E, Han Y, Guo M, Ames S, et al. DNA methylation markers and early recurrence in stage I lung cancer. N Engl J Med 2008;358 (11):1118−28.

[74] Lange CP, Campan M, Hinoue T, Schmitz RF, van der Meulen-de Jong AE, Slingerland H, et al. Genome-scale discovery of DNA-methylation biomarkers for blood-based detection of colorectal cancer. PLoS One 2012;7(11):e50266.

[75] Oh T, Kim N, Moon Y, Kim MS, Hoehn BD, Park CH, et al. Genome-wide identification and validation of a novel methylation biomarker, SDC2, for blood-based detection of colorectal cancer. J Mol Diagn 2013;15(4):498−507.

[76] Lasseigne BN, Burwell TC, Patil MA, Absher DM, Brooks JD, Myers RM. DNA methylation profiling reveals novel diagnostic biomarkers in renal cell carcinoma. BMC Med 2014;12(1):235.

[77] Christman JK. 5-Azacytidine and 5-aza-2'-deoxycytidine as inhibitors of DNA methylation: mechanistic studies and their implications for cancer therapy. Oncogene 2002;21(35):5483—95.

[78] Plumb JA, Strathdee G, Sludden J, Kaye SB, Brown R. Reversal of drug resistance in human tumor xenografts by 2'-deoxy-5-azacytidine-induced demethylation of the hMLH1 gene promoter. Cancer Res 2000;60(21):6039—44.

[79] Daskalakis M, Nguyen TT, Nguyen C, Guldberg P, Kohler G, Wijermans P, et al. Demethylation of a hypermethylated P15/INK4B gene in patients with myelodysplastic syndrome by 5-Aza-2'-deoxycytidine (decitabine) treatment. Blood 2002;100(8):2957—64.

[80] Ginder GD, Whitters MJ, Pohlman JK. Activation of a chicken embryonic globin gene in adult erythroid cells by 5-azacytidine and sodium butyrate. Proc Natl Acad Sci USA 1984;81(13):3954—8.

[81] Chiu CP, Blau HM. 5-Azacytidine permits gene activation in a previously noninducible cell type. Cell 1985;40(2):417—24.

[82] Jin B, Tao Q, Peng J, Soo HM, Wu W, Ying J, et al. DNA methyltransferase 3B (DNMT3B) mutations in ICF syndrome lead to altered epigenetic modifications and aberrant expression of genes regulating development, neurogenesis and immune function. Hum Mol Genet 2008;17(5):690—709.

[83] Klein CJ, Botuyan MV, Wu Y, Ward CJ, Nicholson GA, Hammans S, et al. Mutations in DNMT1 cause hereditary sensory neuropathy with dementia and hearing loss. Nat Genet 2011;43(6):595—600.

[84] Klein CJ, Bird T, Ertekin-Taner N, Lincoln S, Hjorth R, Wu Y, et al. DNMT1 mutation hot spot causes varied phenotypes of HSAN1 with dementia and hearing loss. Neurology 2013;80(9):824—8.

[85] Tatton-Brown K, Seal S, Ruark E, Harmer J, Ramsay E, Del Vecchio Duarte S, et al. Mutations in the DNA methyltransferase gene DNMT3A cause an overgrowth syndrome with intellectual disability. Nat Genet 2014;46(4):385—8.

[86] Dunn HG, MacLeod PM. Rett syndrome: review of biological abnormalities. Can J Neurol Sci 2001;28(1):16—29.

[87] Amir RE, Van den Veyver IB, Wan M, Tran CQ, Francke U, Zoghbi HY. Rett syndrome is caused by mutations in X-linked MECP2, encoding methyl-CpG-binding protein 2. Nat Genet 1999;23(2):185—8.

[88] Wiedemann HR. [The EMG-syndrome: exomphalos, macroglossia, gigantism and disturbed carbohydrate metabolism]. Zeitschrift fur Kinderheilkunde 1969;106(3):171—85.

[89] Weksberg R, Shuman C, Beckwith JB. Beckwith—Wiedemann syndrome. Eur J Hum Genet 2010;18(1):8—14.

[90] Reik W, Brown KW, Slatter RE, Sartori P, Elliott M, Maher ER. Allelic methylation of H19 and IGF2 in the Beckwith—Wiedemann syndrome. Hum Mol Genet 1994;3(8):1297—301.

[91] Brown KW, Villar AJ, Bickmore W, Clayton-Smith J, Catchpoole D, Maher ER, et al. Imprinting mutation in the Beckwith—Wiedemann syndrome leads to biallelic IGF2 expression through an H19-independent pathway. Hum Mol Genet 1996;5(12):2027—32.

[92] Ogawa O, Eccles MR, Szeto J, McNoe LA, Yun K, Maw MA, et al. Relaxation of insulin-like growth factor II gene imprinting implicated in Wilms' tumour. Nature 1993;362(6422):749—51.

[93] Weksberg R, Shen DR, Fei YL, Song QL, Squire J. Disruption of insulin-like growth factor 2 imprinting in Beckwith—Wiedemann syndrome. Nat Genet 1993;5(2):143—50.

[94] Lee MP, DeBaun MR, Mitsuya K, Galonek HL, Brandenburg S, Oshimura M, et al. Loss of imprinting of a paternally expressed transcript, with antisense orientation to KVLQT1, occurs frequently in Beckwith—Wiedemann syndrome and is independent of insulin-like growth factor II imprinting. Proc Natl Acad Sci USA 1999;96(9):5203—8.

[95]   Horsthemke B, Wagstaff J. Mechanisms of imprinting of the Prader−Willi/Angelman region. Am J Med Genet A 2008;146a(16):2041−52.

[96]   Gibson WT, Hood RL, Zhan SH, Bulman DE, Fejes AP, Moore R, et al. Mutations in EZH2 cause Weaver syndrome. Am J Hum Genet 2012;90(1):110−18.

[97]   Williams SR, Aldred MA, Der Kaloustian VM, Halal F, Gowans G, McLeod DR, et al. Haploinsufficiency of HDAC4 causes brachydactyly mental retardation syndrome, with brachydactyly type E, developmental delays, and behavioral problems. Am J Hum Genet 2010;87(2):219−28.

[98]   Kerr B, Gedeon A, Mulley J, Turner G. Localization of non-specific X-linked mental retardation genes. Am J Med Genet 1992;43(1-2):392−401.

[99]   Walker FO. Huntington's disease. Lancet 2007;369(9557):218−28.

[100] Sadri-Vakili G, Cha JH. Mechanisms of disease: histone modifications in Huntington's disease. Nat Clin Pract Neurol 2006;2(6):330−8.

[101] McFarland KN, Das S, Sun TT, Leyfer D, Xia E, Sangrey GR, et al. Genome-wide histone acetylation is altered in a transgenic mouse model of Huntington's disease. PLoS One 2012;7(7):e41423.

[102] Nucifora Jr. FC, Sasaki M, Peters MF, Huang H, Cooper JK, Yamada M, et al. Interference by huntingtin and atrophin-1 with cbp-mediated transcription leading to cellular toxicity. Science 2001;291(5512):2423−8.

[103] Gusella JF, MacDonald ME. Molecular genetics: unmasking polyglutamine triggers in neurodegenerative disease. Nat Rev Neurosci 2000;1(2):109−15.

[104] Minucci S, Pelicci PG. Histone deacetylase inhibitors and the promise of epigenetic (and more) treatments for cancer. Nat Rev Cancer 2006;6(1):38−51.

[105] West AC, Johnstone RW. New and emerging HDAC inhibitors for cancer treatment. J Clin Invest 2014;124(1):30−9.

[106] Steffan JS, Bodai L, Pallos J, Poelman M, McCampbell A, Apostol BL, et al. Histone deacetylase inhibitors arrest polyglutamine-dependent neurodegeneration in *Drosophila*. Nature 2001;413(6857):739−43.

[107] Ferrante RJ, Kubilus JK, Lee J, Ryu H, Beesen A, Zucker B, et al. Histone deacetylase inhibition by sodium butyrate chemotherapy ameliorates the neurodegenerative phenotype in Huntington's disease mice. J Neurosci 2003;23(28):9418−27.

[108] Hockly E, Richon VM, Woodman B, Smith DL, Zhou X, Rosa E, et al. Suberoylanilide hydroxamic acid, a histone deacetylase inhibitor, ameliorates motor deficits in a mouse model of Huntington's disease. Proc Natl Acad Sci USA 2003;100(4):2041−6.

[109] Jia H, Morris CD, Williams RM, Loring JF, Thomas EA. HDAC inhibition imparts beneficial transgenerational effects in Huntington's disease mice via altered DNA and histone methylation. Proc Natl Acad Sci USA 2015;112(1):E56−64.

[110] Glaser KB, Staver MJ, Waring JF, Stender J, Ulrich RG, Davidsen SK. Gene expression profiling of multiple histone deacetylase (HDAC) inhibitors: defining a common gene set produced by HDAC inhibition in T24 and MDA carcinoma cell lines. Mol Cancer Ther 2003;2(2):151−63.

[111] Lopez-Atalaya JP, Ito S, Valor LM, Benito E, Barco A. Genomic targets, and histone acetylation and gene expression profiling of neural HDAC inhibition. Nucleic Acids Res 2013;41(17):8072−84.

[112] Jirtle RL, Skinner MK. Environmental epigenomics and disease susceptibility. Nat Rev Genet 2007;8(4):253−62.

[113] Evans RM. The steroid and thyroid hormone receptor superfamily. Science 1988;240 (4854):889−95.

[114] Tsai MJ, O'Malley BW. Molecular mechanisms of action of steroid/thyroid receptor superfamily members. Annu Rev Biochem 1994;63:451−86.

[115] Ribeiro RC, Kushner PJ, Baxter JD. The nuclear hormone receptor gene superfamily. Annu Rev Med 1995;46:443−53.

[116] Gross KL, Cidlowski JA. Tissue-specific glucocorticoid action: a family affair. Trends Endocrinol Metab 2008;19(9):331−9.

[117] John S, Sabo PJ, Johnson TA, Sung MH, Biddie SC, Lightman SL, et al. Interaction of the glucocorticoid receptor with the chromatin landscape. Mol Cell 2008;29(5):611−24.

[118] John S, Sabo PJ, Thurman RE, Sung MH, Biddie SC, Johnson TA, et al. Chromatin accessibility pre-determines glucocorticoid receptor binding patterns. Nat Genet 2011;43 (3):264−8.

[119] He HH, Meyer CA, Chen MW, Jordan VC, Brown M, Liu XS. Differential DNase I hypersensitivity reveals factor-dependent chromatin dynamics. Genome Res 2012;22(6):1015−25.

[120] Burd CJ, Archer TK. Chromatin architecture defines the glucocorticoid response. Mol Cell Endocrinol 2013;380(1−2):25−31.

[121] Gertz J, Savic D, Varley KE, Partridge EC, Safi A, Jain P, et al. Distinct properties of cell-type-specific and shared transcription factor binding sites. Mol Cell 2013;52(1):25−36.

[122] Magnani L, Lupien M. Chromatin and epigenetic determinants of estrogen receptor alpha (ESR1) signaling. Mol Cell Endocrinol 2014;382(1):633−41.

[123] Chen H, Lin RJ, Xie W, Wilpitz D, Evans RM. Regulation of hormone-induced histone hyperacetylation and gene activation via acetylation of an acetylase. Cell 1999;98(5):675−86.

[124] Ito K, Barnes PJ, Adcock IM. Glucocorticoid receptor recruitment of histone deacetylase 2 inhibits interleukin-1beta-induced histone H4 acetylation on lysines 8 and 12. Mol Cell Biol 2000;20(18):6891−903.

[125] Sharma D, Fondell JD. Ordered recruitment of histone acetyltransferases and the TRAP/ mediator complex to thyroid hormone-responsive promoters in vivo. Proc Natl Acad Sci USA 2002;99(12):7934−9.

[126] Gadaleta RM, Magnani L. Nuclear receptors and chromatin: an inducible couple. J Mol Endocrinol 2014;52(2):R137−49.

[127] Roseboom T, de Rooij S, Painter R. The Dutch famine and its long-term consequences for adult health. Early Hum Dev 2006;82(8):485−91.

[128] Schulz LC. The Dutch Hunger Winter and the developmental origins of health and disease. Proc Natl Acad Sci USA 2010;107(39):16757−8.

[129] Dolinoy DC, Weidman JR, Jirtle RL. Epigenetic gene regulation: linking early developmental environment to adult disease. Reprod Toxicol 2007;23(3):297−307.

[130] McGowan PO, Sasaki A, D'Alessio AC, Dymov S, Labonte B, Szyf M, et al. Epigenetic regulation of the glucocorticoid receptor in human brain associates with childhood abuse. Nat Neurosci 2009;12(3):342−8.

[131] Feil R, Fraga MF. Epigenetics and the environment: emerging patterns and implications. Nat Rev Genet 2011;13(2):97−109.

[132] Heijmans BT, Tobi EW, Stein AD, Putter H, Blauw GJ, Susser ES, et al. Persistent epigenetic differences associated with prenatal exposure to famine in humans. Proc Natl Acad Sci USA 2008;105(44):17046−9.

[133] Tobi EW, Lumey LH, Talens RP, Kremer D, Putter H, Stein AD, et al. DNA methylation differences after exposure to prenatal famine are common and timing- and sex-specific. Hum Mol Genet 2009;18(21):4046−53.

[134] Weaver IC, Cervoni N, Champagne FA, D'Alessio AC, Sharma S, Seckl JR, et al. Epigenetic programming by maternal behavior. Nat Neurosci 2004;7(8):847−54.

[135] Jaenisch R, Bird A. Epigenetic regulation of gene expression: how the genome integrates intrinsic and environmental signals. Nat Genet 2003;33(Suppl):245−54.

[136] Dolinoy DC, Huang D, Jirtle RL. Maternal nutrient supplementation counteracts bisphenol A-induced DNA hypomethylation in early development. Proc Natl Acad Sci USA 2007;104(32):13056−61.

[137] Reamon-Buettner SM, Mutschler V, Borlak J. The next innovation cycle in toxicogenomics: environmental epigenetics. Mutat Res 2008;659(1-2):158−65.

[138] Toledo-Rodriguez M, Lotfipour S, Leonard G, Perron M, Richer L, Veillette S, et al. Maternal smoking during pregnancy is associated with epigenetic modifications of the brain-derived neurotrophic factor-6 exon in adolescent offspring. Am J Med Genet B Neuropsychiatr Genet 2010;153b(7):1350−4.

[139] Fraga MF, Agrelo R, Esteller M. Cross-talk between aging and cancer: the epigenetic language. Ann NY Acad Sci 2007;1100:60−74.

[140] Fraga MF, Esteller M. Epigenetics and aging: the targets and the marks. Trends Genet 2007;23(8):413−18.

[141] Horvath S. DNA methylation age of human tissues and cell types. Genome Biol 2013;14 (10):R115.

[142] Gertz J, Varley KE, Reddy TE, Bowling KM, Pauli F, Parker SL, et al. Analysis of DNA methylation in a three-generation family reveals widespread genetic influence on epigenetic regulation. PLoS Genet 2011;7(8):e1002228.

[143] Holliday R, Grigg GW. DNA methylation and mutation. Mutat Res 1993;285(1):61−7.

[144] Morgan HD, Sutherland HG, Martin DI, Whitelaw E. Epigenetic inheritance at the agouti locus in the mouse. Nat Genet 1999;23(3):314−18.

[145] Skinner MK. Environmental epigenomics and disease susceptibility. EMBO Rep 2011;12(7):620−2.

[146] Painter RC, Osmond C, Gluckman P, Hanson M, Phillips DI, Roseboom TJ. Transgenerational effects of prenatal exposure to the Dutch famine on neonatal adiposity and health in later life. BJOG 2008;115(10):1243−9.

[147] Anway MD, Cupp AS, Uzumcu M, Skinner MK. Epigenetic transgenerational actions of endocrine disruptors and male fertility. Science 2005;308(5727):1466−9.

[148] Pembrey ME, Bygren LO, Kaati G, Edvinsson S, Northstone K, Sjostrom M, et al. Sex-specific, male-line transgenerational responses in humans. Eur J Hum Genet 2006;14 (2):159−66.

[149] Waterland RA, Dolinoy DC, Lin JR, Smith CA, Shi X, Tahiliani KG. Maternal methyl supplements increase offspring DNA methylation at Axin Fused. Genesis 2006;44(9):401−6.

[150] Dolinoy DC, Das R, Weidman JR, Jirtle RL. Metastable epialleles, imprinting, and the fetal origins of adult diseases. Pediatr Res 2007;61(5 Pt 2):30r−7r.

[151] Li E. Chromatin modification and epigenetic reprogramming in mammalian development. Nat Rev Genet 2002;3(9):662−73.

[152] Mardis ER. Next-generation DNA sequencing methods. Annu Rev Genomics Hum Genet 2008;9:387−402.

[153] Shendure J, Ji H. Next-generation DNA sequencing. Nat Biotechnol 2008;26(10):1135−45.

[154] Schones DE, Zhao K. Genome-wide approaches to studying chromatin modifications. Nat Rev Genet 2008;9(3):179−91.

[155] Lister R, Pelizzola M, Dowen RH, Hawkins RD, Hon G, Tonti-Filippini J, et al. Human DNA methylomes at base resolution show widespread epigenomic differences. Nature 2009;462(7271):315−22.

[156] Bell JT, Pai AA, Pickrell JK, Gaffney DJ, Pique-Regi R, Degner JF, et al. DNA methylation patterns associate with genetic and gene expression variation in HapMap cell lines. Genome Biol 2011;12(1):R10.

[157] Fraser HB, Lam LL, Neumann SM, Kobor MS. Population-specificity of human DNA methylation. Genome Biol 2012;13(2):R8.

[158] Banovich NE, Lan X, McVicker G, van de Geijn B, Degner JF, Blischak JD, et al. Methylation QTLs are associated with coordinated changes in transcription factor binding, histone modifications, and gene expression levels. PLoS Genet 2014;10(9): e1004663.

[159] Kundaje A, Meuleman W, Ernst J, Bilenky M, Yen A, Heravi-Moussavi A, et al. Integrative analysis of 111 reference human epigenomes. Nature 2015;518(7539):317−30.

[160] Heintzman ND, Hon GC, Hawkins RD, Kheradpour P, Stark A, Harp LF, et al. Histone modifications at human enhancers reflect global cell-type-specific gene expression. Nature 2009;459(7243):108−12.

[161] Leung D, Jung I, Rajagopal N, Schmitt A, Selvaraj S, Lee AY, et al. Integrative analysis of haplotype-resolved epigenomes across human tissues. Nature 2015;518(7539):350−4.

[162] Robertson KD. DNA methylation and human disease. Nat Rev Genet 2005;6(8):597−610.

[163] Rakyan VK, Down TA, Balding DJ, Beck S. Epigenome-wide association studies for common human diseases. Nat Rev Genet 2011;12(8):529−41.

[164] Michels KB, Binder AM, Dedeurwaerder S, Epstein CB, Greally JM, Gut I, et al. Recommendations for the design and analysis of epigenome-wide association studies. Nat Methods 2013;10(10):949−55.

[165] Consortium TEP. An integrated encyclopedia of DNA elements in the human genome. Nature 2012;489(7414):57−74.

[166] Karlic R, Chung HR, Lasserre J, Vlahovicek K, Vingron M. Histone modification levels are predictive for gene expression. Proc Natl Acad Sci USA 2010;107(7):2926−31.

[167] Farh KK, Marson A, Zhu J, Kleinewietfeld M, Housley WJ, Beik S, et al. Genetic and epigenetic fine mapping of causal autoimmune disease variants. Nature 2015;518(7539):337−43.

[168] Zhou VW, Goren A, Bernstein BE. Charting histone modifications and the functional organization of mammalian genomes. Nat Rev Genet 2011;12(1):7−18.

Chapter 3

The State of Whole-Genome Sequencing

Andrew Dervan and Jay Shendure
University of Washington, Seattle, WA, United States

Chapter Outline

INTRODUCTION

We are now over a decade past the completion of the Human Genome Project (HGP), in which a canonical "reference" genome for the human species was assembled [1], with a length of approximately 3 gigabases (Gb) across 22 autosomes and 2 sex chromosomes. The two haploid copies of the human genome present in each of us are highly related to this reference, as well as to all other human genomes in our species. Genetic variation occurs as: (1) single-nucleotide variants (SNVs) occurring every approximately 1000 bases (several million SNVs per genome); (2) small (a few base pairs) insertions or deletions (indels) (several hundred thousand per genome); and (3) larger structural variants, such as deletions, duplications, and inversions of contiguous segments of DNA. This inherited variation, i.e., "genotype," underlies a substantial component of our individual risk for both rare and common human diseases, i.e., "phenotype." Identifying the links between genotype and phenotype is the *raison d'etre* for human genetics.

Since its inception, the ascertainment of genotype—that is, genetic differences between individuals—has been the rate-limiting step of human genetics. However, this situation has changed with the emergence of a new generation of DNA sequencing technologies [1]. Whereas the cost of sequencing an individual human genome to a reasonable level of quality and completeness was around $100 million in 2001, it is now a few thousand dollars and continues to fall [2]. Affordable sequencing opens up tremendous opportunities for the field of human genetics in both research and clinical settings, while at the same time highlighting the difficulty of deciphering the biological impact of genetic variation.

In this chapter, we review the current state of whole-genome sequencing (WGS). We focus on the most cost-effective platforms, but also discuss emerging technologies that may displace or complement these. We also briefly summarize the practical challenges of generating and analyzing WGS data, and anticipate how this might evolve over time. Finally, we review how WGS illuminates new strategies for discovering the genes and variants that underlie both rare and common human diseases, and consider the current challenges facing the field of human genetics.

The First Human Genome

From its inception in the late 1970s through the early 2000s, DNA sequencing was primarily achieved by methods pioneered in the laboratories of Fred Sanger and Wally Gilbert, respectively termed Sanger sequencing and Maxam–Gilbert sequencing. Sanger sequencing proved to be more automatable and therefore more scalable and was the method by which essentially all of the sequence data for the HGP was generated. In the context of the HGP, high-throughput Sanger sequencing consisted of: (1) cloning genomic DNA fragments to a bacterial vector; (2) a primer extension reaction on individual clones, with stochastic chain termination by fluorescently labeled dideoxynucleotides, yielding a ladder of fragments wherein the identity of the dye correlates with the identity of each fragment's terminal base; (3) capillary gel electrophoresis to separate fragments with single base-pair resolution; (4) automated detection and recording of fluorescence at each length to yield a chromatogram. By the end of the HGP, nearly all aspects of this workflow could be automated and achieved reads of up to 1 kilobase (kb) at very high accuracies (> 99.99%) and low costs (around $1 per base pair read) [3].

Although there was intense competition between public (HGP) and private (Celera) groups during that period, the reference human genome that is broadly used today descends directly from the public (HGP) project [4,5]. But whose genome is "the" human genome, and how does its availability impact our approach to the sequencing of additional human genomes? The public reference human genome was generated by the HGP using genomic DNA from 20 anonymous individuals from Buffalo, NY, although it mostly derives from just one of these individuals [6]. A key point in this context is that the primary challenge for the HGP was the *assembly* of a human genome, which is a very different task from the ascertainment of genetic variation within or between individuals. The frequency of repetitive elements and low-complexity sequences (in addition to inevitable heterozygosity due to the diploid nature of human DNA samples) render the genome extraordinarily difficult to assemble from Sanger sequence reads alone. Consequently, an immense amount of information *in addition* to pure sequence was required (e.g., paired ends, physical maps, genetic maps, radiation hybrid maps) to create the high-quality human genome assembly that the HGP produced [7].

More Human Genomes?

The primary motivation for sequencing additional human genomes is the ascertainment of genetic variation within an individual or a population, termed resequencing. The critical performance parameters with respect to sequencing technology differ for genome assembly versus resequencing. With assembly, read length, accuracy, and long-range contiguity information are critical. However, given that any two human genomes are overwhelmingly more similar to one another than they are different, resequencing simply requires the mapping of sequence reads to the high-quality reference genome, which is readily possible with much shorter reads than are necessary for high-quality *de novo* genome assembly.

On the other hand, the overall *amount* of sequencing required may be greater for resequencing than for assembly. With "shotgun" sequencing of genomic DNA from a diploid human, one is essentially sampling short fragments derived from one or the other allele/haplotype. In order to ensure that both alleles are adequately sampled for accurate and comprehensive identification of heterozygous variants, redundant coverage of each position is required. Specifically, at least 15-fold, and preferably 30-fold or greater, coverage (i.e., redundancy) is necessary (e.g., sensitively detecting heterozygous variants in a 3 Gb genome may require 90 Gb of generated sequence for 30-fold coverage). In contrast, the HGP completed the human reference genome with less than 30 Gb of sequence data.

NEXT-GENERATION DNA SEQUENCING

Progress toward the development of alternatives to Sanger and Maxam−Gilbert sequencing has been continuous for several decades. However, a major milestone came around 2005, when several alternative approaches became cost-competitive to Sanger sequencing [8,9]. These platforms, as well as the next-generation DNA sequencing platforms most widely used today, follow a shared paradigm which we term "cyclic array sequencing" [10]. In brief, this paradigm involves interrogation of a dense, two-dimensional array of DNA "features" by iterative cycles of enzymatic manipulation and imaging-based data collection. Initial DNA library construction is accomplished by random fragmentation of genomic (source) DNA, followed by in vitro ligation of platform-specific double-stranded adaptor sequences to these DNA fragments to generate a shotgun library. Single-molecule fragments from the library are then hybridized to a suitable detection surface (e.g., a primer-coated flat array or micron-scale bead) and amplified/replicated in a spatially clustered fashion. Each target fragment sequence, which is now present as an "*in vitro* colony," is then interrogated base-by-base by a polymerase (and fluorescently labeled nucleotides) or ligase (and fluorescently labeled oligonucleotides). After each elongation step, the surface to which the DNA colonies are tethered is imaged, the labels cleaved, and the process repeated. The fluorescence pattern exhibited by a given colony over the course of the full experiment (captured in a series of images) is used to determine a contiguous sequence read. For further details, see Refs. [10,11].

Other reviews detail the specifics of each cyclic array sequencing platform [10,11], and that is beyond the scope of this chapter. However, it is important to note the key characteristic that underlies the cost-effectiveness of cyclic array sequencing relative to Sanger sequencing. With Sanger sequencing, each read derives from a single physical compartment, necessitating its own microliter scale reaction. With cyclic array sequencing, in contrast, a single microliter scale reaction volume is used to manipulate millions to billions of immobilized DNA colonies in parallel. Whereas Sanger sequencing produces its ladder in a single reaction, tens to hundreds of independent cycles of biochemical manipulation (and imaging) are required with cyclic array sequencing (proportional to the read length). However, this is entirely offset and outdone by massive parallelization and miniaturization. Related advantages are the parallelization of library construction (*in vivo* cloning and colony picking vs *in vitro* shotgun libraries) and data acquisition (one detector per Sanger read vs a single camera capturing composite images of millions to billions of DNA colonies at each cycle).

Limitations of cyclic array sequencing platforms include the introduction of errors during *in vitro* nucleic acid manipulations and the need for clonally amplified molecules because of limitations of single-molecule sensitivity for luminescence- or fluorescence-based detection. Also, read lengths are limited by the propensity for clonal strands to get out of phase with one another as sequencing progresses due to biochemistry inefficiencies at each cycle. Consequently, each base of sequence produced by cyclic array sequencing is markedly less expensive than with Sanger sequencing, but the read lengths are also generally shorter. However, as discussed above, our present task is not to assemble additional human genomes, and technologies yielding greater depth of coverage with shorter reads may be entirely appropriate for genome *resequencing*. These same performance characteristics are also highly appropriate for applications that effectively use sequencing instruments as a molecular counter, e.g., to quantify the transcriptome, to assess chromatin accessibility, or to investigate other biological phenomena [12].

WGS on the Illumina Platform

The most widely used cyclic array sequencing platform today is that of Illumina, Inc., which achieves DNA colony amplification by bridge PCR (in which fragments are amplified upon primers attached to a solid surface), sequencing-by-synthesis (polymerase extensions with reversibly labeled, reversibly terminating deoxynucleotides), and total internal reflection fluorescence imaging. There are several types of Illumina sequencing instruments currently available, including the Illumina MiSeq, HiSeq, NextSeq, and HiSeq X Ten. The HiSeq X Ten instruments offer the lowest cost per base, but their adoption has been limited to a few locations because of capital cost and a requirement that they only be used for WGS. The minimal "fleet" of HiSeq X Ten instruments includes 10 instruments (costing ~$10 million altogether) that collectively can produce over 18,000 genomes per year. Although the precise costs of consumables are not available, Illumina claims that the consumables cost less than $1000 to achieve approximately 30-fold coverage of a human genome in pairs of 150 bp reads in less than 3 days. Consistent with these claims, several vendors that were early adopters of the HiSeq X Ten systems have begun to offer 30-fold WGS for about $1500 per individual (which presumably covers consumables, labor, instrument depreciation, etc.). As of the end of 2014, at least 16 such fleets had been purchased [13]. Although it may be some time before these are fully operationalized, the collective capacity when they are will be a staggering 288,000 human genomes per year.

A direct consequence of the release of the HiSeq X Ten systems is that although many academic and commercial groups continue to access Illumina sequencing through NextSeq or HiSeq instruments for diverse applications, at the time of this writing, sequencing "in house" is no longer a cost-effective approach for human WGS. Rather, we anticipate that both medical resequencing of individual patients and large-scale sequencing of population-based cohorts will increasingly (and eventually, perhaps exclusively) be conducted at a handful of such centralized facilities.

WGS on the Complete Genomics and Ion Torrent Platforms

It is beyond the scope of this chapter to consider all next-generation DNA sequencing technologies in depth, but there are at least two platforms that are commercially available now and potentially cost-competitive with Illumina for WGS in the very near term. The first of these, Complete Genomics, is a cyclic array sequencing technology that relies on amplified "nanoballs" that self-assemble to an ordered array and are sequenced by iterative cycles of sequencing-by-ligation and fluorescence-based imaging [14]. From its inception, Complete Genomics was unique in offering WGS solely as a service, rather than selling instrumentation and consumables. Now owned by the Beijing Genomics Institute, Complete Genomics continues to offer WGS under a service model at prices on par with centers operating the Illumina HiSeq X Ten. A limitation is that because of the format of the raw sequence (short, discontinuous reads) and the service model, publically available bioinformatics tools are not as widely available for Complete Genomics sequence data as they are for Illumina sequence data, and the data are less well understood by the community. Nonetheless, the Complete Genomics platform continues to be used for WGS by various projects with success, such as in a recent WGS study of severe intellectual disability [15]. Furthermore, the underlying biases may be distinct for Complete Genomics versus Illumina sequencing, such that the data may be complementary [16–18].

Another competitor, Ion Torrent (acquired by Life Technologies, which is now part of Thermo Fisher Scientific), is similar to cyclic array platforms in that it amplifies DNA colonies on beads, which are then immobilized within wells on a patterned surface. However, Ion Torrent systems utilize a novel, nonoptical detection step, measuring the release of a hydrogen ion when a new base is incorporated during an elongation cycle. This is accomplished using an ultrasensitive pH meter on a semiconductor chip. Given this pH change can be measured simultaneously in every well where these elongation reactions occur, the lengthy process of capturing and storing light images is eliminated [19]. One caveat, however, is when there are homopolymeric regions, i.e., two or more of the same base, the pH detection becomes less precise as to exactly how many bases of the same type were added. That criticism notwithstanding, the rapidity of sequence generation this platform may enable Ion Torrent to improve quickly, in principle exploiting advances in semiconductor technology. Although it is presently significantly more expensive to conduct WGS on an Ion Torrent than with an Illumina X Ten, WGS is possible on the Ion Torrent [19], and the technology may become more competitive for WGS upon its further development.

The Pacific Biosciences Platform

A key limitation of the platforms described above is that the read lengths that they produce are relatively short, i.e., presently a few hundred bases at best. As discussed, this is generally sufficient for uniquely mapping reads to the reference human genome for ascertainment of genetic variation. However, there are portions of the human genome that are not accurately mapped to (segmental duplications, long stretches of nearly identical sequence), and forms of genetic variation that are not reliably detected by sequence reads of this length (modestly sized structural variants, rearrangements mediated by highly identical sequences, and variation within segmental duplications). As it is plausible that genetic variation in these regions is more likely to be associated with human disease than other regions, this missed genetic variation may have disproportionate consequences. In the end, improving sensitivity for routinely detecting genetic variation in challenging regions of the genome may simply require much longer read lengths. Although currently not cost-competitive for routine WGS, single-molecule real-time (SMRT) sequencing is the first "next-generation" platform to offer markedly longer read lengths than cyclic array platforms at costs that are at least much lower than Sanger sequencing. SMRT sequencing is based on sequencing-by-synthesis and real-time detection of the incorporation of fluorescent labels by directly monitoring DNA polymerase activity at the single-molecule level [20]. Nucleotide incorporation is detected through fluorescence resonance energy transfer (FRET) with zero-mode waveguides (ZMWs) [21]. ZMWs are essentially very small holes in a metal film on a silica surface that enable detection of FRET in the immediate vicinity of the hole, where a single polymerase and primed template are tethered. Because the incorporation of labeled nucleotides occurs at a much slower timescale than diffusion of free nucleotides in and out of the illuminated volume, incorporation events are readily detectable, albeit with a relatively high error rate (15−20% per base) [21]. Presently, a single sequencing run interrogates an array of 150,000 ZMWs in parallel, only about one-third of which are loaded with polymerases due to Poisson statistics. The read lengths are geometrically distributed, with half the data in reads more than 20 kb, and the longest reads at around 60 kb.

Presently, human WGS on the Pacific Biosciences platform is considerably more expensive than alternatives (i.e., >$50,000). However, there are several reasons to keep a close eye on this technology for large-scale WGS. First, the combination of long reads and minimal GC bias (i.e., GC-rich sequences being consistently under-sequenced with most other platforms) allows access to a substantially larger portion of human genetic variation [22], an intrinsic and important advantage over short read sequencing technologies. Second, the platform is steadily improving in terms of read lengths and costs—it is easy to forget it has only been a few years since WGS on the Illumina platform cost more than $50,000. Third, although the raw error rate seems unlikely to greatly improve, the fact that it appears random rather than systematic means that high accuracies are readily achievable with sufficient coverage.

Oxford Nanopore and Other Technologies

A distinct real-time approach, also yielding long reads, involves driving nucleic acids through a nanopore (a biological membrane protein such as MspA (*Mycobacterium smegmatis* porin A) or a synthetic pore) [23]. Fluctuations in

DNA conductance through the pore, affected by the nucleotides within the pore at any given moment, are used to infer the nucleotide sequence. Nanopore technologies are noteworthy because of the potential for long reads and of using low quantities of input nucleic acids that do not need manipulation. Many years in the making [24–27], both academic [28] and commercial [29] efforts to reduce this concept to practice have now come to fruition. The first commercial devices implementing nanopore sequencing are now available in a limited access program from Oxford Nanopore. However, many challenges remain, including an error rate that is even greater than that of the Pacific Biosciences system [29]. Provided that these challenges can be overcome, nanopores offer the exciting prospect of "bedside" genomics, i.e., portable devices that can perform WGS with data immediately available for analysis.

There are many other companies that continue to develop advanced next-generation DNA sequencing technologies. These include Genia (synthetic nanopore), Nabsys (hybridization-assisted nanopore), Qiagen (sequencing-by-synthesis), Stratos Genomics, GnuBIO, and many others (summarized at http://allseq.com/knowledgebank). As such, it is difficult to predict how we will be performing WGS beyond just a few years from now, as entirely new methods and platforms will likely become available.

EXOME SEQUENCING

The focus of this chapter is on WGS. However, it seems important to give at least some consideration to exome sequencing, which remains the predominant approach by which human genomic DNA samples have been analyzed in large-scale sequencing projects. Exome sequencing—the targeted capture and sequencing of the approximately 1% of the human genome that is protein-coding—was developed at a time when WGS of anything more than a handful of samples was prohibitively expensive for most groups [30]. Many or most of the sequencing-based discoveries of the past few years have been through exome sequencing, rather than WGS. However, the cost of WGS has declined at a pace faster than that of exome sequencing, simply because of the fixed per-sample costs (e.g., hybrid capture). Today, the cost ratio is around 3:1, i.e., exomes are available from service providers for about one-third of the cost of WGS (~$1500 vs ~$500). We anticipate that as the cost of WGS continues to fall and as this gap further narrows, the field will increasingly shift to WGS, both in the context of medical resequencing of individual patients as well as large-scale sequencing of population-based cohorts.

HAPLOTYPE-RESOLVED GENOME SEQUENCING

Haplotypes—constellations of genetic variants that co-occur along single chromosomes—are essential to the complete description and interpretation of an individual human genome. For example, haplotype information is necessary for knowing the *cis* (on the same haplotype) versus *trans* (on opposing haplotypes) relationship of potentially causal recessive variation (e.g., where two pathogenic mutations are detected but they may be present on the same allele, and thus not lead to the recessive condition). Haplotype information is also necessary for noninvasive inference of a fetal genome [31]. However, this aspect of genetic information, almost completely lost through the process of shotgun sequencing of short fragments, has been largely ignored in the course of "complete" WGS of individual genomes in the past few years. Although common variant haplotypes can be partly obtained by population-based inference, and complete haplotypes obtained by sequencing of pedigrees, a better solution is sequencing technologies that directly recover haplotype information from an individual sample. In recent years, we and others have developed technologies for direct molecular measurement of haplotypes, either to dense blocks spanning hundreds of kilobases [32−35] or sparsely across entire chromosomes [36,37]. Such methods are only beginning to become commercially available, typically requiring on the order of twofold more sequencing than shotgun WGS. Although this premium is limiting its uptake, we anticipate that the added cost will come down with time and that the recovery of haplotype information will soon become a routine component of WGS [78].

ANALYSIS AND INTERPRETATION OF WGS DATA

The bioinformatics challenges in handling WGS data are daunting but surmountable. The key steps of human WGS data analysis include quality control, mapping of reads to a reference human genome, post-mapping processing including the removal of PCR duplicates and recalibration of quality scores, and finally, the identification of genetic variants (SNVs, indels, and structural variants), with the last task carried out on either a single sample (single-sample calling) or on many samples simultaneously for added sensitivity (multi-sample calling).

Although a full consideration of WGS bioinformatics tools is beyond the scope of this chapter, we emphasize the following points:

1. The bioinformatics tools and data formats for WGS have matured considerably over the past few years [38]. Several tools are currently routinely used in academics, including the Burrows−Wheeler aligner software for short read mapping against the large reference genome [39] and the genome analysis toolkit to analyze high-throughput sequencing data for variant calling [40]. There are "best practices" guidelines for using these tools [41,42], as well as reviews of the broader set of available tools [43], and an increasingly large community of expert users. Nonetheless, reproducibility of variant calling, particularly when different sequencing platforms are used, currently remains imperfect [44−46].

2. There are also platforms such as the Galaxy toolbox, which enable these and many other tools to be used on the cloud, even by individuals with relatively limited bioinformatics skillsets.
3. It is likely that many processing-intensive aspects of WGS will increasingly be handled by the centralized facilities that are generating the data, both with respect to analysis and storage, alleviating many end users from these challenges. Thus, although a "data flood" of hundreds of thousands of human whole-genome sequences will be generated in the next few years, the challenge of properly analyzing them is surmountable with the standardization of workflows and data formats, the increasing ubiquity of cloud-based computing and storage, and the growing community of WGS data users.

The process of annotating genetic variation in an individual human genome includes gathering allele frequencies (as the vast majority of variation in any single human is *common*, overwhelmingly nonpathogenic variation), assessing conservation information across species (i.e., the less variation between species, theoretically, the more critical that sequence is to the homologous gene product), predicting the actual effect on a gene product via *in silico* computer models, or though model systems or organisms if such data are available, and investigating whether the variant has been previously annotated in the literature as being associated with a condition. Although these aspects can seem straightforward, the definitive interpretation of individual sequence variants found in individual genomes for their relevance to phenotype or disease risk remains an immense challenge. In the long run, we anticipate that large WGS databases of ethnically diverse, deeply characterized (phenotyped) individuals will be necessary for improving genome interpretation, and these will likely come from a combination of public consortia (like the Exome Aggregation Consortium, an aggregate data set from 61,000 individuals from a variety of large-scale sequencing projects) and private efforts (like the Regeneron Pharmaceuticals and Geisinger Health System collaboration, where these organizations plan to link more than 100,000 patient medical records to individual WGS data) to generate and/or aggregate data. However, even with such databases in hand, our ability to interpret genetic variants will ultimately depend on our underlying knowledge with respect to the gene's underlying contribution to disease risk. Furthermore, even when a gene is known to confer risk, the interpretation of newly observed variants is not always straightforward (e.g., missense variants of uncertain significance in *BRCA1*, an established and well-studied breast cancer risk gene).

WGS AND DISEASE GENE DISCOVERY

In this last section, we summarize how WGS (or whole-exome sequencing) has unlocked new strategies for discovering genes and variants underlying both rare and common human diseases. We anticipate that these paradigms will continue to advance our understanding of pathophysiology with the increasing availability and cost-effectiveness of WGS.

Rare Variants, Rare Diseases

The first several decades of medical genetics were focused on uncovering the genetic underpinnings of rare Mendelian disorders, wherein highly penetrant mutations, typically in a single gene, underlie each disorder. These genes were discovered usually by family-based linkage or candidate gene analysis. In the past few years, exome and genome sequencing has led to a renaissance in monogenic disease gene discovery by enabling gene discovery for disorders that proved recalcitrant the conventional strategies, e.g., *de novo* mutations [47,48]. The genetic basis for hundreds of Mendelian disorders has been identified by exome or genome sequencing in the past few years, and sequencing is increasingly used for the diagnosis of Mendelian disorders in a clinical setting. These diagnoses can lead to meaningful changes in clinical management, not to mention peace of mind and the opportunity for prospective guidance [49,50]. Moreover, given that 1% of new births are affected by a Mendelian disorder requiring specialized medical attention, the ability to cost-effectively sequence prospective parents' genomes may allow them to know their carrier status for hundreds of severe recessively inherited disorders prior to conception [51]. In addition, after conception, fetal cell−free DNA detected in the maternal blood can potentially be used, along with the parents' haplotype-resolved genomes, to construct the inherited fetal genome [52]. Such approaches may eventually allow for the noninvasive, prenatal diagnosis of thousands of Mendelian disorders.

Common Variant, Common Disease

Through genome-wide association studies (GWAS), more is being understood about the genetic contribution to, and the complex heritability of, common diseases [53]. Historically, these studies have been performed with genome-wide genotyping arrays, which capture a large fraction of common variation in a human population by virtue of linkage disequilibrium. However, it is likely that WGS will increasingly be used to comprehensively ascertain variation in case-control cohorts, for concurrently associating both rare and common variants with common diseases. It is likely that enormous cohorts will be necessary to drive new discovery [54], and WGS cannot, by itself, solve the challenge of identifying the precise variant(s) in the region of a genetic association that are causal for conferring the risk.

Rare Variants, Common Disease

While GWAS have greatly improved our understanding of the genetic basis of common disease, they have failed in their current form to pinpoint most presumed genetic risk (genetic variance) [55,56]. This "missing heritability" may correspond to poorly captured forms of common genetic variation, epistatic effects, scores of small effect common variants, or other sources [57−59]. That said, it is incontrovertible that rare variants contribute meaningfully to common disease and the study of rare variation is a powerful lens though which to understand the molecular basis of disease. For example, in 2003, two families with autosomal dominant elevated low-density lipoprotein (LDL) levels (and an increase in coronary heart disease) were found to have rare gain-of-function mutations in *PCSK9* [60]. Subsequent studies linked rare loss-of-function mutations with low LDL and reduced incidence of coronary heart disease [61]. This has led several major pharmaceutical companies to pursue antibodies against PCSK9 as a target to reduce heart disease [62,63]. While rare variants in the LDL receptor have been associated with increased cardiovascular events for decades [64], the elucidation of the impact of rare variants on heart disease continues at a brisk pace (e.g., *APOA5* and additional LDL receptor variants increase the risk for early heart attacks [65], and inactivating mutations in *NPC1L1* protect from heart disease [66]).

With exome and genome sequencing of large cohorts, the pace of this type of discovery is accelerating. Examples of additional high-penetrance rare variants include: *GPR120* and obesity [67], *SLC30A8* and type 2 diabetes [68], *PLS3* and osteoporosis [69], *PTEN* and insulin resistance [70], *TREM2* and Alzheimer's disease [71], and *ATG16L1* and Crohn's disease [72]. Similar success is found in the neuropsychiatric field (e.g., autism, intellectual disability) where an excess of rare, moderate to high penetrance *de novo* mutations has been described in affected individuals, suggesting that a substantial proportion of these neurodevelopmental disorders may be effectively monogenic [73,74]. In autism, disease is likely driven (from a genetic perspective) by a combination of both oligogenic familial risk alleles and *de novo* mutations [75,76].

THE FUTURE OF WGS

Looking both forward and backward, we see the following reasons for optimism with respect to WGS:

1. In a single decade, next-generation DNA sequencing technologies have risen from obscurity to ubiquity. It seems likely that well in excess of one million whole-genome sequences will be generated in the next 10 years as WGS becomes increasingly routine in both genetics research and clinical medicine.

2. Although platforms may come and go, there is an expanding community of producers and users of WGS data with a shared language and common tools. Future transitions with respect to the underlying sequencing technologies are likely to be less challenging than the inception of WGS has been for many groups.

3. As we move toward sequencing millions of genomes and implementing effective data sharing across research institutions and hospital networks, there will be unprecedented opportunities for new discovery. Ideally, this will be a virtuous cycle, where increasing discovery drives more widespread adoption of WGS in a clinical setting, which in turn further accelerates discovery.

The ideal technology for human WGS will produce a gapless, errorless, end-to-end, haplotype-resolved diploid genome sequence. It is important to recognize both that such capability does not exist yet and that further investment in sequencing technology development is necessary to reach this goal. As the ascertainment of genotype ceases to be the rate-limiting factor in both research and clinical genetics, what are the new bottlenecks? There are three current prominent limitations. First, the interpretation of genetic variants in genes with established disease risk remains an immense challenge, and one that is not necessarily solved by simply more sequencing. Second, new approaches to phenotyping, e.g., phenotyping more deeply or of specific systems based on knowledge of genotype, are needed. For example, the extreme genetic heterogeneity of neurodevelopmental disorders such as autism may yield genotype-dependent subtypes upon deep phenotyping [77]. Finally, as we move toward WGS of millions of individuals within diverse health networks, the ability to seamlessly share and integrate genotypic and phenotypic information will be critical for extracting maximal value from our collective efforts.

GLOSSARY TERMS, ACRONYMS, AND ABBREVIATIONS

Gigabase 1×10^9 bases

Kilobase 1×10^6 bases

Paired ends sequencing both ends of a fragment of DNA in opposite directions; this facilitates mapping of reads back to a reference genome, as well as the detection of various classes of structural variation.

Physical maps using various techniques (e.g., restriction enzyme digestion, fluorescent in situ hybridization, and sequence-tagged site and others) to characterize the content of DNA molecules and from that information construct maps or orderings of the relative positions in a genome from which the DNA molecules derive.

Genetic maps using genetic techniques, e.g., breeding or pedigree analysis, to construct maps or orderings of the relative positions of genes and other sequence features of a genome.

Radiation hybrid maps using high-dose X-rays to fragment DNA, and then the relative frequency with which pairs of known DNA markers are separated from one another by this process to infer their linkage and/or distance.

Shotgun sequencing the sequencing of a long DNA molecule by breaking it into short, random fragments, determining the sequence of each fragment, and then either mapping these to a reference (resequencing) or using overlaps to guide assembly.

Diploid a cell or an organism with paired chromosomes, i.e., one from each parent.

Allele one form of a gene (i.e., with one or a particular set of variations) at a given location along a chromosome.

Haplotype a set of variants, alleles, or polymorphisms that reside on the same chromosome and thus tend to be inherited together.

Missense variant a substitution variant that results in a nonsynonymous amino acid change. Missense variants can have a full range of effects on protein function, ranging from no effect to rendering it nonfunctional.

REFERENCES

[1] International Human Genome Sequencing Consortium, Finishing the euchromatic sequence of the human genome. Nature 2004;431(7011):931–45.

[2] Wetterstrand KA. DNA sequencing costs: data from the NHGRI Genome Sequencing Program (GSP). Available from: http:www.genome.gov/sequencingcostsdata, 2016 [accessed 16.10.16].

[3] Shendure J, Mitra RD, Varma C, Church GM. Advanced sequencing technologies: methods and goals. Nat Rev Genet 2004;5(5):335–44.

[4] Waterston RH, Lindblad-Toh K, Birney E, Rogers J, Abril JF, Agarwal P, et al. Initial sequencing and comparative analysis of the mouse genome. Nature 2002;420(6915):520–62.

[5] Church DM, Schneider VA, Graves T, Auger K, Cunningham F, Bouk N, et al. Modernizing reference genome assemblies. PLoS Biol 2011;9(7):e1001091.

[6] Osoegawa K, Mammoser AG, Wu C, Frengen E, Zeng C, Catanese JJ, et al. A bacterial artificial chromosome library for sequencing the complete human genome. Genome Res 2001;11(3):483–96.

[7] Waterston RH, Lander ES, Sulston JE. On the sequencing of the human genome. Proc Natl Acad Sci USA 2002;99(6):3712–16.

[8] Shendure J, Porreca GJ, Reppas NB, Lin X, McCutcheon JP, Rosenbaum AM, et al. Accurate multiplex polony sequencing of an evolved bacterial genome. Science 2005;309 (5741):1728–32.

[9] Margulies M, Egholm M, Altman WE, Attiya S, Bader JS, Bemben LA, et al. Genome sequencing in microfabricated high-density picolitre reactors. Nature 2005;437 (7057):376–80.

[10] Shendure J, Ji H. Next-generation DNA sequencing. Nat Biotechnol 2008;26 (10):1135–45.

[11] Mardis ER. Next-generation sequencing platforms. Annu Rev Anal Chem (Palo Alto, Calif) 2013;6:287–303.

[12] Shendure J, Lieberman Aiden E. The expanding scope of DNA sequencing. Nat Biotechnol 2012;30(11):1084–94.

[13] Allseq. Web. http://allseq.com/x-ten, 2014 [accessed 16.10.16].

[14] Drmanac R, Sparks AB, Callow MJ, Halpern AL, Burns NL, Kermani BG, et al. Human genome sequencing using unchained base reads on self-assembling DNA nanoarrays. Science 2010;327(5961):78–81.

[15] Gilissen C, Hehir-Kwa JY, Thung DT, van de Vorst M, van Bon BW, Willemsen MH, et al. Genome sequencing identifies major causes of severe intellectual disability. Nature 2014;511(7509):344−7.

[16] Wall JD, Tang LF, Zerbe B, Kvale MN, Kwok PY, Schaefer C, et al. Estimating genotype error rates from high-coverage next-generation sequence data. Genome Res 2014;24 (11):1734−9.

[17] Rieber N, Zapatka M, Lasitschka B, Jones D, Northcott P, Hutter B, et al. Coverage bias and sensitivity of variant calling for four whole-genome sequencing technologies. PLoS One 2013;8(6):e66621.

[18] Lam HY, Clark MJ, Chen R, Chen R, Natsoulis G, O'Huallachain M, et al. Performance comparison of whole-genome sequencing platforms. Nat Biotechnol 2012;30(1):78−82.

[19] Rothberg JM, Hinz W, Rearick TM, Schultz J, Mileski W, Davey M, et al. An integrated semiconductor device enabling non-optical genome sequencing. Nature 2011;475 (7356):348−52.

[20] Eid J, Fehr A, Gray J, Luong K, Lyle J, Otto G, et al. Real-time DNA sequencing from single polymerase molecules. Science 2009;323(5910):133−8.

[21] Levene MJ, Korlach J, Turner SW, Foquet M, Craighead HG, Webb WW. Zero-mode waveguides for single-molecule analysis at high concentrations. Science 2003;299 (5607):682−6.

[22] Chaisson MJ, Huddleston J, Dennis MY, Sudmant PH, Malig M, Hormozdiari F, et al. Resolving the complexity of the human genome using single-molecule sequencing. Nature 2015;517(7536):608−11.

[23] Deamer DW, Akeson M. Nanopores and nucleic acids: prospects for ultrarapid sequencing. Trends Biotechnol 2000;18(4):147−51.

[24] Soni GV, Meller A. Progress toward ultrafast DNA sequencing using solid-state nanopores. Clin Chem 2007;53(11):1996−2001.

[25] Cockroft SL, Chu J, Amorin M, Ghadiri MR. A single-molecule nanopore device detects DNA polymerase activity with single-nucleotide resolution. J Am Chem Soc 2008;130 (3):818−20.

[26] Meller A, Nivon L, Brandin E, Golovchenko J, Branton D. Rapid nanopore discrimination between single polynucleotide molecules. Proc Natl Acad Sci USA 2000;97(3):1079−84.

[27] Healy K, Schiedt B, Morrison AP. Solid-state nanopore technologies for nanopore-based DNA analysis. Nanomedicine 2007;2(6):875−97.

[28] Laszlo AH, Derrington IM, Ross BC, Brinkerhoff H, Adey A, Nova IC, et al. Decoding long nanopore sequencing reads of natural DNA. Nat Biotechnol 2014;32(8):829−33.

[29] Ashton PM, Nair S, Dallman T, Rubino S, Rabsch W, Mwaigwisya S, et al. MinION nanopore sequencing identifies the position and structure of a bacterial antibiotic resistance island. Nat Biotechnol 2015;33(3):296−300.

[30] Ng SB, Turner EH, Robertson PD, Flygare SD, Bigham AW, Lee C, et al. Targeted capture and massively parallel sequencing of 12 human exomes. Nature 2009;461 (7261):272−6.

[31] Snyder MW, Simmons LE, Kitzman JO, Santillan DA, Santillan MK, Gammill HS, et al. Noninvasive fetal genome sequencing: a primer. Prenat Diagn 2013;33(6):547−54.

[32] Kitzman JO, Mackenzie AP, Adey A, Hiatt JB, Patwardhan RP, Sudmant PH, et al. Haplotype-resolved genome sequencing of a Gujarati Indian individual. Nat Biotechnol 2011;29(1):59−63.

[33] Kuleshov V, Xie D, Chen R, Pushkarev D, Ma Z, Blauwkamp T, et al. Whole-genome haplotyping using long reads and statistical methods. Nat Biotechnol 2014;32(3):261−6.

[34] Peters BA, Kermani BG, Sparks AB, Alferov O, Hong P, Alexeev A, et al. Accurate whole-genome sequencing and haplotyping from 10 to 20 human cells. Nature 2012;487 (7406):190−5.

[35] Amini S, Pushkarev D, Christiansen L, Kostem E, Royce T, Turk C, et al. Haplotype-resolved whole-genome sequencing by contiguity-preserving transposition and combinatorial indexing. Nat Genet 2014;46(12):1343−9.

[36] Selvaraj S, R Dixon J, Bansal V, Ren B. Whole-genome haplotype reconstruction using proximity-ligation and shotgun sequencing. Nat Biotechnol 2013;31(12):1111−18.

[37] Fan HC, Wang J, Potanina A, Quake SR. Whole-genome molecular haplotyping of single cells. Nat Biotechnol 2011;29(1):51−7.

[38] Suzuki T, Tsurusaki Y, Nakashima M, Miyake N, Saitsu H, Takeda S, et al. Precise detection of chromosomal translocation or inversion breakpoints by whole-genome sequencing. J Hum Genet 2014;59(12):649−54.

[39] Li H, Durbin R. Fast and accurate short read alignment with Burrows−Wheeler transform. Bioinformatics 2009;25(14):1754−60.

[40] McKenna A, Hanna M, Banks E, Sivachenko A, Cibulskis K, Kernytsky A, et al. The Genome Analysis Toolkit: a MapReduce framework for analyzing next-generation DNA sequencing data. Genome Res 2010;20(9):1297−303.

[41] DePristo MA, Banks E, Poplin R, Garimella KV, Maguire JR, Hartl C, et al. A framework for variation discovery and genotyping using next-generation DNA sequencing data. Nat Genet 2011;43(5):491−8.

[42] Van der Auwera GA, Carneiro MO, Hartl C, Poplin R, Del Angel G, Levy-Moonshine A, et al. From FastQ data to high confidence variant calls: the Genome Analysis Toolkit best practices pipeline. Curr Protoc Bioinformatics 2013;11(1110), 11.10.1−11.10.33.

[43] Bao R, Huang L, Andrade J, Tan W, Kibbe WA, Jiang H, et al. Review of current methods, applications, and data management for the bioinformatics analysis of whole exome sequencing. Cancer Inform 2014;13(Suppl. 2):67−82.

[44] Brownstein CA, Beggs AH, Homer N, Merriman B, Yu TW, Flannery KC, et al. An international effort towards developing standards for best practices in analysis, interpretation and reporting of clinical genome sequencing results in the CLARITY Challenge. Genome Biol 2014;15(3):R53.

[45] Dewey FE, Grove ME, Pan C, Goldstein BA, Bernstein JA, Chaib H, et al. Clinical interpretation and implications of whole-genome sequencing. JAMA 2014;311 (10):1035−45.

[46] O'Rawe J, Jiang T, Sun G, Wu Y, Wang W, Hu J, et al. Low concordance of multiple variant-calling pipelines: practical implications for exome and genome sequencing. Genome Med 2013;5(3):28.

[47] Bamshad MJ, Ng SB, Bigham AW, Tabor HK, Emond MJ, Nickerson DA, et al. Exome sequencing as a tool for Mendelian disease gene discovery. Nat Rev Genet 2011;12 (11):745−55.

[48] Boycott KM, Vanstone MR, Bulman DE, MacKenzie AE. Rare-disease genetics in the era of next-generation sequencing: discovery to translation. Nat Rev Genet 2013;14 (10):681−91.

[49] Bainbridge MN, Wiszniewski W, Murdock DR, Friedman J, Gonzaga-Jauregui C, Newsham I, et al. Whole-genome sequencing for optimized patient management. Sci Transl Med 2011;3(87):87re3.

[50] Berg JS. Genome-scale sequencing in clinical care: establishing molecular diagnoses and measuring value. JAMA 2014;312(18):1865−7.

[51] Bell CJ, Dinwiddie DL, Miller NA, Hateley SL, Ganusova EE, Mudge J, et al. Carrier testing for severe childhood recessive diseases by next-generation sequencing. Sci Transl Med 2011;3(65):65ra4.

[52] Kitzman JO, Snyder MW, Ventura M, Lewis AP, Qiu R, Simmons LE, et al. Noninvasive whole-genome sequencing of a human fetus. Sci Transl Med 2012;4(137):137ra76.

[53] Johnson AD, O'Donnell CJ. An open access database of genome-wide association results. BMC Med Genet 2009;10:6.

[54] Schizophrenia Working Group of the Psychiatric Genomics Consortium, Biological insights from 108 schizophrenia-associated genetic loci. Nature 2014;511(7510):421–7.

[55] Maher B. Personal genomes: the case of the missing heritability. Nature 2008;456 (7218):18–21.

[56] Manolio TA, Collins FS, Cox NJ, Goldstein DB, Hindorff LA, Hunter DJ, et al. Finding the missing heritability of complex diseases. Nature 2009;461(7265):747–53.

[57] Cirulli ET, Goldstein DB. Uncovering the roles of rare variants in common disease through whole-genome sequencing. Nat Rev Genet 2010;11(6):415–25.

[58] Visscher PM, Hill WG, Wray NR. Heritability in the genomics era—concepts and misconceptions. Nat Rev Genet 2008;9(4):255–66.

[59] Eichler EE, Flint J, Gibson G, Kong A, Leal SM, Moore JH, et al. Missing heritability and strategies for finding the underlying causes of complex disease. Nat Rev Genet 2010;11(6):446–50.

[60] Abifadel M, Varret M, Rabes JP, Allard D, Ouguerram K, Devillers M, et al. Mutations in PCSK9 cause autosomal dominant hypercholesterolemia. Nat Genet 2003;34(2): 154–6.

[61] Cohen JC, Boerwinkle E, Mosley Jr. TH, Hobbs HH. Sequence variations in PCSK9, low LDL, and protection against coronary heart disease. N Engl J Med 2006;354 (12):1264–72.

[62] Stein EA, Mellis S, Yancopoulos GD, Stahl N, Logan D, Smith WB, et al. Effect of a monoclonal antibody to PCSK9 on LDL cholesterol. N Engl J Med 2012;366 (12):1108–18.

[63] Mullard A. Cholesterol-lowering blockbuster candidates speed into phase III trials. Nat Rev Drug Discov 2012;11(11):817–19.

[64] Hobbs HH, Brown MS, Goldstein JL. Molecular genetics of the LDL receptor gene in familial hypercholesterolemia. Hum Mutat 1992;1(6):445–66.

[65] Do R, Stitziel NO, Won H, Jorgensen AB, Duga S, Angelica Merlini P, et al. Exome sequencing identifies rare LDLR and APOA5 alleles conferring risk for myocardial infarction. Nature 2015;518(7537):102–6.

[66] Stitziel NO, Won HH, Morrison AC, Peloso GM, Do R, Lange LA, et al. Inactivating mutations in NPC1L1 and protection from coronary heart disease. N Engl J Med 2014;371(22):2072–82.

[67] Ichimura A, Hirasawa A, Poulain-Godefroy O, Bonnefond A, Hara T, Yengo L, et al. Dysfunction of lipid sensor GPR120 leads to obesity in both mouse and human. Nature 2012;483(7389):350–4.

[68] Flannick J, Thorleifsson G, Beer NL, Jacobs SB, Grarup N, Burtt NP, et al. Loss-of-function mutations in SLC30A8 protect against type 2 diabetes. Nat Genet 2014;46 (4):357–63.

[69] van Dijk FS, Zillikens MC, Micha D, Riessland M, Marcelis CL, de Die-Smulders CE, et al. PLS3 mutations in X-linked osteoporosis with fractures. N Engl J Med 2013;369 (16):1529–36.

[70] Pal A, Barber TM, Van de Bunt M, Rudge SA, Zhang Q, Lachlan KL, et al. PTEN mutations as a cause of constitutive insulin sensitivity and obesity. N Engl J Med 2012;367(11):1002−11.

[71] Neumann H, Daly MJ. Variant TREM2 as risk factor for Alzheimer's disease. N Engl J Med 2013;368(2):182−4.

[72] Murthy A, Li Y, Peng I, Reichelt M, Katakam AK, Noubade R, et al. A Crohn's disease variant in Atg16l1 enhances its degradation by caspase 3. Nature 2014;506 (7489):456−62.

[73] Sullivan PF, Daly MJ, O'Donovan M. Genetic architectures of psychiatric disorders: the emerging picture and its implications. Nat Rev Genet 2012;13(8):537−51.

[74] Krumm N, O'Roak BJ, Shendure J, Eichler EE. A de novo convergence of autism genetics and molecular neuroscience. Trends Neurosci 2014;37(2):95−105.

[75] Gaugler T, Klei L, Sanders SJ, Bodea CA, Goldberg AP, Lee AB, et al. Most genetic risk for autism resides with common variation. Nat Genet 2014;46(8):881−5.

[76] Robinson EB, Samocha KE, Kosmicki JA, McGrath L, Neale BM, Perlis RH, et al. Autism spectrum disorder severity reflects the average contribution of de novo and familial influences. Proc Natl Acad Sci USA 2014;111(42):15161−5.

[77] Bernier R, Golzio C, Xiong B, Stessman HA, Coe BP, Penn O, et al. Disruptive CHD8 mutations define a subtype of autism early in development. Cell 2014;158(2):263−76.

[78] Snyder MW, Adey A, Kitzman JO, Shendure J. Haplotype-resolved genome sequencing: experimental methods and applications. Nat Rev Genet 2015;16(6):344−58. Available from: http://dx.doi.org/10.1038/nrg3903. Epub 2015 May 7.

Chapter 4

The Human Microbiome

Jacquelyn S. Meisel and Elizabeth A. Grice
University of Pennsylvania, Philadelphia, PA, United States

Chapter Outline

INTRODUCTION

The human microbiome refers to the communities of microorganisms living in association with our bodies. These topographically diverse and temporally complex microbial populations are largely commensal, providing us with genetic variation and gene functions that human cells have not had to evolve on their own. Development of culture-independent isolation techniques and next-generation DNA sequencing technologies has enabled high-throughput surveys of human microbiota. These studies have linked alterations of both microbial community composition and diversity to various disease states. Although the microbiome has been shown to play an important role in shaping the host immune response, influencing metabolism, and modulating drug interactions, many important questions must be answered before we can fully utilize its prognostic and predictive potential. This chapter highlights the progress of the genomic technologies that drive microbiome research, examines how the microbiome modulates health and contributes to disease, and discusses the future challenges facing this emerging field of study.

16S RIBOSOMAL RNA GENE SEQUENCING

In the late 1800s, Robert Koch developed techniques to cultivate and isolate bacteria cells, which were then identified and characterized by biochemical

Genomic and Precision Medicine. DOI: http://dx.doi.org/10.1016/B978-0-12-800681-8.00004-9

staining, microscopic observation of their morphology, and the use of enrichment cultures. For over a hundred years, these culture-based techniques were the gold standard for classifying microbes. However, these approaches are restricted to the small subset of microbes that are able to survive in isolation and under specific laboratory conditions.

Genomic classification approaches offered a solution to biases of culture-based practices. In the late 1970s, Carl Woese and colleagues [1] generated the first bacterial phylogeny based on the small subunit 16S ribosomal RNA (rRNA). Unique to prokaryotic organisms, the 16S rRNA gene is highly conserved, but contains nine hypervariable regions with species-specific signatures. Soon after bacterial phylogeny was established, Norman Pace and colleagues [2] developed a technique to isolate the 16S rRNA gene from genomic DNA using PCR amplification. Sequences of the 16S rRNA gene could then be compared to the phylogenetic "reference" tree for taxonomic classification. While the 16S rRNA gene is ideal for profiling bacteria, the 18S rRNA and internal transcribed spacer (ITS) regions are similarly used to classify fungal species.

Standard human microbiome studies involve the extraction and sequencing of DNA from a sample containing a heterogeneous mixture of microbes, followed by computational analysis to examine those populations (Fig. 4.1). Rapid advances in DNA sequencing technology have been a key impetus for culture-independent microbiome studies. Early human microbial surveys relied upon fingerprinting techniques or Sanger sequencing of the amplified and cloned 16S rRNA gene. Today, next-generation sequencing platforms offer faster sequencing and vastly increased sampling depths at much lower costs.

The type of sequencing platform used is ultimately determined by the question being asked. In general, shorter reads are sufficient for most microbial community characterization studies, but decrease taxonomic precision. Longer read lengths are beneficial for studies attempting to distinguish between strains or species. Paired-end sequencing is often used to mitigate the problems associated with shorter read lengths by sequencing reads bidirectionally and merging the resulting pairs into a single, longer read.

Upon its introduction, many researchers in the field relied upon the Roche/454 pyrosequencing platform, which produced reads approximately 400−500 bp long. Currently, the Illumina MiSeq benchtop sequencer, which produces reads up to 300 bp, is a popular tool used in 16S rRNA characterization studies. A single run on a MiSeq can generate up to 50 million paired-end 300 bp reads in less than three days. Hundreds of samples can be sequenced on a single run by incorporating sample-specific barcodes into the 5′ primer sequence, in a process known as multiplexing.

A number of open-source software packages exist for computational analysis once microbial samples are sequenced. Two commonly used programs are QIIME [3] and mothur [4], which provide automated scripts for each

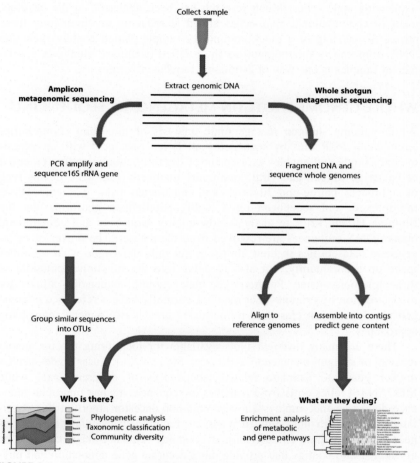

FIGURE 4.1   Microbiome study workflow for sample collection, sequencing, and analysis.

step of their bioinformatics pipelines. Raw DNA sequencing data is first demultiplexed into sample-specific sequences and filtered to remove low-quality sequences that may inflate diversity estimations or falsely suggest the presence of novel organisms. Highly similar sequences are grouped into operational taxonomic units (OTUs), which are compared to reference databases for taxonomic classification and used to calculate within-sample (alpha) and between-sample (beta) diversity. Statistical tests are used to identify significant associations between microbiome components and factors of interest.

General sequencing error, amplification bias introduced by selection of PCR primers or conditions, and the formation of hybrid sequences known as chimeras are just a few potential sources of inaccuracy in amplicon-based sequencing approaches. Much research has been dedicated to the development of computational approaches aimed at reducing or eliminating these errors. Sequencing of a mock community sample, which contains genomic DNA from known microorganisms in specified quantities, alongside experimental samples is one way of estimating sequencing error rates.

WHOLE-GENOME SHOTGUN METAGENOMIC SEQUENCING

Whole-genome shotgun metagenomic analysis of microbial communities circumvents PCR bias by sequencing all DNA associated with an experimental sample and enables assessment of the full genomic coding potential of bacterial, fungal, and viral community members (Fig. 4.1). In this type of approach, paired-end libraries are constructed from extracted DNA, multiplexed, and sequenced on a highly parallelized platform, like the Illumina HiSeq. Prior to analysis, low-quality sequences and contaminant human DNA sequences are removed from the dataset. The power of metagenomic datasets lies in their ability to not only determine what microbes make up a community, but also to delve into the functional potential of these microorganisms. Furthermore, metagenomic sequencing allows for reconstruction of genomes that may not currently have a reference genome and are thereby not classified by culturing or 16S rRNA gene sequencing approaches.

There are many different tools available for identifying the taxonomic makeup of shotgun metagenomic datasets. MetaPhlAn [5] uses clade-specific marker genes to estimate relative abundances of different taxa, while MEGAN [6] relies on BLAST searches of sequences against microbial reference databases and employs a lowest common ancestor algorithm for classification. Although unassembled reads are required to calculate frequencies necessary for sample comparisons, overlapping sequence reads can also be assembled into contigs that provide more accurate gene annotation and phylogeny prediction. Assembly of the various genomes in complex metagenomic datasets is challenging. Toolkits, like IDBA-UD [7] and Ray Meta [8], utilize algorithms to assemble longer contigs with high accuracy. The functional capacity of the metagenome can be determined by comparing predicted protein-coding genes, identified by a BLASTX search, to databases such as the KEGG (Kyoto Encyclopedia of Genes and Genomes) pathway database [9] and/or COG (Clusters of Orthologous Groups of proteins) functional categories database [10].

Metagenomic studies are a computationally intensive undertaking, generating an extremely large volume of sequence data. Subsequent analysis relies on incomplete reference databases that are highly biased toward cultivable organisms and genes with known functions. Thus, development of new methods for cultivating and isolating different organisms is crucial for construction of robust references. Once reference genome sequences are available, additional obstacles to metagenomic sequencing analysis include the annotation of putative open reading frames and functional classification of hypothetical proteins.

CHARACTERIZING THE HEALTHY HUMAN MICROBIOME

In 2007, the NIH funded the Human Microbiome Project (HMP) and one of its key objectives was to define the "normal" human adult microbiome and investigate its role in various diseases [11]. Sampling a cohort of 242 volunteers at 18 diverse sites from five body areas, the HMP found that relative abundances of metabolic and functional pathways identified from the metagenomic data were much more stable than organismal abundances measured by 16S rRNA sequences (Fig. 4.2). Pathogenic organisms were rarely present in these microbial populations, and, as seen in previous microbiome studies, intrapersonal variation between body sites of the same subject was more significant than interpersonal variation between the same body sites of different subjects [12−16]. Because the communities found at each body site are highly specialized, the human microbiome can be considered as a composite of many different microbiomes. In the following sections, we highlight significant findings from individual studies of the gut, oral cavity, lung, urogenital tract, and skin, focusing on the contributions of the microbiota to human health.

Gastrointestinal Tract Microbiome

The gut is one of the first and most well-studied human body habitats regarding microbial communities. Fecal samples are commonly collected and used in microbiome analyses. The MetaHIT (Metagenomics of the Human Intestinal Tract) Consortium has been a key leader in gut microbiome and metagenomics research. Their study of 124 Europeans described a "core" gut metagenome containing genes essential for host−microbe interactions [14]. Analysis of this dataset, in conjunction with others, introduced the idea of "enterotypes", or groups of individuals defined by the composition of their gut microbiota [17]. Three enterotypes were identified, which could not be explained by nationality, body mass index (BMI), age, or gender. The notion that the composition of the human gut microbiota may be stratified, and not continuous, has sparked much debate in the field [18].

Analysis of the human gut virome has drawn attention to the prominence of bacteriophages, viruses that infect bacteria. Metagenomic sequencing of viruses colonizing a single adult gut found that almost 80% of the viral community persisted throughout the 2½-year study [19]. In addition to viral temporal stability, the study also identified high nucleotide substitution rates in certain bacteriophage families. The authors suggest that rapid evolution of

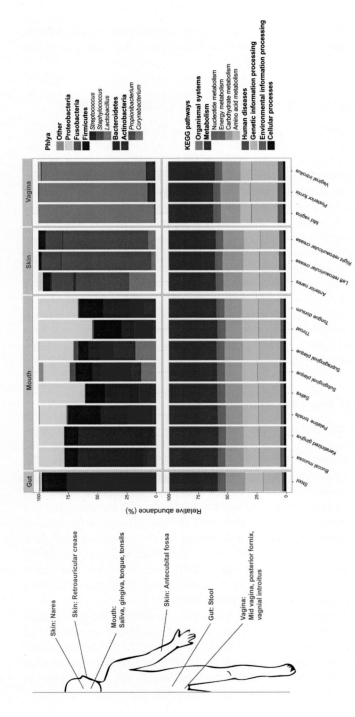

FIGURE 4.2 Taxonomic and functional relative abundance profiles of healthy individuals obtained via shotgun metagenomic sequencing as part of the HMP (MetaPhlAn taxonomic data file downloaded from http://www.hmpdacc.org/HMSMCP/ and KEGG pathway coverage calls downloaded from http://www.hmpdacc.org/HMRC/).

long-term gut residents could give rise to new viral species, which may contribute, in part, to the gut's high interpersonal variability.

The gut microbiome is known to contribute to a variety of human diseases, including cancer and obesity. Colorectal carcinoma is associated with increased abundances of *Fusobacterium*, which is rarely found in the healthy gut [20,21]. Recent work suggests that this correlation is causal. One study found that introducing *Fusobacterium* into mice that develop intestinal tumors accelerated tumor development and induced a proinflammatory response [22]. These findings are supported by a second study that identified a highly conserved *Fusobacterium nucleatum* virulence factor, adhesion FadA, as an inducer of oncogenic and inflammatory responses that promote cancer cell growth [23].

While we do not fully understand human genetic variation associated with obesity, it has been established that the gut microbiota of obese individuals is significantly different from microbiota of lean individuals and carries with it a greater capacity for energy harvest [24]. Born and reared in sterile environments, germ-free mice are not colonized by microorganisms and are often utilized to determine the effects of microbial changes. One such study transplanted gut microbes from twins discordant for obesity into germ-free mice in order to elucidate how interactions between diet and the gut microbiome influence the human host [25]. Ridaura and colleagues saw that mice colonized with bacteria from the obese twin had significantly greater body mass and adiposity than mice colonized with bacteria from the lean twin. These differences in body composition were correlated with metabolic differences. Cohousing the mice not only prevented weight gain in mice colonized with bacteria from obese twins, but also caused their metabolic profiles to shift towards the profile of their lean cage mates. These results were dependent on the diet fed to the mice.

Other studies have similarly shown that diet has a strong influence on gut microbial communities. Wu and colleagues [26] demonstrated that gut enterotypes are strongly correlated with long-term dietary patterns. Gut enterotype identity was not affected by short-term dietary changes. Rapid shifts in both gut microbial community structure and gene expression were observed in volunteers who consumed either an animal- or a plant-based diet for five consecutive days [27]. The animal-based diet had a greater impact on the gut microbiome than the plant-based diet and was associated with decreased levels of *Firmicutes*, which metabolize plant polysaccharides, and increased expression of genes for the degradation of polycyclic aromatic hydrocarbons, compounds produced during the charring of meat.

Significant changes in the gut virome were also observed when the host was placed on a defined diet and these diet-induced changes co-varied with changes in the gut bacterial community [28]. Furthermore, the gut virus populations converged in individuals placed on similar diets. In contrast, another study found the gut virome to be stable over time [29]. The source of the differences observed between these two studies is unknown, however these conflicting results emphasize the need for experimental standardization.

Host genetics has also been shown to influence gut microbial composition and function [30]. Analysis of fecal samples from monozygotic and dizygotic twin pairs identified *Christensenellaceae* as heritable taxa associated with low BMI. Furthermore, the addition of a *Christensenellaceae* species to an obese-associated microbiome reduced weight gain in germ-free mice. The authors suggest that the species not found to be heritable are more heavily influenced by environmental factors, such as diet.

Oral Microbiome

Streptococcus dominates the oral cavity, but other abundant genera include *Veillonella*, *Gamella*, *Rothia*, *Fusobacterium*, and *Neisseria* [31,32]. A recent study that used statistical models to partition human microbiome data into body-site specific community types identified a significant association between gut and oral community types, despite their strong taxonomic differences [33]. One potential explanation for this connection is that oral bacterial populations seed the gut, thereby giving rise to distinct gut community types.

The majority of human oral viruses are bacteriophage, individual-specific, and persist over time [34]. Genome-encoded clustered regularly interspaced short palindromic repeats (CRISPRs) are a form bacterial defense mechanism against mobile genetic elements like bacteriophage and provide a genomic record of phage–bacteria interactions. Streptococcal CRISPR sequences in the oral cavity revealed great diversity within individuals, suggesting that each individual was exposed to unique viral populations [35].

The oral microbiome has been linked to both dental caries (cavities) and periodontitis (gum disease). The complex microbial communities of caries are taxonomically and functionally different from those colonizing healthy oral cavities [36]. In periodontitis, *Porphyromonas gingivalis* is the suspected etiological agent. Small quantities of this bacterium were shown to induce changes in the oral microbiota by exploiting the complement cascade to cause periodontal bone loss [37]. Epidemiological studies have suggested a correlation between periodontitis and atherosclerosis. These two seemingly unrelated diseases may be linked by microbiota, as the types and abundance of bacteria in atherosclerotic plaques correlated with the abundance of those same bacteria in the oral cavity [38]. These studies indicate the potential utility of the microbiome as a clinical biomarker.

Lung Microbiome

Although healthy lungs were once thought to be a sterile environment, recent studies have characterized the lung microbiome and its associations with diseases such as asthma, chronic obstructive pulmonary diseases, and cystic fibrosis (CF). The lung microbiome is especially difficult to study because of its low biomass and the difficulty of sampling only microbiota from the lower respiratory tract without also picking up carryover microbes from the upper respiratory tract. Analysis of six healthy human lungs found that although the lung bacteria were much lower in biomass, they were compositionally similar to bacteria in the upper airways [39].

Infection and bronchiolitis obliterans syndrome are common causes of death after a lung transplant and can be partially attributed to microbial factors. Amplicon-based studies of bacterial and fungal communities have shown that the lungs of transplant subjects are significantly different from healthy subjects in both composition and diversity [40]. Furthermore, the lung microbiome of transplant recipients was less similar to their upper respiratory tract microbial communities and contained lung-enriched bacteria. Longitudinal analysis of lung samples after transplantation also identified significant differences between healthy and transplanted lungs and found that a majority of microbes present were transient colonizers [41].

Lung infection and inflammation is the primary cause of death in patients with CF. As a result, CF lung microbiota have been described at various stages of the disease. One group studied the lung microbiome of three stable and three progressing CF patients for over a decade [42]. They found that the lungs of patients with the progressing disease had decreased microbial diversity, and that antibiotic treatment is a stronger driver of this decrease in diversity than both age and lung function. A more recent study analyzing the daily lung microbiome of four subjects over 25 total days found that bacterial communities remained constant during periods of clinical stability, and microbial shifts were sometimes observed with the onset of CF respiratory exacerbations [43].

Fungal species have also been detected as important players in CF lungs, with *Candida* dominating the relatively stable mycobiome [44]. A metagenomic pilot study analyzing sputum samples from CF lungs identified differences in metabolic profiles of three patients with different responses to antibiotic treatment [45]. Additionally, they identified a reservoir of antibiotic resistance genes that may provide insight into microbial response to treatment.

Urogenital Tract Microbiome

Multiple urogenital diseases, including bacterial vaginosis (BV), yeast infections, sexually transmitted diseases, urinary tract infections, and human immunodeficiency virus (HIV), have been associated with vaginal microbiota. In reproductive-aged women, vaginal bacterial communities generally fall into one of five groups, four of which are dominated by *Lactobacillus* species. Associated with a greater abundance of anaerobic species and increased bacterial diversity, the fifth group is also linked to higher vaginal pH and Nugent scores, both of which are indicators of BV [15]. A longitudinal study indicated that in cases of recurring BV, antibiotic treatments successfully depleted BV-associated bacteria, but the bacteria returned after the treatment ended [46]. The paper also noted the dynamic nature of vaginal microbial communities, finding that *Gardnerella vaginalis* and *Lactobacillus iners* increase in abundance during menstruation, possibly due to the increased availability of iron from menstrual blood. Another study collected daily samples from 135 women over 10 weeks [47]. Initial analysis revealed that vaginal microbiota associated with asymptomatic BV lacked *Lactobacillus* species and was comprised of strict anaerobes prior to symptomatic BV.

While some vaginal communities frequently fluctuate between several of the five different bacterial profiles, others are more stable [48]. During pregnancy, vaginal communities change as a function of gestational age, increasing in *Lactobacillus* species and decreasing in anaerobic species as pregnancy progresses [49]. No differences in microbiota were observed between women who had spontaneous preterm birth and those who delivered at full term [50].

Microbiota colonizing the male genitourinary tract are not as well studied, however they are known to play an important role in sexually transmitted infections. In a longitudinal study of the coronal sulcus microbiome of 77 uncircumcised compared to 79 circumcised African males, circumcision was shown to decrease both bacterial load and overall diversity [51]. In particular, anaerobic bacteria levels decreased, which the authors hypothesize may contribute to the reduced risk of HIV acquisition in circumcised males.

Skin Microbiome

The skin is home to a variety of microorganisms, including bacteria, fungi, viruses, and mites. Studies utilizing 16S rRNA gene sequencing to characterize skin microbial communities have found that microenvironment has the strongest influence on bacterial community composition. Oily microenvironments (such as the back and face) tend to be less diverse and are predominantly populated by *Actinobacteria*, whereas dry sites (arms and legs) harbor *Proteobacteria* and are typically more diverse [12,13]. Alterations in the composition and diversity of skin bacterial communities have been linked to multiple dermatological conditions. Acne is associated with a particularly virulent strain of *Propionibacterium acnes* [52], and atopic dermatitis is characterized by increased colonization of *Staphylococcus aureus* and decreased bacterial diversity [53].

Fungi are known to thrive on the skin and have been implicated in disorders such as toenail infections and athlete's foot. Fungi colonizing healthy human skin have been characterized by amplification and sequencing of the 18S rRNA gene and ITS regions. *Malassezia* species are the predominant community members of most sampled sites, except for sites on the feet that were much more fungally diverse [54–56]. *Demodex* mites, which reside in facial sebaceous glands and hair follicles, are known to increase in abundance as we age and may play a role in disorders such as rosacea [57–59].

Until recently, whole-metagenome shotgun sequencing of skin microbiota was impeded by low microbial burden, preventing collection of the large amounts of DNA required for sequencing, and high quantities of human contamination. Advances in technology have enabled whole-metagenome studies of the skin, which emphasize the importance of biogeography in both taxonomic composition and functional potential [60]. This first metagenomic examination of healthy skin also allowed for identification of strain-level variation in the commensals, *Propionibacterium acnes* and *Staphylococcus epidermidis*, as well as reference-independent analysis of previously uncharacterized species.

CONCLUSIONS

Despite the major advances made over the last decade, human microbiome research is still in its infancy and faces many challenges on the road ahead. One of these challenges will be dealing with the massive volume of sequencing data. While increasingly inexpensive DNA sequencing makes generating data relatively easy, the bioinformatics expertise and computational resources required to store, process, and analyze this data are expensive and hard to come by.

Well-designed studies will produce the greatest advances in understanding the human microbiome. Because the human microbiome is an ecosystem, an important step forward will be integrating strategies and findings from ecology and environmental microbiology into human studies. Furthermore, researchers must take care to collect biologically relevant samples with well-annotated metadata to generate meaningful microbiome datasets.

There are still many unanswered questions regarding the role of the microbiome in human health. How are commensal microbiota regulated and maintained? How does the microbiome educate the immune system to distinguish between threatening pathogens and nonthreatening commensals? Can we manipulate the microbiota or the host response to microbiota to treat, or even prevent, disease? New approaches will be crucial in addressing the questions above, and functional studies will be required to move beyond associations of the microbiome with disease to causation.

REFERENCES

[1] Woese CR, Fox GE. Phylogenetic structure of the prokaryotic domain: the primary kingdoms. Proc Natl Acad Sci USA 1977;74:5088−90.

[2] Lane DJ, Pace B, Olsen GJ, Stahl DA, Sogin ML, Pace NR. Rapid determination of 16S ribosomal RNA sequences for phylogenetic analyses. Proc Natl Acad Sci USA 1985;82:6955−9.

[3] Caporaso JG, Kuczynski J, Stombaugh J, Bittinger K, Bushman FD, Costello EK, et al. QIIME allows analysis of high-throughput community sequencing data. Nat Methods 2010;7:335−6.

[4] Schloss PD, Westcott SL, Ryabin T, Hall JR, Hartmann M, Hollister EB, et al. Introducing mothur: open-source, platform-independent, community-supported software for describing and comparing microbial communities. Appl Environ Microbiol 2009;75:7537−41.

[5] Segata N, Waldron L, Ballarini A, Narasimhan V, Jousson O, Huttenhower C. Metagenomic microbial community profiling using unique clade-specific marker genes. Nat Methods 2012;9:811−14.

[6] Huson DH, Auch AF, Qi J, Schuster SC. MEGAN analysis of metagenomic data. Genome Res 2007;17:377−86.

[7] Peng Y, Leung HC, Yiu SM, Chin FY. IDBA-UD: a de novo assembler for single-cell and metagenomic sequencing data with highly uneven depth. Bioinformatics 2012;28:1420−8.

[8] Boisvert S, Raymond F, Godzaridis E, Laviolette F, Corbeil J. Ray Meta: scalable de novo metagenome assembly and profiling. Genome Biol 2012;13:R122.

[9] Kanehisa M, Goto S. KEGG: Kyoto encyclopedia of genes and genomes. Nucleic Acids Res 2000;28:27−30.

[10] Tatusov RL, Galperin MY, Natale DA, Koonin EV. The COG database: a tool for genome-scale analysis of protein functions and evolution. Nucleic Acids Res 2000;28:33−6.

[11] Peterson J, Garges S, Giovanni M, McInnes P, Wang L, Schloss JA, et al. The NIH Human Microbiome Project. Genome Res 2009;19:2317−23.

[12] Costello EK, Lauber CL, Hamady M, Fierer N, Gordon JI, Knight R. Bacterial community variation in human body habitats across space and time. Science 2009;326:1694−7.

[13] Grice EA, Kong HH, Conlan S, Deming CB, Davis J, Young AC, et al. Topographical and temporal diversity of the human skin microbiome. Science 2009;324:1190−2.

[14] Qin J, Li R, Raes J, Arumugam M, Burgdorf KS, Manichanh C, et al. A human gut microbial gene catalogue established by metagenomic sequencing. Nature 2010;464:59−65.

[15] Ravel J, Gajer P, Abdo Z, Schneider GM, Koenig SS, McCulle SL, et al. Vaginal microbiome of reproductive-age women. Proc Natl Acad Sci USA 2011;108(Suppl. 1):4680−7.

[16] Turnbaugh PJ, Hamady M, Yatsunenko T, Cantarel BL, Duncan A, Ley RE, et al. A core gut microbiome in obese and lean twins. Nature 2009;457:480−4.

[17] Arumugam M, Raes J, Pelletier E, Le Paslier D, Yamada T, Mende DR, et al. Enterotypes of the human gut microbiome. Nature 2011;473:174−80.

[18] Knights D, Ward TL, McKinlay CE, Miller H, Gonzalez A, McDonald D, et al. Rethinking "enterotypes". Cell Host Microbe 2014;16:433−7.

[19] Minot S, Bryson A, Chehoud C, Wu GD, Lewis JD, Bushman FD. Rapid evolution of the human gut virome. Proc Natl Acad Sci USA 2013;110:12450−5.

[20] Kostic AD, Gevers D, Pedamallu CS, Michaud M, Duke F, Earl AM, et al. Genomic analysis identifies association of *Fusobacterium* with colorectal carcinoma. Genome Res 2012;22:292−8.

[21] Castellarin M, Warren RL, Freeman JD, Dreolini L, Krzywinski M, Strauss J, et al. *Fusobacterium nucleatum* infection is prevalent in human colorectal carcinoma. Genome Res 2012;22:299–306.

[22] Kostic AD, Chun E, Robertson L, Glickman JN, Gallini CA, Michaud M, et al. *Fusobacterium nucleatum* potentiates intestinal tumorigenesis and modulates the tumor-immune microenvironment. Cell Host Microbe 2013;14:207–15.

[23] Rubinstein MR, Wang X, Liu W, Hao Y, Cai G, Han YW. *Fusobacterium nucleatum* promotes colorectal carcinogenesis by modulating E-cadherin/beta-catenin signaling via its FadA adhesin. Cell Host Microbe 2013;14:195–206.

[24] Turnbaugh PJ, Ley RE, Mahowald MA, Magrini V, Mardis ER, Gordon JI. An obesity-associated gut microbiome with increased capacity for energy harvest. Nature 2006;444:1027–31.

[25] Ridaura VK, Faith JJ, Rey FE, Cheng J, Duncan AE, Kau AL, et al. Gut microbiota from twins discordant for obesity modulate metabolism in mice. Science 2013;341:1241214.

[26] Wu GD, Chen J, Hoffmann C, Bittinger K, Chen YY, Keilbaugh SA, et al. Linking long-term dietary patterns with gut microbial enterotypes. Science 2011;334:105–8.

[27] David LA, Maurice CF, Carmody RN, Gootenberg DB, Button JE, Wolfe BE, et al. Diet rapidly and reproducibly alters the human gut microbiome. Nature 2014;505:559–63.

[28] Minot S, Sinha R, Chen J, Li H, Keilbaugh SA, Wu GD, et al. The human gut virome: inter-individual variation and dynamic response to diet. Genome Res 2011;21:1616–25.

[29] Reyes A, Haynes M, Hanson N, Angly FE, Heath AC, Rohwer F, et al. Viruses in the faecal microbiota of monozygotic twins and their mothers. Nature 2010;466:334–8.

[30] Goodrich JK, Waters JL, Poole AC, Sutter JL, Koren O, Blekhman R, et al. Human genetics shape the gut microbiome. Cell 2014;159:789–99.

[31] Aas JA, Paster BJ, Stokes LN, Olsen I, Dewhirst FE. Defining the normal bacterial flora of the oral cavity. J Clin Microbiol 2005;43:5721–32.

[32] Bik EM, Long CD, Armitage GC, Loomer P, Emerson J, Mongodin EF, et al. Bacterial diversity in the oral cavity of 10 healthy individuals. ISME J 2010;4:962–74.

[33] Ding T, Schloss PD. Dynamics and associations of microbial community types across the human body. Nature 2014;509:357–60.

[34] Abeles SR, Robles-Sikisaka R, Ly M, Lum AG, Salzman J, Boehm TK, et al. Human oral viruses are personal, persistent and gender-consistent. ISME J 2014;8:1753–67.

[35] Pride DT, Sun CL, Salzman J, Rao N, Loomer P, Armitage GC, et al. Analysis of streptococcal CRISPRs from human saliva reveals substantial sequence diversity within and between subjects over time. Genome Res 2011;21:126–36.

[36] Belda-Ferre P, Alcaraz LD, Cabrera-Rubio R, Romero H, Simon-Soro A, Pignatelli M, et al. The oral metagenome in health and disease. ISME J 2012;6:46–56.

[37] Hajishengallis G, Liang S, Payne MA, Hashim A, Jotwani R, Eskan MA, et al. Low-abundance biofilm species orchestrates inflammatory periodontal disease through the commensal microbiota and complement. Cell Host Microbe 2011;10:497–506.

[38] Koren O, Spor A, Felin J, Fak F, Stombaugh J, Tremaroli V, et al. Human oral, gut, and plaque microbiota in patients with atherosclerosis. Proc Natl Acad Sci USA 2011;108 (Suppl. 1):4592–8.

[39] Charlson ES, Bittinger K, Haas AR, Fitzgerald AS, Frank I, Yadav A, et al. Topographical continuity of bacterial populations in the healthy human respiratory tract. Am J Respir Crit Care Med 2011;184:957–63.

[40] Charlson ES, Diamond JM, Bittinger K, Fitzgerald AS, Yadav A, Haas AR, et al. Lung-enriched organisms and aberrant bacterial and fungal respiratory microbiota after lung transplant. Am J Respir Crit Care Med 2012;186:536–45.

[41] Borewicz K, Pragman AA, Kim HB, Hertz M, Wendt C, Isaacson RE. Longitudinal analysis of the lung microbiome in lung transplantation. FEMS Microbiol Lett 2013;339:57–65.

[42] Zhao J, Schloss PD, Kalikin LM, Carmody LA, Foster BK, Petrosino JF, et al. Decade-long bacterial community dynamics in cystic fibrosis airways. Proc Natl Acad Sci USA 2012;109:5809–14.

[43] Carmody LA, Zhao J, Kalikin LM, LeBar W, Simon RH, Venkataraman A, et al. The daily dynamics of cystic fibrosis airway microbiota during clinical stability and at exacerbation. Microbiome 2015;3:12.

[44] Willger SD, Grim SL, Dolben EL, Shipunova A, Hampton TH, Morrison HG, et al. Characterization and quantification of the fungal microbiome in serial samples from individuals with cystic fibrosis. Microbiome 2014;2:40.

[45] Lim YW, Evangelista 3rd JS, Schmieder R, Bailey B, Haynes M, Furlan M, et al. Clinical insights from metagenomic analysis of sputum samples from patients with cystic fibrosis. J Clin Microbiol 2014;52:425–37.

[46] Srinivasan S, Liu C, Mitchell CM, Fiedler TL, Thomas KK, Agnew KJ, et al. Temporal variability of human vaginal bacteria and relationship with bacterial vaginosis. PLoS One 2010;5:e10197.

[47] Ravel J, Brotman RM, Gajer P, Ma B, Nandy M, Fadrosh DW, et al. Daily temporal dynamics of vaginal microbiota before, during and after episodes of bacterial vaginosis. Microbiome 2013;1:29.

[48] Gajer P, Brotman RM, Bai G, Sakamoto J, Schutte UM, Zhong X, et al. Temporal dynamics of the human vaginal microbiota. Sci Transl Med 2012;4:132ra52.

[49] Romero R, Hassan SS, Gajer P, Tarca AL, Fadrosh DW, Nikita L, et al. The composition and stability of the vaginal microbiota of normal pregnant women is different from that of non-pregnant women. Microbiome 2014;2:4.

[50] Romero R, Hassan SS, Gajer P, Tarca AL, Fadrosh DW, Bieda J, et al. The vaginal microbiota of pregnant women who subsequently have spontaneous preterm labor and delivery and those with a normal delivery at term. Microbiome 2014;2:18.

[51] Liu CM, Hungate BA, Tobian AA, Serwadda D, Ravel J, Lester R, et al. Male circumcision significantly reduces prevalence and load of genital anaerobic bacteria. MBio 2013;4:e00076.

[52] Fitz-Gibbon S, Tomida S, Chiu BH, Nguyen L, Du C, Liu M, et al. *Propionibacterium acnes* strain populations in the human skin microbiome associated with acne. J Invest Dermatol 2013;133:2152–60.

[53] Kong HH, Oh J, Deming C, Conlan S, Grice EA, Beatson MA, et al. Temporal shifts in the skin microbiome associated with disease flares and treatment in children with atopic dermatitis. Genome Res 2012;22:850–9.

[54] Findley K, Oh J, Yang J, Conlan S, Deming C, Meyer JA, et al. Topographic diversity of fungal and bacterial communities in human skin. Nature 2013;498:367–70.

[55] Paulino LC, Tseng CH, Strober BE, Blaser MJ. Molecular analysis of fungal microbiota in samples from healthy human skin and psoriatic lesions. J Clin Microbiol 2006;44:2933–41.

[56] Paulino LC, Tseng CH, Blaser MJ. Analysis of *Malassezia* microbiota in healthy superficial human skin and in psoriatic lesions by multiplex real-time PCR. FEMS Yeast Res 2008;8:460–71.

[57] Lacey N, Delaney S, Kavanagh K, Powell FC. Mite-related bacterial antigens stimulate inflammatory cells in rosacea. Br J Dermatol 2007;157:474—81.

[58] Georgala S, Katoulis AC, Kylafis GD, Koumantaki-Mathioudaki E, Georgala C, Aroni K. Increased density of *Demodex folliculorum* and evidence of delayed hypersensitivity reaction in subjects with papulopustular rosacea. J Eur Acad Dermatol Venereol 2001;15:441—4.

[59] Li J, O'Reilly N, Sheha H, Katz R, Raju VK, Kavanagh K, et al. Correlation between ocular *Demodex* infestation and serum immunoreactivity to *Bacillus* proteins in patients with Facial rosacea. Ophthalmology 2010;117:870—877.e1.

[60] Oh J, Byrd AL, Deming C, Conlan S, Kong HH, Segre JA, et al. Biogeography and individuality shape function in the human skin metagenome. Nature 2014;514:59—64.

Chapter 5

Quantitative Proteomics for Clinical Translation

Zhaohui Chen and Jennifer E. Van Eyk
Cedars Sinai Medical Center, Los Angeles, CA, United States

Chapter Outline

INTRODUCTION

The study of a specific set of proteins expressed under defined biological conditions at a specific time (proteome) is termed proteomics. Proteomics is the set of technologies and approaches that provide quantitative information on protein abundance, variations, and modifications, along with their interacting partners and networks that are involved with and regulate cellular processes. Proteomics, thus, can be defined as a small- or/and large-scale scientific discipline, encompassing the characterization of a single protein or the simultaneous identification and measurement of the thousands of proteins that collectively dictate cellular function.

With respect to precision medicine, the identification of proteins that are reflective of one's personal health status is key to the development and use of circulating biomarkers and also to the identification of the mechanistic pathways involved in disease and the assessment of the effect of therapy. It is clear that to reflect the physiology and/or pathophysiology of an individual will require a panel of protein markers alone or in combination with other measures, such as DNA methylation, mRNA, miRNA, and/or metabolites. In order to achieve this goal, it is essential to accurately and reproducibly measure protein concentrations, their isoforms or polymorphism expression, and the many potential cotranslational and posttranslational modifications.

Over the last several decades, the proteomic field has matured with respect to its approaches, methods, and instrumentation to allow for increased number of quantitative measurements per single analysis. Through the use of the current, state-of-art, high precision, and accuracy mass spectrometry (MS) technology, quantitative proteomics has become increasingly proposed as a method to address biomedical questions. But are MS approaches and instruments ready for the comprehensive proteomics quantitation or use in a clinical chemistry core for diagnosis of patients? If not, what is required? This chapter will examine the question by reviewing the recent advances in quantitative proteomics regarding two directions: large number of analytes with a relatively small scale of samples (numbering in the hundreds) and several key proteins or panels for high-throughput applications (requiring >1000 samples).

PROTEOMICS AND THE ROAD TO ABSOLUTE QUANTIFICATION

Proteomics began with the development of two-dimensional gel electrophoresis coupled to MS in late 1990s, when global changes in protein quantities were measured and visualized in the context of thousands of protein spots representing hundreds of proteins [1]. Simultaneously, the MS-based shotgun approach [2] was developed, in which proteins were digested and then analyzed on an MS instrument, directly or following peptide fractionation. Since then, MS-based methods have become increasingly utilized due to robustness of instrumentation, sample preparation, and informatics over the years. The use of differential stable isobaric chemical labeling with a specific mass tag that is covalently bound to the peptides can be recognized by a mass spectrometer and at the same time provide the basis for fold-change quantification in a multiplexed format [3–7].

In addition, improvements in MS instrument performance and the development of software tools hand in hand have also allowed direct quantification of without tags (referred to as label free [8]). Label-free quantitation is especially appealing because it can be applied to any proteomic sample without the need of introducing isotopes for quantitation. With up-to-date chromatographic and MS technologies and computer algorithms, variability in liquid chromatography (LC)-MS resulting in retention time shifts and errors introduced by slight differences in sample fractionation steps in label-free quantitation is gradually overcome by the matching of thousands of peptides across samples.

It is recognized that global proteomics suffers from limited dynamic range and complicated sampling procedures due to the complexity of biological samples. This has somewhat been addressed by incorporating orthogonal fractionation steps, although each step potentially adds a source of technical error and increases the work load dramatically. The move toward data-independent acquisition (DIA) and in particular SWATH (*Sequential Window Acquisition of all Theoretical Mass Spectra*) may allow a more efficient pathway for the analysis of a larger number of samples [9]. In DIA, most observable peptides can be quantified. The first step is often the generation of a peptide MS spectral library of all of these observable peptides. This MS special library then acts as a reference map in the second step and is used to compare all the subsequence analyses. Each MS peptide library is cell or organ and species-specific, and generation of a high-quality library requires extensive labor and time due to the need for extensive fractionation of the representative sample with subsequent MS analysis of each fraction. However, subsequent analysis of individual samples can be carried out without or with minimal fractionation, saving time and effort while achieving approximately the same proteome coverage as traditional shotgun-based methods. On the other hand, the potential of this method is the scalability of the method and ability for the same cost and time. What used to allow someone to analyze 10 samples can now be expanded to hundreds of samples by SWATH. This opens the door to running larger cohorts and providing an ability to determine an individual's biological variability in context to a specific disease or treatment setting. This is fundamental for the discovery and quantification of new biomarkers and their proteome among the hundreds of individuals who have high clinical utility and most importantly, allows for personalization.

Another important trend to highlight is the recent evolution of MS for more precise and accurate protein measurements of a few proteins, with each protein represented by particular peptides that are unique to the target proteins. Multiple reaction monitoring (MRM, also referred to as selected reaction monitoring (SRM)) can be carried out on large number of samples even in samples comprising thousands of proteins (Fig. 5.1). MRM can be

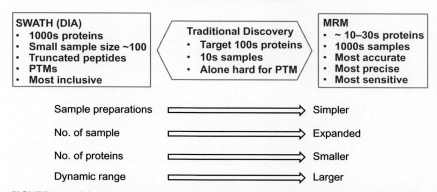

Current Quantitative Proteomics

SWATH (DIA)
- 1000s proteins
- Small sample size ~100
- Truncated peptides
- PTMs
- Most inclusive

Traditional Discovery
- Target 100s proteins
- 10s samples
- Alone hard for PTM

MRM
- ~ 10–30s proteins
- 1000s samples
- Most accurate
- Most precise
- Most sensitive

Sample preparations	⟹	Simpler
No. of sample	⟹	Expanded
No. of proteins	⟹	Smaller
Dynamic range	⟹	Larger

FIGURE 5.1   Schematic Diagram of Current Quantitative Proteomics: Discovery (SWATH or DIA) and Target Proteomics (MRM).

used to determine protein ratio changes in a relative manner or absolute quantification using spiked-in labeled standards. Thus, MRM can be used to generate quantitative data equal to that which is generated by enzyme-linked immunosorbent assay (ELISA).

ELISAs are based on immuno-based quantitation that uses antibody pairs to different amino acid sequences of the target protein. ELISA methodology has existed before the birth of proteomics on both small-scale and large-scale automated platforms and today still plays important role in the field. While immunoassays enjoy high specificity and high throughput and thus are used in point of care and core clinical chemistry laboratories for protein biomarker quantification, their use constantly faces technical challenges. Since immunoassays depend critically on highly specific antibodies, unfortunately, the development can be long and costly, the batches' quality may not be consistent, and ELISAs can display cross-reactivity with biological matrix and analytes related interferences [10]. On the other hand, an MRM assay could target one to five peptides per protein and obtains independent measurements for each peptide. Accordingly, each protein will have multiple quantifications within a given sample. The implication is that interferences or modifications and alterations to the protein analytes will be noticed. Thus, MRM has analytic specificity above that of an ELISA and an ELISA may miss this critical information in patients [11] (Table 5.1).

MRM has become the gold standard for proteomics MS-based quantitation [12]. MRM is a tandem MS technique based on the molecular specificity

---

TABLE 5.1 Characteristics of ELISA Versus MRM

- Relative and absolute quantitation
- Primarily antibody-free assay
- Multiple protein quantified simultaneously
- Differentiation between protein isoforms (splice variants/gene)
- Multiple PTM quantified simultaneously
- Any body fluid and tissue/cells

	Quantitative Western	ELISA	MRM
Standards	Protein	Protein	Protein or peptide
Coefficient of variation (%)	>20%	2–20%	2–20%
Lower LOQ	mg	ng-pg/mL	10 amol−nmol/peptide
			mg-pg/mL* (unless enriched)
Multiplex	Not common	<10	<40
Measure PTMs/variants	Challenging	Challenging	Feasible (constrained MRM)

and unlike shotgun or SWATH, the MS instrument acts as two mass filters, eliminating all peptides except the ones targeted in the MRM assay. The first filter (MS) is for selecting the peptide molecular ion mass, normally digested peptides from the target protein (or endogenous peptides or intact small proteins) and the second filter (MS/MS) is for the smaller cleaved peptide fragments, which independently confirm the amino acid sequence (these peptide-specific fragments are called transitions).

MRM has the flexibility to target specific amino acid sequences, allowing one to differentiate protein isoforms (e.g., TGF beta 1, 2, and 3 [13]) or disease-induced posttranslational modifications (PTMs) (e.g., cardiac troponin I phosphorylation [14] or citrullination of GFAP [15]). With these MRM assays, the signature peptide selection can be constrained, as the peptide must include the amino acid residue(s) of polymorphism, PTM, or the unique isoform sequence variation. This can force the selection of a signature peptide that has a weak MRM response or that is confounded by high background, but there are strategies that can be used to get around this potential limitation [13]. MRM is ideal for monitoring multiple proteins and/or proteins with specific forms to correlate to a unique biological state to predict the progression of a chronic disease or biochemical processes.

As each peptide being monitored has a different hydrophobicity and retention time and thus enters the MS instrument at a different time, it is straightforward to build a multiplex assay comprising of the various peptides in a single run with wide dynamic linear range and high throughput [16]. In many cases MRM enables one to simplify sample preparation by avoiding the initial sample pretreatments or fractionations, although one can enrich the protein or specific peptides (SISCAPA [17]), if needed. With MRM the incorporation of stable isotope-labeled internal standards for every peptide within the multiplex allows the measurement of absolute protein abundances to become possible. As an internal standard, a known concentration of an isotopic labeled (e.g., C^{13}, N^{15}) synthetic peptide composed of the same amino acid sequence as the endogenous targeted peptide is added to the sample. By measuring the levels of native peptide in relation to a chemically identical but labeled heavy reference peptide (a standard curve could be generated using the labeled peptides), the exact concentration of the endogenous nonlabeled can be determined and further, the original protein will be identified and quantified in different normal and diseased stages.

TRANSITIONING TARGETED MRM ASSAY TO THE CLINIC

As mentioned above, compared to immunoassay, MRM is molecule-specific and peptide-dependent, and thus, the quantitation for a multiplex of different proteins [18,19] or PTMs [20,21] or isoforms [22] that reflect the disease process can be employed. Although conceptually straightforward, there are numerous technical challenges to move MRM assays into the clinical situation, either a Clinical Laboratory Improvement Amendments (CLIA) or a core clinical chemistry laboratory. This includes (1) robust pipelines for the development of MRM assays, which involves efficient identification of the proteotypic peptides or the constrained peptides that are robust within a diverse patient population and that have minimal interferences; (2) efficient and robust sample preparation that is capable of being scaled depending on the application; (3) MS instruments, which are easily maintained and suitable for the core laboratories; and finally, (4) a financial structure that allows the business model to be successful.

Clinical MRM assays must provide excellent selectivity and specificity regardless of the complexity of the sample systems. In many cases, internal peptide or protein standards perform adequately in simple phosphate-buffered saline and/or in matrix when spiked in as proteins or peptides. However, results from endogenous proteins may behave differently and be inaccurate due to the added complexity and variability of the biological (clinical) samples. In addition, the normal reference ranges (the concentration range) of different target proteins can vary greatly in plasma and serum (also referred as a large dynamic range) and can add challenges with respect to the accuracy of measurement in which proteins can be multiplexed together. Proteomics analyses in clinical applications not only require highly accurate and reproducible measurements, but also high-sample throughput. Therefore, both issues need to be addressed from the viewpoint of structure-related (analytical chemistry) and sample-related (biology) aspects.

For any MRM assay to be useful clinically, there needs to be a true evaluation of its accuracy. To accomplish this, one can compare the results of the method with results from an established reference (gold standard) analytical method, in this case, ELISA. As an example, an analysis of correlations between an LC-MS/MS assay and an FDA-approved immunoassay platform using 89 unidentified patient discards across the insulin concentration range revealed good agreement [23]. This approach assumes that the uncertainty of the reference method is known and that the ELISA is the "true and accurate" concentration, which it may not be. Accuracy can also be evaluated by analyzing samples with known concentrations (e.g., a control sample or certified reference material) and comparing the measured value with the true value as supplied. Alternatively, a blank sample matrix of interest can be spiked with a known concentration by weight or volume and the recovery can be determined to assess the effectiveness of sample preparations. In the case of insulin MRM measurement, the assay was also assessed by spike and recovery experiments in patient serum and was found to have an overall recovery of 94–113% across the measured range [23].

Sample matrix also plays important role to affect the sensitivity of the detection as well. The limit of quantitation (LOQ) is applied to assess the sensitivity and usually determined by multiple measuring the target protein at concentration levels close to the expected LOQ. The lowest concentration whose CV is less than 20% will be the LOQ. Although the reported lower end for protein measurement has been able to reach as low as 10 amol, the development of sensitive MRM tests for low-concentration proteins in biofluids is always challenging. One of the major hurdles is the complexity of the matrix. Without any sample fractionation or immuno-enrichment, MRM has been repeatedly reported to reach an LOQ around the low microgram per milliliter or high nanogram per milliliter level in serum and plasma [17−20]. However, with careful experimental designs and quality controls, without the assist of antibodies, the LOQ could reach 0.03 ng/mL on peptide angiotensin in plasma [24] and 104 pg/mL for small protein insulin [23]. Another recently developed strategy without antibody capture called PRISM (high-pressure, high-resolution separations with intelligent selection and multiplexing) effectively enriches target peptides for conventional LC-SRM analysis and achieve high-sensitivity measurements for low-abundance proteins at sub-nanogram per milliliter level in plasma or serum (or any body fluid) [25].

For all candidate and established biomarkers, it must be recognized that the target protein may exist in multiple forms and that the MRM assay should be developed to a consistent and stable region of the protein (i.e., using a peptide sequence that is present in all circulating forms) or should exploit these differences. In the latter case, MRM assays can be developed to quantify each form. For example, circulating cardiac TnI can be a dimer (cTnI plus either troponin T or C, which make up the protein complex troponin that regulates muscle contraction) or tetramer (cTnI, troponin T and troponin C) or modified via proteolysized or phosphorylated [21]. Once the protocol is established, the correlation has to be tested by expanding from small number of sample ($n < 10$) to large scale (better $n > 100$) to achieve persuasive statistic significant results. The correlation between MS and ELISA for multiple proteins faces complications while the recovery experiments are relatively less challenging to pursue.

Finally, the transfer of peptide-based MRM assays specifically for protein quantification into a clinical setting will require that there is an assay(s) or multiplex(s) to have clinical utility. This means a clinically relevant menu. Preferably, the assays will be best performed using MS rather than ELISA technology. There are potentially a number of situations where MRM assays are advantageous, including multiplexing where panels of proteins can be quantified in a single analysis (single run). As well, MRM is advantageous when measuring a protein PTM, such as cTnI, where you can obtain both the total protein and the modification ratio in the same analysis. Having such assays may allow a push to adopt MS more widely into clinical setting. Furthermore, if the move is into a CRO or CLIA laboratory then the assay should target chronic diseases where there is sufficient time to allow for the individual's sample to be sent and analysis completed. It will be more challenging to have MRM move into the acute diagnosis initially, as this will require rapid response at the clinical chemistry core lab. Lastly, the technical issues need to be addressed such as the automation of the sample preparation, development of robust MS instruments, adequate service and maintenance contracts for immediate repair, and assay kits being available with adequate menu.

CONCLUSIONS

The human proteome is complex. Proteins can be present in a cell as different isoforms and/or with various cotranslational and posttranslational modifications. The exact composition of the cellular proteome dictates biology function and thus should reflect an individual's physiological or pathological status. The ability to quantify the diverse forms of proteins can be done using ELISA or MS based technologies (such as MRM assays). MS-based targeted MRM has the advantages of specifically quantifying a protein panel as well as individual protein isoforms and PTMs in single analysis. If precision medicine takes advantage of the diversity of the proteome within the clinical setting, it will be necessary to translate the MS workflows into a CLIA or clinical chemistry core laboratory setting. This is feasible especially since MS instrument for quantification of small molecules is present in some core laboratories. As clinically valued MRM assays are developed, this will further the adaption and development of emerging and cutting edge technologies and open up new ways of assessing an individual's disease status.

REFERENCES

[1] Blackstock WP, Weir MP. Proteomics: quantitative and physical mapping of cellular proteins. Trends Biotechnol 1999;17:121−7.

[2] McDonald WH, Yates III JR. Shotgun proteomics: integrating technologies to answer biological questions. Curr Opin Mol Ther 2003;5:302−9.

[3] Tao WA, Aebersold R. Advances in quantitative proteomics via stable isotope tagging and mass spectrometry. Curr Opin Biotechnol 2003;14:110−18.

[4] Aebersold R. Quantitative proteome analysis: methods and applications. J Infect Dis 2003;187(Suppl. 2):S315−20.

[5] Ong SE, Foster LJ, Mann M. Mass spectrometric-based approaches in quantitative proteomics. Methods 2003;29:124−30.

[6] Ong SE, Mann M. Mass spectrometry-based proteomics turns quantitative. Nat Chem Biol 2005;1:252−62.

[7] Panchaud A, Affolter M, Moreillon P, Kussmann M. Experimental and computational approaches to quantitative proteomics: status quo and outlook. J Proteomics 2008;71: 19−33.

[8] Schiess R, Mueller LN, Schmidt A, Mueller M, Wollscheid B, Aebersold R. Analysis of cell surface proteome changes via label-free, quantitative mass spectrometry. Mol Cell Proteomics 2009;8:624−38.

[9] Gillet LC, Navarro P, Tate S, Rost H, Selevsek N, Reiter L, et al. Targeted data extraction of the MS/MS spectra generated by data-independent acquisition: a new concept for consistent and accurate proteome analysis. Mol Cell Proteomics 2012;11(O111):016717.

[10] Hoofnagle AN, Roth MY. Clinical review: improving the measurement of serum thyro-globulin with mass spectrometry. J Clin Endocrinol Metab 2013;98:1343–52.

[11] Schoenhoff FS, Fu Q, Van Eyk JE. Cardiovascular proteomics: implications for clinical applications. Clin Lab Med 2009;29:87–99.

[12] Bantscheff M, Schirle M, Sweetman G, Rick J, Kuster B. Quantitative mass spectrometry in proteomics: a critical review. Anal Bioanal Chem 2007;389:1017–31.

[13] Liu X, Jin Z, O'Brien R, Bathon J, Dietz HC, Grote E, et al. Constrained selected reaction monitoring: quantification of selected post-translational modifications and protein iso-forms. Methods 2013;61:304–12.

[14] Zhang P, Kirk JA, Ji W, dos Remedios CG, Kass DA, Van Eyk JE, et al. Multiple reac-tion monitoring to identify site-specific troponin I phosphorylated residues in the failing human heart. Circulation 2012;126:1828–37.

[15] Jin Z, Fu Z, Yang J, Troncosco J, Everett AD, Van Eyk JE. Identification and characteri-zation of citrulline-modified brain proteins by combining HCD and CID fragmentation. Proteomics 2013;13:2682–91.

[16] Liebler DC, Zimmerman LJ. Targeted quantitation of proteins by mass spectrometry. Biochemistry 2013;52:3797–806.

[17] Anderson L, Hunter CL. Quantitative mass spectrometric multiple reaction monitoring assays for major plasma proteins. Mol Cell Proteomics 2006;5:573–88.

[18] Jia Y, Wu T, Jelinek CA, Bielekova B, Chang L, Newsome S, et al. Development of pro-tein biomarkers in cerebrospinal fluid for secondary progressive multiple sclerosis using selected reaction monitoring mass spectrometry (SRM-MS). Clin Proteomics 2012;9:9.

[19] Kuzyk MA, Smith D, Yang J, Cross TJ, Jackson AM, Hardie DB, et al. Multiple reaction monitoring-based, multiplexed, absolute quantitation of 45 proteins in human plasma. Mol Cell Proteomics 2009;8:1860–77.

[20] Barnidge DR, Goodmanson MK, Klee GG, Muddiman DC. Absolute quantification of the model biomarker prostate-specific antigen in serum by LC-MS/MS using protein cleavage and isotope dilution mass spectrometry. J Proteome Res 2004;3:644–52.

[21] McDonough JL, Van Eyk JE. Developing the next generation of cardiac markers: disease-induced modifications of troponin I. Prog Cardiovasc Dis 2004;47:207–16.

[22] Zhou H, Hoek M, Yi P, Rohm RJ, Mahsut A, Brown P, et al. Rapid detection and quanti-fication of apolipoprotein L1 genetic variants and total levels in plasma by ultra-performance liquid chromatography/tandem mass spectrometry. Rapid Commun Mass Spectrom 2013;27:2639–47.

[23] Chen Z, Caulfield MP, McPhaul MJ, Reitz RE, Taylor SW, Clarke NJ. Quantitative insu-lin analysis using liquid chromatography-tandem mass spectrometry in a high-throughput clinical laboratory. Clin Chem 2013;59:1349–56.

[24] Bystrom CE, Salameh W, Reitz R, Clarke NJ. Plasma renin activity by LC-MS/MS: development of a prototypical clinical assay reveals a subpopulation of human plasma samples with substantial peptidase activity. Clin Chem 2010;56:1561–9.

[25] Shi T, Fillmore TL, Sun X, Zhao R, Schepmoes AA, Hossain M, et al. Antibody-free, targeted mass-spectrometric approach for quantification of proteins at low picogram per milliliter levels in human plasma/serum. Proc Natl Acad Sci USA 2012;109:15395–400.

Chapter 6

From Data to Knowledge: An Introduction to Biomedical Informatics

Philip R.O. Payne

Washington University in St. Louis, St. Louis, MO, United States

Chapter Outline

INTRODUCTION

The field of Biomedical Informatics (BMI) has emerged over the last several decades as a driving force behind the ability of the biological and healthcare research and delivery communities relative to the ability to harness and make sense of every increasing volumes and complexities of incumbent data. For the purposes of clarity in the remainder of this chapter, we define BMI as follows (per the conventions put forth by the American Medical Informatics Association):

> *Biomedical informatics (BMI) is the interdisciplinary field that studies and pursues the effective uses of biomedical data, information, and knowledge for scientific inquiry, problem solving, and decision making, driven by efforts to improve human health. [1]*

Genomic and Precision Medicine. DOI: http://dx.doi.org/10.1016/B978-0-12-800681-8.00006-2

It is important to note that the scientific field of BMI as defined above is both different from and complementary to the areas of Computer and Quantitative Science. Concretely, the aforementioned fields emphasize theories and methods focused upon mechanisms to collect, manage, analyze, and report upon data, while BMI emphasized the broader issues of how we generate and use such data-driven insight given any number of driving problems, yielding information (e.g., contextualized data) and actionable knowledge (e.g., information that is delivered in the right format, place, and time to enable decision making or analogous processes). The complementarity of these constituent areas is often encountered when BMI practitioners use methods and technology solutions derived from the Computer and Quantitative Science domains in order to collect, store, assess, and present large-scale and/or heterogeneous data, as are frequently encountered throughout the biomedical and healthcare settings. Building upon these foundations, BMI practitioners are able to use domain-specific theories and methods in order to interpret and reason upon those data and generate and deliver actionable knowledge at multiple end points, such as the laboratory, point-of-care, or population health settings. In this chapter, we present a broad model for the critical evaluation and understanding of how such methods are selected and applied. In doing so, we hope to equip readers with the ability to navigate an ever-evolving armamentarium of such methods and approaches in a thoughtful and systematic manner.

A PRIMER ON THE ROLE OF BMI IN THE ERA OF PRECISION APPROACHES TO RESEARCH, HEALTHCARE DELIVERY, AND POPULATION HEALTH

Across a spectrum from basic science investigation, to clinical research, to healthcare delivery, and ultimately to population health, it is becoming increasingly important to apply data-driven and precision approaches that leverage the best available scientific knowledge [2−4]. This evolving model is predicated on the establishment of close and well-instrumented interconnections among research, healthcare delivery, and population health investigators, practitioners, and their respective methodologies. At the present time, unfortunately, prevailing approaches to the aforementioned domains are usually not well integrated, resulting in data analytic and decision-making processes that do not take advantage of up-to-date scientific knowledge or well-validated techniques or approaches to reasoning, sense making, and delivery of information or knowledge in a manner consistent with predisposing or enabling sociocultural frameworks [3−5].

Fortunately, there exists a constantly growing body of multimethod and knowledge synthesis techniques incumbent to the field of BMI that can overcome the preceding challenges—for example, exploring potential linkages been biomolecular and clinical phenotypes as well as drug targeting

information so as to design and delivery highly tailored and precision therapeutics for a number of important disease states in a partially or fully in-
silico manner [2,5−11]. Examples of the type of problems and corresponding
BMI theories and methods that can be used to address such information and
knowledge needs are provided in Table 6.1. It is of note that evolving and
increasingly commonly utilized BMI methods and techniques seek to bridge
the gap between humans, available domain knowledge, and large-scale or
heterogeneous data sets, so as to enable the efficient, timely, and empirically
defensible conduct of research, healthcare delivery, and/or population-health
interventions [7−9,12−14]. Such approaches and methods are regularly
being recognized as serving a central role in the conduct of studies or clinical
care delivery paradigms predicated on a genomic medicine approach to
health and wellness [2,5,8,10,11,15−18]. These gaps in knowledge and practice, as well as the promise afforded by BMI theories and methods in such
contexts, serve as the motivation for the review and recommendations made
in the remainder of this discussion.

A FRAMEWORK FOR SELECTING, UNDERSTANDING, AND ASSESSING BMI METHODS

As was introduced in the preceding section, BMI theories and methods are
broadly aligned with a central dogma for the field in which data are contextualized so as to produce information, and that information is in turn delivered
in an actionable manner as knowledge. In this dogma, the process of moving
from data to information involves the *augmentation* of data with contextual-
izing information—for example, in the case of clinical data corresponding to
a given patient, such augmentation could involve the coupling of individual
data points with metadata that identifies the patient, the measurement methods used to generate said data, and pertinent temporal characteristics of that
measure. Similarly, the process of rendering such information as actionable
knowledge involves the *delivery* of information to the right end users in the
right format and time, so as to support the intended decision making or otherwise data-driven process. For each of the phases of the preceding central
dogma, a range of methods can be employed. This conceptual model is illustrated in Fig. 6.1 and explained below. At a high level, these methods can be
aligned with the end points of: (1) data generation, (2) information generation, or (3) knowledge generation. In addition, a broad class of crosscutting
(or core) methods can be used across this entire spectrum of activities.

Methods Aligned With Data Generation

- *Data Discovery and Mining*: One of the initial steps in almost any biomedical or health-focused data analysis project is the identification,
  extraction, and normalization, or quality assurance of source data sets

TABLE 6.1 Examples of Application Domains and Problem Areas Addressed by BMI Theories and Methods

Application Domain	Related Problem Areas Addressed by BMI Theories and Methods
Biomedical Data Science	• Data collection, storage, management, and dissemination • Data integration, harmonization, and sharing • Syntactic and semantic standards development and application (basic theories and methods) • Knowledge-based systems design and application such as those related to the use of artificial intelligence and/or cognitive computing in biomedical settings (basic theories and methods)
Bioinformatics and Computational Biology	• Analysis and interpretation of the output of biomolecular phenotyping instruments (e.g., genomics, proteomics, metabolomics) • Derivation and evaluation of biological networks • Annotation and enrichment of biological data sets using public data resources • Biomolecular data visualization, exploration, and hypothesis generation/testing in the preceding problem areas • Syntactic and semantic standards development and application (as applied to biological data types) • Knowledge-based systems design and application such as those related to the use of artificial intelligence and/or cognitive computing in biomedical settings (as applied to biological data types)
Clinical and Translational Informatics	• Design, implementation, and management of data, information, and knowledge management systems for use in both patient-focused research and care settings (e.g., Clinical Research Management Systems, Electronic Health Records, Personal Health Records, Data Warehouses) • Clinical decision support systems design and evaluation • Syntactic and semantic standards development and application (as applied to clinical data and research data types) • Knowledge-based systems design and application such as those related to the use of artificial intelligence and/or cognitive computing in biomedical settings (as applied to clinical data and research data types)
Imaging Informatics	• Design, implementation, and management of data, information, and knowledge management systems for use in the collection, storage, transaction, and delivery of data generated via imaging instruments

(Continued)

TABLE 6.1 (Continued)

Application Domain	Related Problem Areas Addressed by BMI Theories and Methods
	• Computer-aided interpretation and feature extraction from biomedical images (for hypothesis discovery and clinical decision-making purposes) • Syntactic and semantic standards development and application (as applied to imaging data types)
Public Health and/or Population Health Informatics	• Epidemiological surveillance using patient level and higher-order data sets combined across care and population settings • Design, delivery, and evaluation of tailored health communication and intervention measures at a population level • Geographic and/or temporal reasoning across population-level data types for hypothesis generation and/or testing purposes • Syntactic and semantic standards development and application (as applied to population-level data types) • Knowledge-based systems design and application such as those related to the use of artificial intelligence and/or cognitive computing in biomedical settings (as applied to population-level data types)

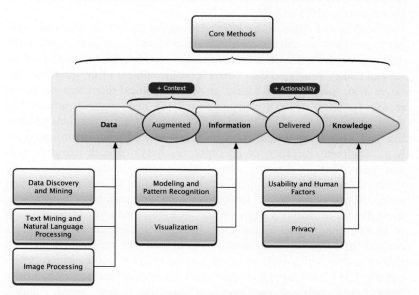

FIGURE 6.1 Overview of the central dogma of the field of BMI, concerned with the generation of information and knowledge from foundational data, and the alignment of methodologies that support or enable the constituent activities comprising that dogma.

that exhibit some type of definable structure. Such structured source data can be distributed across any number of resources such as (but not limited to): (1) structured or semi-structured databases (e.g., relational, hierarchical, or object-oriented data structures), (2) linked open data, (3) metadata repositories, and/or (4) transactional or otherwise operational data repositories. A number of critical methods can be employed to address the identification and downstream processing of such data-centric targets so as to support follow-on analyses, such as (but not limited to): (1) logical and platform-independent data modeling and the linkage of such models to physical repositories via approaches such as object-relational mapping; (2) the use of physical and metadata mining algorithms to derive the structure and content of heretofore informally or undefined data repositories; (3) the application of semantic reasoning algorithms to link data resources to formal syntactic and/or semantic definitions of common data elements, controlled terminologies, or more complex and semantically rich ontologies; and (4) the use of data mining techniques such as column-level indexing and graph-based data linkage invocation, so as to support the structural and/or content-level integration of heterogeneous but related data resources. All of these methods, and many others that apply to the fundamental problem area of data discovery and mining, involve the platform independent modeling, linkage, and integration of data based upon the generation or curation of some type of syntactic and/or semantic model that serves to define the fundamental characteristics of said data sets [10,11,18−21].

- *Text Mining and Natural Language Processing (NLP)*: In a similar manner to the methods and approaches used to identify and extract structured data resources, there exist a body of techniques that can be used to generate and codify structured data based upon the context of unstructured narrative text, such as that found in the biomedical literature or health records (to name a few of many such sources). Such methods usually rely upon statistical and/or rule-based artificial intelligence algorithms to identify and assess the lexical, semantic, and other higher-order meanings (e.g., sentiment, negation of concepts, certainty) of such narrative text, and then codify its meaning using highly expressive data structures and semantic annotations (the later often involving the annotation of text using formal terminology or ontology-defined concepts). Ultimately, these classes of methods allow for the creation of structured data where it is absent, thus enabling the cross-linkage of those data with others derived from structured resources, so as to facilitate integrative follow-on analysis. That being said, such text mining or NLP techniques are infrequently 100% accurate, given their artificial intelligence and probabilistic "roots," and therefore can introduce potential biases, "noise," and/or errors into such downstream analyses. As such, they should be used carefully and with full awareness of such potential impacts on the data analytics process they are incumbent to [19,22].

- *Image Processing*: Again, and as was the case in the preceding discussion of text mining and/or NLP approaches, there are additional sources of data to be found in the unstructured data that comprise a variety of imaging modalities. Such imaging resources can span multiple granularities, from biomolecules to human anatomy and physiology, with each such granularity and level presenting its own unique challenges. However, regardless of such scales, the basic principles that define image processing involve: (1) the conversion of what is often an analog or otherwise continuous scale measurement at a discrete juncture (such as a pixel or otherwise spatially oriented data point) into a structured and actionable data element; (2) the application of search-space constraints to render further analysis of image contents computationally tractable or otherwise understandable from a spatial or structural standpoint; and (3) the identification of patterns in such data using statistical or machine learning methods; ultimately resulting in the generation of discrete measures of features of interest that are definitional to the initial image. For example, such a process could be used to evaluate a clinical image, via which an initial analog signal derived from an imaging instrument is converted into a spatially aligned matrix of pixel-level measures, that matrix is then segmented to isolate an anatomical area of interest, and then pattern recognition techniques are used to identify aberrations in the area of interest as compared to some predefined reference standard image, all for the purposes of diagnostic decision making. Again and mirroring the prior discussion of text mining or NLP techniques, image processing algorithms are also infrequently 100% accurate, and therefore can introduce potential biases, "noise," and/or errors into downstream analyses. As such, they too should be used carefully and with full awareness of such potential impacts on the data analytics process involved [19,23].

Methods Aligned With Information Generation

- *Modeling and Pattern Recognition*: As was noted in a number of the preceding methods aligned with data generation, there is a frequent need in the broad biomedical and healthcare data analytics domain to extract structured data from otherwise unstructured sources so as to enable integrative analyses [14,17,18]. These same types of methods can also be employed to make sense of and contextualize resulting data by cross-linking it with complementary resources (e.g., metadata, an understanding of the biological or clinical basis for a given set of measurements) through a process of modeling and higher-order pattern recognition. In a broad stroke, such methods can be thought of as focusing on multimodeling. These multimodeling and pattern recognition approaches are designed to overcome the limitations of reductionist approaches to scientific discovery, replacing decomposition focused problem solving with integrative network-based modeling and analysis techniques [5,16]. Systems-level analysis of complex problem domains ultimately enables the study of critical interactions that influence health and wellness across a scale from molecules to populations, and that are not observable when such systems are broken down into constituent components. The use of systems-level analysis methodologies is well supported by the foundational theory of vertical reasoning first proposed by Blois [24]. This theory holds that effective decision making in the biomedical domain is predicated on the vertical integration of multiple scales and levels of reasoning. This fundamental premise is the basis for the correlative framework put forth by Tsafnat and colleagues [25] that the ability to replicate expert reasoning relative to complex biomedical problems using computational agents requires the replication of such multiscale and integrative decision making. In order to achieve such an outcome, Tsafnat posits that multiscale decision making in an in-silico context requires both: (1) the generation of component decision-making models at multiple scales and (2) the similar generation of interchange layers that define important pair-wise connections between entities situated in two or more component models, often referred to as vertical linkages [25]. Of note, this type of approach is extremely reliant upon graph-theoretic reasoning and representational models, using a network paradigm that allows for the application of logical reasoning operations spanning the entities and relationships that make up a multimodel [16].

- *Visualization*: While all of the preceding methods, spanning a spectrum from data discovery and integration to multimodeling and pattern recognition, have great promise in terms of enabling the analysis and understanding of complex and high-throughput biomedical and healthcare data, they cannot address every use case or need. In fact, humans possess certain and currently non-computationally-reproducible cognitive strengths in the areas of pattern recognition and multiscale reasoning across and between data sets. As such, there remains a strong need for the use of visualization methods to extract, present, and deliver data in a manner that leverages such unique cognitive strengths so as to support hybrid human—computer analytical processes. These visualization methods can involve simple approaches such as the generation of conventional data graphics, as well as approaches with increasing complexity such as the use of multiscale and —resolution delivery of complex data in immersive and potentially three-dimensional environments (employing advanced computational and user-interaction technologies) [20,21,26—28].

Methods Aligned With Knowledge Generation

- *Usability and Human Factors*: Once data have been identified and contextualized as information, the next step in the BMI dogma is to deliver
it to the right stakeholder in the right format and temporality so as
to yield actionable knowledge. Doing so requires that we understand the
core usability and human-centric dimensions that serve to define
the information needs and environmental factors that will influence or
predispose the actionability of said information. The domains of
human–computer interaction, user experience (UX) design, workflow
assessment and modeling, and cognitive science (to name a few of many)
provide a variety of methods to be employed in this capacity. Such
approaches can include the formal modeling and instrumentation of
"real-world" workflows so as to understand how technologies are to be
used and what environmental factors influence such utilization, to the
detailed assessment of the usability of technology-based interventions
intended to deliver necessary information to targeted recipients, to the
qualitative and quantitative evaluation of internalized decision-making
models employed by human beings when acting upon said information
(and how that may impact their ability to do so, given the preceding
workflow and technical issues that may be elucidated during the course
of a study or an implementation effort). Many of these areas represent
the "fuzzy" science that exists at the implementation level of BMI,
wherein a combination of quantitative and qualitative methods must be
used to triangulate an otherwise difficult to measure "ground truth" and
understand how to achieve actionability relative to a given information
resource [26,27].
- *Privacy*: Finally (and importantly given the increasing integration of
biomolecular and clinical phenotyping in translational science or medicine
contexts), it is imperative that the delivery of actionable knowledge
account for or otherwise incorporate methods to ensure appropriate patient
confidentiality or privacy. These types of methods can involve the use of
de-identification algorithms that can be applied to patient-derived data
sets to reduce and/or eliminate re-identification potential and therefore
enable research using that data in those cases where direct patient
consent is infeasible, or to support the application of rule-based or
knowledge-based constraints at the point-of-care so as to ensure that
actionable knowledge is only delivered to individuals who have a justifiable and ethically defensible reason for accessing that information.
These types of methods remain one of the most emergent in the field of
BMI, and simultaneously, are perhaps some of the most important given
the need to bridge basic science and clinical practice in the era of
genomic medicine [29,30].

Crosscutting (Core) Methods

While the methods introduced in the preceding discussion are organized to roughly align with the major steps in the BMI dogma introduced at the outset of this section, there also exist a number of crosscutting (or core) methods that can support or enable a full spectrum of data analytics, presentation, and delivery needs. These methods can include but are not limited to:

- Data structure and algorithm design
- Multilevel conceptual data modeling
- Knowledge engineering and management
- Probabilistic modeling and analyses
- Implementation and application of leaning systems such as classifiers

A comprehensive and pragmatic treatment of these methodologies can be found in the excellent text provided by Sarkar and colleagues [27].

Additional Readings

A further exploration of the methods and approaches enumerated in the preceding discussion can be found in the selected articles indicated in Table 6.2.

THE RELATIONSHIP BETWEEN PROBLEM SOLVING AND METHODS SELECTION

When selecting appropriate (and often times multimethod) approaches to biomedical problem solving, it is helpful to position such decision making in the broader context of biomedical or healthcare problem solving. Of note, such a general problem-solving model applies equally to research and operational scenarios. This approach is shown in Fig. 6.2 and briefly explained below.

Problem-Solving Process

The biomedical and healthcare related problem-solving process initiates with the definition of a problem and associated use case (e.g., who are the stakeholders involved, what are the end points those stakeholders wish to achieve, and the metrics or other measurements of success therein). Such use cases easily apply across a variety of problem spaces from research (e.g., hypothesis generation and testing) to clinical care (e.g., diagnostic decision making and therapy planning) to population health (e.g., intervention planning and evaluation of outcomes). Subsequently the problem-solving process involves the identification and engagement of necessary data sets, the concomitant identification and engagement of contextual resources such as pertinent domain knowledge or prior work, and the derivation of appropriate delivery

TABLE 6.2 Selected Readings Focusing Upon Common or Emergent BMI Methodologies, Organized by Thematic and Application Focus

Author	Thematic Focus	Application Focus	Title	Publication	Year
Sarkar IN	Core Methods	Broad	*Methods in biomedical informatics: a pragmatic approach*	Methods in Biomedical Informatics: A Pragmatic Approach	2013
Holzinger A, Jurisica I	Data Discovery and Mining	Broad	*Knowledge discovery and data mining in biomedical informatics: the future is in integrative, interactive machine learning solutions*	Interactive Knowledge Discovery and Data Mining in Biomedical Informatics	2014
Hersh WR, Cimino J, Payne PRO, Embi P, Logan J, Weiner M, et al.	Data Discovery and Mining	Healthcare Delivery	*Recommendations for the use of operational electronic health record data in comparative effectiveness research*	eGEMs (Generating Evidence and Methods to improve patient outcomes)	2013
Embi PJ, Hebert C, Cordillo G, Kelleher K, Payne PRO	Knowledge Engineering	Clinical Research	*Knowledge management and informatics considerations for comparative effectiveness research: a case-driven exploration*	Medical Care	2013
Holmes JH	Modeling and Pattern Recognition	Basic Science	*Methods and applications of evolutionary computation in biomedicine*	Journal of Biomedical Informatics	2014
Jiang X, Cai B, Xue D, Lu X, Cooper GF, Neapolitan RE	Modeling and Pattern Recognition	Clinical Research	*A comparative analysis of methods for predicting clinical outcomes using high-dimensional genomic datasets*	Journal of the American Medical Informatics Association	2014

(Continued)

TABLE 6.2 (Continued)

Author	Thematic Focus	Application Focus	Title	Publication	Year
Liu M, Hinz ERM, Matheny ME, Denny JC, Schildcrout JS, Miller RA, et al.	Modeling and Pattern Recognition	Healthcare Delivery	*Comparative analysis of pharmacovigilance methods in the detection of adverse drug reactions using electronic medical records*	Journal of the American Medical Informatics Association	2013
Gkoulalas-Divanis A, Lokides G, Xiong L, Sun J	Privacy	Healthcare Delivery	*Informatics methods in medical privacy*	Journal of Biomedical Informatics	2014
Ohno-Machado L, Nadkarni P, Johnson K	Text Mining and Natural Language Processing	Broad	*Natural language processing: algorithms and tools to extract computable information from EHRs and from the biomedical literature*	Journal of the American Medical Informatics Association	2013
Boland MR, Rusanov A, So Y, Lopez-Jimenez C, Busacca L, Steinman RC, et al.	Usability and Human Factors	Broad	*From expert-derived user needs to user-perceived ease of use and usefulness: a two-phase mixed-methods evaluation framework*	Journal of Biomedical Informatics	2013
Turkay C, Jeanquartier F, Holzinger A, Hauser H	Visualization	Broad	*On computationally-enhanced visual analysis of heterogeneous data and its application in biomedical informatics*	Interactive Knowledge Discovery and Data Mining in Biomedical Informatics	2014

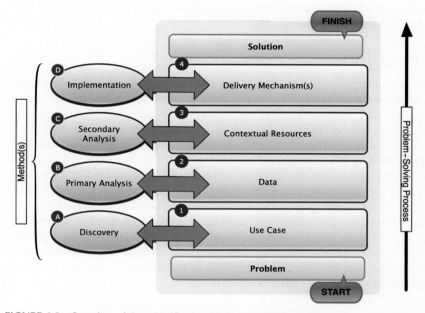

FIGURE 6.2 Overview of the scientific or operational problem-solving process, comprised of: (1) use case definition, (2) data discovery, (3) contextualization of said data, and (4) identification and use of appropriate delivery mechanisms so as to arrive at a solution to the underlying problem. In this same model, core BMI theories and methods can be aligned with: (A) discovery activities, supporting step [1]; (B) primary analyses, supporting step [2]; (C) secondary analyses, supporting step [3]; and (D) implementation approaches, supporting step [4].

mechanisms for ensuing actionable knowledge generated via the problem-solving process (combining the prior components) so as to achieve success given the underlying use case (e.g., a solution to the initial problem).

Alignment With BMI Methods

Given the preceding problem-solving process model, the methods introduced previously can be assigned to one or more categories, corresponding with: (1) the initial discovery phase for the identification of the definitional characteristics underlying the use case; (2) the primary analysis phase for the discovery, integration, and assessment of source data; (3) the secondary analysis phase for the contextualization of initial data and analytics results so as to generate information; and (4) the implementation phase, via which appropriate delivery mechanisms, aligned with the underlying use case and success criteria can be selected and evaluated in a meaningful manner. For each of the methods we have introduced, alignment can be induced with none or more than one of these phases, depending on problem and use-case specific needs and requirements.

CONCLUSIONS

As was introduced at the outset of this chapter, the field of BMI focuses first and foremost on multimethod approaches to generating contextualized information and actionable knowledge from a variety of biological and healthcare-relevant data types. In order to achieve such outcomes, BMI practitioners adopt and adapt methods drawn from the computational, quantitative, and qualitative sciences. In this chapter, we have reviewed a selection of such methods and the rationale for how they can be applied to address driving biological and clinical problems drawn from the "real world." These types of use cases span a range from the biomolecular characterization of disease states to the comprehensive phenotyping of patients to the promotion of population health. Ultimately, this review provides the basis for critically understanding and evaluating the ways in which such multimethod approaches can be optimally utilized so as to advance biomedical research, clinical practice, and population-level health.

GLOSSARY AND ABBREVIATIONS

Biomedical Informatics (BMI) "Biomedical informatics (BMI) is the interdisciplinary field that studies and pursues the effective uses of biomedical data, information, and knowledge for scientific inquiry, problem solving, and decision making, driven by efforts to improve human health." [1]

Data facts or information used usually to calculate, analyze, or plan something (source: Merriam Webster Dictionary)

Information knowledge that you get about someone or something; facts or details about a subject; knowledge obtained from investigation, study, or instruction (source: Merriam Webster Dictionary)

Knowledge the fact or condition of knowing something with familiarity gained through experience or association; acquaintance with or understanding of a science, art, or technique; the sum of what is known, the body of truth, information, and principles acquired by humankind (source: Merriam Webster Dictionary)

Methodology a set of methods, rules, or ideas that are important in a science or art, a particular procedure or set of procedures (source: Merriam Webster Dictionary)

REFERENCES

[1] Kulikowski CA, Shortliffe EH, Currie LM, Elkin PL, Hunter LE, Johnson TR, et al. AMIA Board white paper: definition of biomedical informatics and specification of core competencies for graduate education in the discipline. J Am Med Inform Assoc 2012;19 (6):931−8.

[2] Hood L, Perlmutter RM. The impact of systems approaches on biological problems in drug discovery. Nat Biotechnol 2004;22(10):1215−17.

[3]   Auffray C, Chen Z, Hood L. Systems medicine: the future of medical genomics and healthcare. Genome Med 2009;1(1):2.1−2.11.

[4]   Hood L, Friend SH. Predictive, personalized, preventive, participatory (P4) cancer medicine. Nat Rev Clin Oncol 2010;8:184−7.

[5]   Ahn AC, Tewari M, Poon CS, Phillips RS. The limits of reductionism in medicine: could systems biology offer an alternative? PLoS Med 2006;3(6):709−13.

[6]   Schadt EE, Bjorkegren JL. Network-enabled wisdom in biology, medicine, and health care. Sci Transl Med 2012;4(115):115rv1.

[7]   Payne PR, Johnson SB, Starren JB, Tilson HH, Dowdy D. Breaking the translational barriers: the value of integrating biomedical informatics and translational research. J Investig Med 2005;53(4):192−200.

[8]   Payne PR, Embi PJ, Sen CK. Translational informatics: enabling high throughput research paradigms. Physiol Genomics 2009;39(3):131−40.

[9]   Pattin KA, Greene AC, Altman RB, Cohen KB, Wethington E, Görg C, et al. Training the next generation of quantitative biologists in the era of big data. InPac Symp Biocomput 2015;20:488−92.

[10] Regan K., Abrams Z., Sharpnack M., Srivastava A., Huang K., Shah N., et al. Discovery of Molecularly Targeted Therapies. InPacific Symposium on Biocomputing. Pacific Symposium on Biocomputing 2016 (Vol. 21, p. 1). NIH Public Access.

[11] Regan K, Payne PRO. From molecules to patients: the clinical applications of translational bioinformatics. Yearb Med Inform 2015;10(1):164.

[12] Embi PJ, Payne PRO. Advancing methodologies in Clinical Research Informatics (CRI): foundational work for a maturing field. J Biomed Inform 2014;52:1−3.

[13] Han D, Wang S, Jiang C, Jiang X, Kim H-E, Sun J, et al. Trends in biomedical informatics: automated topic analysis of JAMIA articles. J Am Med Inform Assoc 2015;22 (6):1153−63.

[14] Payne PRO, Embi PJ. An introduction to translational informatics and the future of knowledge-driven healthcare. Translational informatics. Springer; 2015. p. 3−19.

[15] Butcher EC, Berg EL, Kunkel EJ. Systems biology in drug discovery. Nat Biotechnol 2004;22(10):1253−9.

[16] Barabasi AL, Oltvai ZN. Network biology: understanding the cell's functional organization. Nat Rev Genet 2004;5(February):101−13.

[17] Hripcsak G, Duke JD, Shah NH, Reich CG, Huser V, Schuemie MJ, et al. Observational Health Data Sciences and Informatics (OHDSI): opportunities for observational researchers. MEDINFO 2015;15.

[18] Payne PRO, Embi PJ. Driving clinical and translational research using biomedical informatics. Translational informatics. Springer; 2015. p. 99−117.

[19] Bellazzi R, Gabetta M, Leonardi G. Engineering principles in biomedical informatics. Methods in biomedical informatics: a pragmatic approach 2013;313.

[20] Embi PJ, Hebert C, Gordillo G, Kelleher K, Payne PRO. Knowledge management and informatics considerations for comparative effectiveness research: a case-driven exploration. Med Care 2013;51(S38-S44).

[21] Holzinger A, Jurisica I. Knowledge discovery and data mining in biomedical informatics: the future is in integrative, interactive machine learning solutions. Interactive knowledge discovery and data mining in biomedical informatics. Springer; 2014. p. 1−18.

[22] Ohno-Machado L, Nadkarni P, Johnson K. Natural language processing: algorithms and tools to extract computable information from EHRs and from the biomedical literature. J Am Med Inform Assoc 2013;20(5):805.

[23] Najarian K, Splinter R. Biomedical signal and image processing. CRC Press; 2012.

[24] Blois M. Medicine and the nature of vertical reasoning. N Engl J Med 1988;381 (13):847−51.

[25] Tsafnat G, Coiera EW. Computational reasoning across multiple models. J Am Med Inform Assoc 2009;16(6):768−74.

[26] Boland MR, Rusanov A, So Y, Lopez-Jimenez C, Busacca L, Steinman RC, et al. From expert-derived user needs to user-perceived ease of use and usefulness: A two-phase mixed-methods evaluation framework. J Biomed Inform 2014;52:141−50.

[27] Sarkar IN. Methods in biomedical informatics: a pragmatic approach. Academic Press; 2013.

[28] Turkay C, Jeanquartier F, Holzinger A, Hauser H. On computationally-enhanced visual analysis of heterogeneous data and its application in biomedical informatics. Interactive knowledge discovery and data mining in biomedical informatics. Springer; 2014. p. 117−40.

[29] Gkoulalas-Divanis A, Loukides G, Xiong L, Sun J. Informatics methods in medical privacy. J Biomed Inform 2014;50:1−3.

[30] Hersh WR, Cimino J, Payne PRO, Embi P, Logan J, Weiner M, et al. Recommendations for the use of operational electronic health record data in comparative effectiveness research. eGEMs (Generating Evidence & Methods to improve patient outcomes) 2013;1(1).

Chapter 7

Local and Global Challenges in the Clinical Implementation of Precision Medicine

Robyn Ward[1] and Geoffrey S. Ginsburg[2]
[1]*University of Queensland, Brisbane, QLD, Australia,* [2]*Duke University, Durham, NC, United States*

Chapter Outline

The past decade has witnessed significant advances in genomics directly relevant to disease diagnosis, treatment, and prevention as well as a decline in the cost of detection of genomic variation leading to the use of genomic technologies in routine clinical care [1,2]. However, despite this progress and growing momentum, there remain significant challenges and barriers to the broad implementation of genomic and precision medicine (GPM) and its integration in medical practice globally. These include:

1. Evidence generation
2. Implementation into clinical care
3. Data ownership, data sharing, and data infrastructure
4. Participant engagement and trust

Evidence Generation: Widespread implementation of genomic medicine is significantly hampered by the lack of evidence of its clinical utility or impact on health outcomes. In particular, evidence is needed to demonstrate that GPM approaches deliver improved health outcomes in a cost-effective and affordable manner. Evidence demonstrating the potential of GPM to

Genomic and Precision Medicine. DOI: http://dx.doi.org/10.1016/B978-0-12-800681-8.00007-4

improve care quality and cost-effectiveness will take time to develop, and mechanisms must be put into place to ensure ongoing evidence generation and assessment.

With the availability of ever-increasing numbers of genomic tools and other technologies for clinical decision making, the pace of discovery far outweighs our ability to systematically measure the value of each technology, using traditional approaches such as randomized controlled clinical trials (RCTs). The question thus arises as to when an RCT is required and when other types of evidence, such as from observational studies or analyses of large datasets, is sufficient for clinical adoption, regulatory approval, and reimbursement by payers. Advancing the field of GPM will likely will be best served by drawing upon a range of evidence sources including RCTs, observational and electronic health records. Irrespective of the evidence base, this will require new approaches to the analysis and interpretation of the "big data" generated by genomic analyses.

One strategy for developing the evidence base is to use continuous data collection among diverse populations after a drug or diagnostic test has entered the market place. Thus an evidence framework that guides implementation and ensures ongoing evidence generation either once these products have been approved by regulators and entered the market or across a dedicated "genomic medicine implementation network" (such as IGNITE, http://ignite-genomicmedicine.org/) could ultimately enhance the pace of adoption. Postmarket ongoing data capture and evidence development is a paradigm that until now has been broadly underutilized. Collecting evidence after implementation is similar to the concept of "adaptive licensing" that has been used in clinical drug development [3] and is a strategy that is gaining interest in some health systems. For instance, the UK Government has recently launched the independent "Accelerated Access Review," which has announced its intention to explore this model of evidence generation and approval, with focus on continuous data sharing between the scientific community and regulators [4]. Data collected *after* implementation also helps generate the evidence necessary to assess the health economic impact of GPM approaches. As longitudinal economic data might be difficult to generate within the timeframe of a clinical trial, economic modeling could be applied leaving final regulatory and payer approval made contingent on a postmarket health economic impact analyses.

Finally, it is important to recognize that generating these types of data will only be meaningful if the data collected can be effectively analyzed and compared at scale. In order to ensure that outcomes data can be effectively utilized, it is imperative to address the issue of standardizing data measurement and optimization across institutions and nations. Interoperable data— data that are compatible and can be easily exchanged—allow researchers to create large datasets, so that GPM approaches can be studied and better evaluated at scale. Amassing larger and more diverse datasets and sharing these

---

BOX 7.1 Evidence Generation: Key Points

- A flexible framework is required to balance the use of evidence generated from RCTs, the analysis from large datasets, and from observational studies.
- An evidence framework for implementation and ensures ongoing evidence generation once these products have been approved by regulators and entered the market could enhance the pace of adoption.
- Final regulatory and payer approval might be contingent on the inclusion of a health economic impact analyses.

---

data between health care systems and institutions, or with smaller or resource-limited countries and populations, will be critical for their inclusion in health care strategies locally (Box 7.1).

IMPLEMENTATION: INTEGRATING GPM TECHNOLOGIES SUCCESSFULLY INTO HEALTH CARE SYSTEMS

Achieving the necessary, robust, and continually evolving evidence base for GPM will require overcoming many practical and logistical challenges for clinical staff and healthcare providers to adopt and integrate GPM. Globally, several such barriers include the absence of supporting information technology (IT) infrastructure, a lack of data standards and interoperability, insufficient decision support technology, and insufficient funding for implementation and translational research.

Electronic health records (EHRs), a critical component of the IT infrastructure, can enable linkage between genetic information and other personal information to produce new understandings of health and disease. For health systems with the resources to develop them, EHRs have emerged as a possible solution for delivering actionable genomic information to providers. However, even where EHRs are available, they are largely unprepared to handle genomic information [5]. Compounding this problem is a lack of standards for how such data should be stored and coded. This is referred to as "interoperability" or the ability of IT programs to communicate with other systems both in terms of the data and codes they contain and the operating systems they use [6].

Greater coordination and interoperability will maximize the knowledge that would be obtained from integrating genomic information into health care systems and to accelerate scientific discovery and research globally. As larger datasets are developed, flexible and interoperable IT systems will enable a more rapid and systematic accumulation of evidence compared to current siloed efforts. As such, standards for structured data, or common data elements, to be captured in EHRs must be developed. Having these structured data will also allow datasets to be aggregated across health systems and research cohorts to be created across institutions. Both would improve the statistical power of analyses. Currently, work is being done by the US National Academies of Sciences, Engineering, and Medicine's DIGITizE Action Collaborative to address the standards for integration of genomic information into the EHR (http://www.nationalacademies.org/hmd/Activities/Research/GenomicBasedResearch/Innovation-Collaboratives/EHR.aspx).

Integrating GPM into clinical practice will require an adequate and appropriately trained workforce, equipped with the essential clinical decision support technology and resources. Clinical decision support is becoming more relevant as the sheer volume of information providers need to have "just-in-time" increases and also as clinical conditions continue to be subdivided and treated differently [7,8]. Care complexity and treatment options continue to expand rapidly making physician support tools very important in ensuring patient care is optimized. Clinical decision support has the power to use genetic information to guide decision making, reduce cognitive load, and potentially limit medication errors. Further, decision support tools can be used to facilitate a learning health system (Fig. 7.1) in which patient data builds on itself and generates a continuous feedback loop that enhances learning [9,10].

IT systems including EHRs can transform stored genomic data into something useful that can inform prevention, therapies, and frontline clinical decision making. Health care data is already being integrated into tools for providers. Clinical decision support, empowered by a robust IT infrastructure, represents an important approach for ensuring adherence to evidence-based GPM in clinical practice as well as a platform from which to share clinical success and cost-savings information. A centralized global resource, including optimized and accessible decision support and other analytical tools, would be a major step forward toward uniform advancement of implementation of GPM.

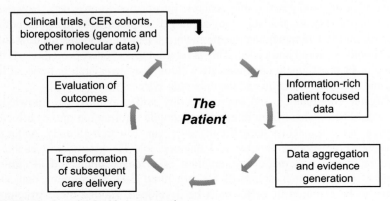

FIGURE 7.1 Learning health system continuum.

BOX 7.2 Implementation: Key Points

- Standardized structured data and common data elements should be captured in EHRs.
- A centralized global genomic medicine resource—or toolbox might be developed and include optimized decision support and other analytical tools.
- Increased investment in translational health research—particularly in the implementation science and diffusion, as well as the outcomes and public health impact phases of translation—will facilitate implementation and integration into clinical practice.
- Incentives, such as "pay for performance" or quality will encourage the adoption of GPM best practices by clinicians and health care delivery systems.

It is important to note that decision support systems will need to be rooted in emerging best practices as defined by experts, likely in the form of clinical practice guidelines. Historically, producing reliable clinical practice guidelines has been laborious and time-consuming. This lag time between the publication of promising research and its adoption into a practice guideline is problematic for the rapidly evolving field of GPM. As such, whenever possible, attempts should be made to modify the guideline generation process to include promising, emerging research. Greater investment in dissemination research and implementation science would benefit this translation. Thus greater investment made in translational health research, particularly in the implementation science and diffusion (so called T3) [11], as well as the outcomes and public health impact (T4) phases of translation. In addition, consideration should be given to incentives, such as "pay for performance" or quality to encourage the adoption of best practices—including genomic medicine—by clinicians and health care delivery systems (Box 7.2).

DATA OWNERSHIP, PRIVACY, AND SHARING

Data sharing among all stakeholders—health care systems, industry, researchers, clinicians, and patients—is integral to evidence generation and to advance our understanding of health and disease.

Yet, currently, prerequisite data sharing platforms and infrastructure are not developed. However, prototypes for sharing genomic and clinical data on a centralized platform are being developed [12]. It is important to acknowledge that economic diversity worldwide inhibits some health care systems and industries from participating in data sharing. Thus many countries may be left out, and their populations either underrepresented or completely absent in the internationally shared datasets. Amassing larger and more diverse datasets and sharing these data with smaller or resource-limited health settings will be critical for their inclusion in GPM strategies locally.

It is also important that regulatory frameworks around genomic and clinical data protect participants address concerns related to data ownership and privacy. Given the intrinsic identifiable nature of genomic data, patients have clear and discernible concerns about their privacy and rights. Because obtaining patient consent is a necessary first step to gathering research data, informed consent process could be made more standardized and streamlined globally including implementing and standardizing nontraditional electronic consent tools and to make it easier on patients and researchers. For example, the US Office of the National Coordinator for Health Information Technology developed eConsent [13], a toolkit designed to address patient questions surrounding consent provides a way for patients to exercise their consent decisions electronically. Apple Inc. has also introduced ResearchKit [14], an open source framework with customizable modules, including visual consent templates to explain the details of a research study and obtain participant signatures.

Ownership of data is a key concern. In a purely research context, ownership of data generally belongs to the research organization, per conditions of the informed consent documents. However, in the context clinical care as a source of research data, a patient's clinical information has historically belonged to them. In addition, concerns about privacy and data breaches might be made a required discussion with participants; this not only helps establish the necessary trust for a long-term research partnership, but also conveys the degree to which researchers are responsible for safeguarding these data, and actions to be taken in the event that data are compromised. As mentioned above, federated query models of data sharing (in which data are stored at individual institutions, but queries of the data are processed through a central server) and the deployment of data use agreements (to preempt malfeasance) may serve to bolster confidence in sharing of data by both patients and institutions and help pave the way for a broader culture of data sharing to support GPM initiatives.

As part of the consenting process, patients could receive information on the benefits of sharing their data, both in terms of the personal benefit to be gained and promotion of the public good. At the same time, clinicians and researchers need to ensure that they inquire about and listen to the concerns of the patient and respond accurately and effectively with the relevant privacy information and protections. However, consent and privacy measures may not be sufficient to adequately address patient concerns about privacy and rights. Countries may consider legislation prohibiting discrimination based on genetic information, such as the Genetic Information Nondiscrimination Act of 2008 [15] in the United States. The informed consent process might be enabled by a framework that (1) addresses the risks to and concerns of patients and the general public, (2) provides information on what privacy protections are currently in place, and (3) provides information on how data sharing promotes the public good (Box 7.3).

BOX 7.3 Data Ownership, Privacy, and Sharing: Key Points

- Data sharing need to be incentivized (among the public, private industry, academics, and government).
- Data governance and ownership frameworks will allow for individuals to customize levels of sharing according to their wishes and values.
- Clarity should be provided on privacy measures and risks associated with data breaches to participants.

PARTICIPANT ENGAGEMENT: TRUST IS A PREREQUISITE

Participants are the heart of GPM. It is their genomic and personal information that contributes to research and eventually to evidence generation. It is their appetite for data sharing that will encourage implementation of genomics into health care. It is their trust that personal data will be used effectively and ethically that will enable data sharing. Given this context, initiatives that pursue deeper engagement between patients and the clinical care and research components of their various health care systems will enable the field to progress. Engagement is being enhanced through several channels, including traditional education approaches (for providers, patients, and the general public), and with patient and health advocacy organizations.

As a key step toward achieving and maintaining trust, participants need to be engaged as partners in the process and their concerns surrounding data privacy and security need to be acknowledged as legitimate. To help secure an ongoing research partnership with patients and their families, and due to the longitudinal nature of evidence generation in GPM, continuous efforts should be made to engage and update patients on the existing and/or potential risks and benefits of participating in GPM research.

Engagement strategies with the participant should pay attention to the diversity of health literacy and numeracy (such that risk information, for example, can be understood) of the population. The degree to which individuals engage with research and clinical care depends upon their ability to access and understand their health care information, relevant consent documents, and the potential impact of testing and treatment on their clinical outcomes. Efforts could be made to not only ensure information is provided to patients (perhaps through patient portals connected to GPM-enabled EHRs, as discussed above), but also that such materials are context-sensitive and culturally appropriate. GPM may exacerbate existing health disparities [16], therefore additional efforts will be required to ensure that diverse ethnic and cultural groups are informed, engaged, and represented to ensure maximum impact of GPM to improve health outcomes.

Finally, employing both traditional and innovative educational approaches to informing and engaging providers, patients, and the general public will be

> **BOX 7.4 Patient and Public Engagement: Key Points**
> - A research agenda is needed to understand (the national and/or local context for) participant preferences, beliefs, literacy, and values in GPM research and clinical care, as well as barriers to public engagement.
> - Programs should target underserved populations with the aim of building understanding and trust for GPM initiatives.
> - Educational programs, beginning at early educational stages, might emphasize GPM and related concepts, including the benefits of data sharing.

important in achieving effective engagement among these groups. In reaching the general public, educational programs (e.g., similar to Science Technology Engineering and Math in the United States) could be developed and initiated, beginning at early educational stages, describing and explaining GPM and related concepts, including data sharing. Innovative efforts to educate the general public about genomics have been implemented in the United States: the National Human Genome Research Institute of the National Institutes of Health partnered with the Smithsonian's National Museum of Natural History to develop a traveling exhibition, "Genome: Unlocking Life's Code." Regarding health care providers, researchers, and clinicians, in both medical school and continuing medical education curricula, training in patient communication should be emphasized, particularly with regard to conversations concerning informed consent and privacy. Resources, such as the Genetic/Genomic Competency Center, a web-based repository of competency-based curricular materials for genetics/genomics education, can help facilitate health care provider education (Box 7.4).

ACHIEVING GLOBAL REACH OF GPM

The genesis of several national implementation strategies for genomic medicine around the globe reflects the growing level of discovery and understanding in this area [17,18], but many such efforts are being conducted in relative isolation. International collaborative projects in clinical implementation are the only way to efficiently generate clinical evidence, evaluate the effectiveness and cost-effectiveness of genomic medicine interventions, and harness relevant information from genomically diverse populations.

Multinational collaborations in genomic medicine implementation are needed. Recognizing the important differences among countries culture, public perceptions, governance structures, health care systems, resources, and infrastructure—and notwithstanding some clear biologic differences in allele frequencies and prevalent diseases—there is so much to be learned and the potential for unnecessary duplication is so great that some degree of coordination and sharing of results is critical [19].

Generating evidence of the value of genomics for patients, clinicians, and health care systems is among the most expensive of potential international collaborations and is already ongoing. Despite differences in health care delivery systems, international collaborations have amply shown the speed with which multinational consortia can answer questions that few countries can tackle on their own, as demonstrated for global burdens of disease [20] and HIV/AIDS [21]. Identifying countries and health care systems willing to permit access to patient data, within appropriate constraints of policy, privacy, and consent, will be a key step in sharing data to fill evidence gaps. Governments and health care providers should agree on the evidentiary standards required for implementation of a genomic medicine intervention. A sufficient body of evidence once available will facilitate professional practice guidelines suitable to a specific setting or country. A framework and principles to facilitate the translation of genomic-based tests from discovery to health care may be modeled on that being developed by Australia's National Health and Medical Research Council [22].

Health information technology is a critical component of genomic medicine implementation, given the vast and complex nature of genome sequence data. Truly global resources for actionable clinical genomic variants are urgently needed and should build on current efforts such as the Clinical Genome Resource (ClinGen) [23,24]. Use of available and widely accepted controlled vocabularies (ontologies) for phenotypes and avoidance of proliferation of local or regional ontologies will be essential to interpretation of variants and sharing of information. The Innovative Medicine Initiative project "ETRIKS," funded jointly by the European Union and industry, aims to create and run an open, sustainable research informatics and analytics platform for sharing data and supporting translational research in GPM [25].

An *educated clinical workforce* will be critical to effective implementation, given the dearth of such expertise worldwide. Competencies for health care professionals at multiple levels within a given system will need to be defined and appropriate educational programs developed [26]. Integration of genomics into health professional curricula will become increasingly necessary. As materials developed in one part of the world are shared globally, translation for language and cultural appropriateness will be needed, however effective training paradigms and best practices should be shared rather than invented (or reinvented) de novo. Relying increasingly on distance learning and other online tools [27] will facilitate rapid implementation and global spread.

Pharmacogenomics represents a probable "early win" ripe for transnational sharing of best practices and lessons learned, given the multiple pharmacogenomic applications that have already been widely implemented in the United States and elsewhere [1,28]. Effective international collaborations have been formed to study the genomics of adverse drug reactions [29,30], but actual implementation efforts have been more isolated.

Guidelines from the Clinical Pharmacogenetics Implementation Consortium [31], which provides recommendations on drug selection and dosing based on an individual's genotypic data, are a notable exception and are increasingly used clinically. Application of whole-genome sequencing in pharmacogenomics could eventually define fully an individual's personalized pharmacogenomics profile. Customizing such an approach in a targeted sequencing effort of the several hundred pharmacogenes involved in drug metabolism and transport, or the smaller subset of clinically actionable pharmacogenes, would reduce costs and make this application more immediately affordable than more comprehensive sequencing efforts.

Multiple international initiatives are addressing *policy recommendations to facilitate data sharing* in genomic research, particularly the Canadian-led Public Population Project in Genomics [32] and the Global Alliance for Genomics and Health [33]. Such efforts are quite relevant to genomic medicine implementation and, as with the evidence realm, an assessment of current activities along with a gap analysis would be important initial steps. *Harmonizing national ethical guidelines and regulatory frameworks* as feasible is essential for successful international collaborations, as is a more complete understanding of regional laws relevant to genomics governing research, privacy, and confidentiality. In evaluating costs, risks, and benefits of genomic interventions, identifying conditions for which genomic tools could have the greatest impact on patient and population outcomes—such as childhood cancers, metabolic disorders, HIV therapy, or cystic fibrosis— would be a useful first step. By *integrating economic assessments into translational research*, not only can the utility of genomic interventions be determined, but also the relative value of such interventions can be assessed to inform health care decision-makers. Expanding single-country studies of cost-effectiveness and financial affordability to multiple health care systems may help identify key underpinning structural components that promote favorable cost−benefit ratios [34]. Multinational collaborations may be particularly valuable for examining different systems and models, such as those with one or a few centralized payers that can provide a more unified and systematic examination of the decision-making process.

GLOBAL COLLABORATION TO IMPLEMENT GENOMIC MEDICINE

Numerous nascent genomic medicine programs are being effectively implemented around the globe demonstrating the potential for genomic medicine to improve our ability to individualize and deliver health care [33]. To facilitate such collaborations, we initiated the Global Genomic Medicine Collaborative (G2MC) hosted by the US National Academy of Medicine as part of its Roundtable on Genomic and Precision Health [35]. Goals of the G2MC are to serve as a nexus for genomic medicine activities globally, develop opportunities for genomic medicine implementation and outcomes research, and capture and disseminate best practices for genomic medicine implementation worldwide. It grew from a gathering of 90 leaders in genomic medicine from the United States and 25 other countries in January 2014 [33]. A second meeting of the group in November 2015 demonstrated growing opportunities for pilot partnerships in the areas of policy, IT, and genomic sequencing and variant interpretation. As we work toward realizing our common interests in the appropriate implementation of genomic medicine, efforts to combine forces in generating and assessing evidence of its impact, and in disseminating best practices for effective implementation, will enhance the use of genomics to improve clinical care worldwide.

ABBREVIATIONS

EHR electronic health record
GPM genomic and precision medicine
G2MC Global Genomic Medicine Collaborative
IT information technology
RCT randomized controlled clinical trial

REFERENCES

[1] McCarthy JJ, McLeod HL, Ginsburg GS. Genomic medicine: a decade of successes, challenges, and opportunities. Sci Transl Med 2013;5(189):189sr4.

[2] Manolio TA, Chisholm RL, Ozenberger B, Roden DM, Williams MS, Wilson R, et al. Implementing genomic medicine in the clinic: the future is here. Genet Med 2013;15 (4):258–67.

[3] Eichler H-G, Baird L, Barker R, Bloechl-Daum B, Børlum-Kristensen F, Brown J, et al. From adaptive licensing to adaptive pathways: delivering a flexible life-span approach to bring new drugs to patients. Clin Pharmacol Ther 2015;97:234–46.

[4] AAR Interim Report. https://www.gov.uk/government/uploads/system/uploads/attachment_data/file/471562/AAR_Interim_Report_acc.pdf; [accessed 29.08.16].

[5] Starren J, Williams MS, Bottinger EP. Crossing the omic chasm: a time for omic ancillary systems. JAMA 2013;309(12):1237–8.

[6] Garde S, Knaup P, Hovenga E, Heard S. Towards semantic interoperability for electronic health records. Methods Inf Med 2007;46(3):332–43.

[7] Castanada C, Nalley K, Mannion C, Bhattacharyya P, Blake P, Pecora A, et al. Clinical decision support systems for improving diagnostic accuracy and achieving precision medicine. J Clin Bioinforma 2015;5(4).

[8] Welch BM, Kawamoto K. The need for clinical decision support integrated with the electronic health record for the clinical application of whole genome sequencing information. J Pers Med 2013;3(4):306–25.

[9] Institute of Medicine (IOM). Best care at lower cost: the path to continuously learning health care in America. Washington, DC: The National Academies Press; 2013.

[10] Ginsburg G. Medical genomics: gather and use genetic data in health care. Nature 2014;508:451–3.

[11] Khoury MJ, Gwinn M, Bowen MS, Dotson WD. Beyond base pairs to bedside: a population perspective on how genomics can improve health. Am J Public Health 2012;102 (1):34−7.

[12] Aronson S, Rehm N. Building the foundation for genomics in precision medicine. Nature 2015;526:336−42.

[13] U.S. Department of Health and Human Services. Patient consent for eHIE—eConsent Toolkit 2014. https://www.healthit.gov/providers-professionals/econsent-toolkit; [accessed 29.08.16].

[14] Apple Inc. ResearchKit. Available from: http://researchkit.org/; 2016 [accessed 29.08.16].

[15] https://www.eeoc.gov/laws/statutes/gina.cfm; [accessed 29.08.16].

[16] National Academies of Sciences, Engineering, and Medicine. Biomarker tests for molecularly targeted therapies: key to unlocking precision medicine. Washington, DC: The National Academies Press; 2016.

[17] Human Genomics Strategy Group. Building on our inheritance: genomic technology in healthcare, https://www.gov.uk/government/uploads/system/uploads/attachment_data/file/213705/dh_132382.pdf; [accessed 29.08.16].

[18] GenomeCanada. Large-scale applied research project competition in genomics and personalized health. Available from: http://www.genomecanada.ca/en/project_search?field_project_category_tid=All&field_project_status_value_i18n_1=All&field_project_sector_tid=All&field_fiscal_year_fund_value=All&field_project_lead_genome_centre_tid=All&field_competition_target_id=24htt; 2012 [accessed 29.08.16].

[19] Manolio TA, Abramowicz M, Al-Mulla F, Anderson W, Balling R, Berger AC, et al. Global implementation of genomic medicine: we are not alone. Sci Transl Med 2015;7.

[20] Murray CJ, Ezzati M, Flaxman AD, Lim S, Lozano R, Michaud C, et al. GBD 2010: a multi-investigator collaboration for global comparative descriptive epidemiology. Lancet 2012;380(9859):2055−8.

[21] IeDEA and ART Cohort Collaborations, Avila D, Althoff KN, Mugglin C, Wools-Kaloustian K, Koller M, et al. Immunodeficiency at the start of combination antiretroviral therapy in low-, middle-, and high-income countries. J Acquir Immune Defic Syndr 2014;65(1):e8−16.

[22] National Health and Medical Research Council of Australia. Principles for the translation of 'omics'-based tests from discovery to health care, https://consultations.nhmrc.gov.au/public_consultations/omics-based-tests; [accessed 29.08.16].

[23] Overby CL, Kohane I, Kannry J, Williams MS, Starren JB, Bottinger E, et al. Opportunities for genomic clinical decision support interventions. Genet Med 2013 Oct;15 (10):817−23.

[24] Ramos EM, Din-Lovinescu C, Berg JS, Brooks LD, Duncanson A, Dunn M, et al. Characterizing genetic variants for clinical action. Am J Med Genet C Semin Med Genet 2014 Mar;166C(1):93−104.

[25] ETRIKS, European Translational Information and Knowledge Management Services. http://www.etriks.org/; [accessed 29.08.16].

[26] Korf BR, Berry AB, Limson M, Marian AJ, Murray MF, O'Rourke PP, et al. Framework for development of physician competencies in genomic medicine: report of the Competencies Working Group of the Inter-Society Coordinating Committee for Physician Education in Genomics. Genet Med 2014;16(11):804−9.

[27] Coursera and University of California, San Francisco. Genomic and precision medicine, https://www.coursera.org/course/genomicmedicine; [accessed 29.08.16].

[28] Carr DF, Alfirevic A, Pirmohamed M. Pharmacogenomics: current state-of-the-art. Genes (Basel) 2014;5(2):430−43.

[29] Holden AL, Contreras JL, John S, Nelson MR. The international serious adverse events consortium. Nat Rev Drug Discov 2014;13(11):795—6.

[30] Pirmohamed M, Burnside G, Eriksson N, Jorgensen AL, Toh CH, Nicholson T, et al. A randomized trial of genotype-guided dosing of warfarin. N Engl J Med 2013 12;369 (24):2294—303.

[31] Relling MV, Klein TE. CPIC: Clinical Pharmacogenetics Implementation Consortium of the pharmacogenomics research network. Clin Pharmacol Ther 2011;89(3):464—7.

[32] Knoppers BM, Fortier I, Legault D, Burton P. The Public Population Project in Genomics (P3G): a proof of concept? Eur J Hum Genet 2008;16(6):664—5.

[33] Global Alliance for Genomics and Health (GA4GH). http://genomicsandhealth.org/; [accessed 29.08.16].

[34] Snyder SR, Mitropoulou C, Patrinos GP, Williams MS. Economic evaluation of pharma-cogenomics: a value-based approach to pragmatic decision-making in the face of com-plexity. Public Health Genomics 2014;17:256—64.

[35] National Academy of Medicine. Roundtable on translating genomic-based research for health. http://www.nationalacademies.org/hmd/Activities/Research/GenomicBasedResearch. aspx; [accessed 26.11.14].

Chapter 8

From Biobanking to Precision Medicine: The Estonian Experience

Liis Leitsalu and Andres Metspalu
University of Tartu, Tartu, Estonia

Chapter Outline

INTRODUCTION

In the mid-1990s, when the Human Genome Project [1] was well underway, people began to think of ways to use the new technologies and information generated in human genetics. In our view, one important step forward was the paper by Risch and Merikangas [2] on genome-wide association studies conducted to understand the genetic underpinnings of complex human diseases. They pointed out that besides the main goal—sequencing of the entire human genome—the Human Genome Project will lead to the identification of polymorphisms for all genes. They also recognized that sequencing technology used to discover and analyze hundreds of thousands of polymorphisms is a limiting factor. Luckily, this problem was soon solved by the Single-Nucleotide Polymorphism (SNP) Consortium [3] and the HapMap Project [4], which lead to the introduction of array technology [5,6]. In another seminal paper published in 1999, Fears and Poste [7] pointed out the potential benefits of genomics and molecular medicine and drafted the roadmap for the future applications of genetics and genomics to health care, to which we can refer today as "precision medicine."

Genomic and Precision Medicine. DOI: http://dx.doi.org/10.1016/B978-0-12-800681-8.00008-6

Early large cohort studies were designed by epidemiologists. However, even one of the largest prospective cohort studies in Europe did not involve the direct storage of DNA. Instead, blood samples were aliquoted into plasma, serum, white blood cells, and erythrocytes [8]. The need for large biobanks with stored DNA and linked medical information was soon recognized, and international efforts were made to organize biobanks into collaborative networks. One of the first commercial activity in this area was the deCode Genetics Inc., founded by K. Stefansson in 1996 in Iceland. The company published many pioneering papers in complex disease genomics and risk factors and was finally sold to Amgen in 2012 (www.decode.com). Genome Canada organized one global consortium in 2002. Four different, but complementary, population genomics research projects involving whole populations—Quebec's CARTaGENE, the GenomEUtwin project (involving eight countries), Estonia's genome project, and the UK Biobank—created an international consortium called the Public Population Project in Genomics (www.p3g.org). A broader European effort soon followed, and 16 member states and one international organization joined forces to establish the Biobanking and BioMolecular resources Research Infrastructures—European Research Infrastructure Consortium (BBMRI-ERIC) (www.bbmri-eric.eu), one of the largest health research infrastructures in Europe today. BBMRI-ERIC aims primarily to establish, operate, and develop a pan-European distributed research infrastructure of biobanks and biomolecular resources. This effort will facilitate access to biological resources and biomedical facilities and support high-quality biomolecular and medical research [9]. The Estonian Biobank (www.biobank.ee) was a founding member of both organizations.

THE ESTONIAN BIOBANK

In 1999, several arguments favored the creation of a large, prospective, population-based cohort in Estonia: The country, population ~1.3 million, sought to become more competitive in the international research community, develop infrastructure for genomic and molecular medicine, and use the results of this research to improve public health. After widespread public debate, the Estonian Parliament (*Riigikogu*) passed the Estonian Human Genes Research Act (HGRA) in 2000 [10]. The act stated clearly the three objectives of the chief processor of the Estonian Biobank, the University of Tartu, where the Estonian Genome Center was founded for this purpose:

1. to promote the development of genetic research;
2. to collect information on the health and genetics of the Estonian population;
3. to use the results of genetic research to improve public health.

The ultimate goal of improving public health, as set out in the third objective, served as the foundation for 15 subsequent years of work leading to the initiation of the Estonian Precision Medicine Pilot Project in 2015 [11].

The Estonian Biobank cohort is a volunteer-based sample of the Estonian resident adult population (aged ≥ 18 years, as they are able to consent) [12]. It was started as a public–private partnership, however the core funding has been provided by the Estonian Government since 2007. The total investment for recruitment and research from year 2001 to 2015 was close to 33 million euros from which 15% was from the private sources, 75% is from the Estonian Government, and 10% from the European Union. The current cohort of almost 52,000 participants represents 5% of the Estonian adult population, making it ideally suited for population-based studies. General practitioners (GPs) and medical personnel in special recruitment offices have recruited participants throughout the country. At baseline, participants provide informed consent (see below) and undergo standardized health examinations performed by GPs; donated blood samples for the analysis of DNA, white blood cells, and plasma; and filled out a thorough questionnaire on health-related topics, such as lifestyle, diet, and clinical diagnoses described in the World Health Organization's *International Classification of Diseases*, 10th Revision [13]. Whole genome sequencing ($n = 2400$) and exome sequencing ($n = 2500$, ongoing) have been completed for significant proportions of the cohort, and the generation of genome-wide SNP array data for all 52,000 participants is in progress. Phenotype data are updated regularly through periodic linking to national health registries and databases (Table 8.1) [14]. A portion of the cohort has been recontacted for follow-up and resampling, and targeted invitations can be distributed for specific purposes, such as the collection and analysis of data from people with a specific diagnosis [12].

Several aspects covered by the HGRA are crucial for the development of the Estonian Genome Center of the University of Tartu (EGCUT) database, as well as its application in a variety of research projects [10,14]. For instance, whereas funding for research projects conducted by the EGCUT is competitively based, the state maintains the specimen collection stored in the biobank and the database, as stated in the HGRA. This arrangement may have contributed to the public's strong support of the biobank, as shown by annual telephone polls conducted over the past decade (Fig. 8.1). Tissue samples and genetic and health information have unique codes and are anonymized (see discussion below on information system for storage of this data). However, decoding is permitted for specific purposes, such as the updating of health descriptions or to contact participants. Data decoding enables linkage of participants' genetic and health data, which is essential for genome-based health research. The HGRA permits recontacting of participants to update and supplement data available in the EGCUT. Recontacting also allows researchers to obtain new tissue samples when

TABLE 8.1 Proportions of EGCUT Database Participants Represented per Data Source

Data Source	Date(s) of Linkage	No. of Participants Found[a]	Percentage of EGCUT Participants Represented
Population register	03-2010, 11-2010, 10-2011, 04-2012, 01-2013, 03-2014, 03-2015	51,800	99.8
Estonian HIF	06-2014	51,607	99.5
ENHIS	02-2014	39,880	76.9
Tartu University Hospital	08-2013	22,492	43.4
Northern Estonian Medical Center	05-2014	21,202	40.9
Estonian Cancer Registry	12-2010, 06-2012, 11-2013, 03-2015	2644	5.1
Estonian Causes of Death Registry	09-2010, 05-2011, 01-2012, 12-2012, 08-2013, 06-2014	2349	4.5
Myocardial Infarction Registry	02-2014	945	1.8
Estonian Tuberculosis Registry	12-2012	260	0.5

[a]As of December 2014.

original samples have been destroyed or do not contain sufficient DNA, or when other types of tissue sample are necessary for specific projects (e.g., epigenetics, expression studies). The act permits the receipt of additional information from other databases to supplement the description of participants' health status.

Several aspects of EGCUT's gene donor consent form (www.biobank.ee), signed by all participants, are crucial for the development of its database and its application in diverse research projects. Participants provide "broad consent," which consists of permission to use the samples collected for future research purposes in general (with the objectives of studying links between genes, environmental factors, and lifestyle and physical characteristics, health, and disease), without identifying specific projects. This type of consent enables scholars to conduct a wide range of research, which would have

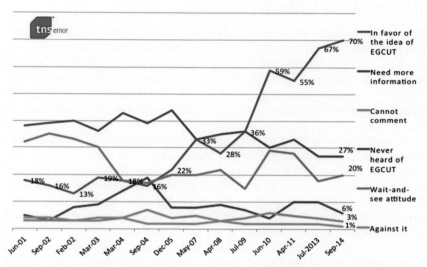

FIGURE 8.1 Results of polling agency www.emor.ee surveys conducted by computer-assisted telephone interviewing reflecting public opinion on and awareness of the Estonian Biobank over the period of 2001–14. Sample = 1000 individuals aged 15–74 years, with distribution proportional to population structure with respect to age, gender, religion, and nationality.

been difficult, if not impossible, to foresee at the time of participant recruitment. Research projects are conducted only with the approval of the Research Ethics Committee of the University of Tartu. The broad consent form also secures permission to receive additional information about participants from other databases. Linkage with other databases and registries enables the updating and enhancement of the EGCUT database on a regular basis without recontacting participants [14]. The participants have the right to be aware of which of their data are available in the database and to receive genetic counseling upon accessing those data. *In other words, participants have the right to obtain research results and to grant their physicians access to that information.* Finally, the consent form stipulates that "I [the participant] may not demand a fee for providing a tissue sample, for the description of my state of health or genealogy, or for the use of the research results. I am aware of the fact that my tissue sample may have some commercial value and research and development institutions as well as commercial enterprises may receive anonymous data about gene donors. The right of ownership of the tissue sample, of the description of my state of health and of other personal data and genealogy shall be transferred to the University of Tartu, the chief processor of the Estonian Genome Project." This type of consent enables the conducting of research and developmental projects with industrial partners and projects, whereby personal genome information is introduced into medical practice for patients' benefit [12,15].

THE ESTONIAN E-HEALTH SYSTEM

Several nationwide information technology development initiatives in the Estonian health sector have had major impacts on the growth of the EGCUT database, as well as the potential to implement the results from genetic research in medical care. Key elements include the introduction of the national electronic identification (ID) card, the launch of the Estonian National Health Information System (ENHIS), and the implementation of digital prescriptions [16–18].

A nationwide governmental technical infrastructure called "X-Road" allows secure communication among various registries and databases [19] (Fig. 8.2). Medical data from hospitals, primary care facilities, and pharmacies can thus be accessed in a regulated manner. The Estonian Ministry of Social Affairs is governing the Estonian Health Information System. After obtaining the approval for a project from the Research Ethical Committee researchers have to apply in order to get access to data.

X-Road currently hosts more than 4000 services and 1000 connected organizations, public registries, and governmental databases [19]. Secure e-services in Estonia are made possible through the use of national electronic ID cards, implemented in 2002 and made compulsory for all Estonians. Digital signatures given with these cards are legally equivalent to handwritten signatures. Authorized medical personnel and citizens use the ID cards to access various services, including electronic health records (EHRs). As of 2015, 1,234,774 ID cards were active, representing more than 95% of the Estonian population, and more than 140 million legally binding electronic signatures had been provided [20].

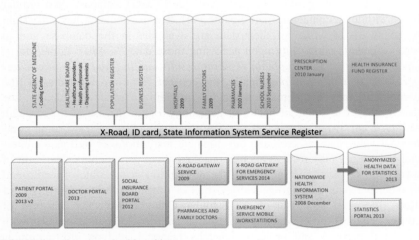

FIGURE 8.2 Estonian e-health architecture.

Healthcare providers have been obliged to forward medical data to the ENHIS electronically since 2008. As of January 2015, the ENHIS contains more than 14 million medical documents for more than 1.3 million individuals in the database [20]. The system integrates data from various healthcare providers, which may use different systems, and generates a common electronic record for each person in a standardized format.

Projects like the ENHIS facilitate the everyday work of healthcare providers and improve patient services. Physicians can provide medical services based on complete health information for patients, provided by the ENHIS. The efficiency of services has also increased. For instance, the digital prescription project was launched in 2010, and by 2013 more than 95% of prescriptions were being issued electronically through X-Road. Patients are more involved and informed, as they are able to access their records through a patient portal, review records of their visits and prescriptions, and control access to their EHRs. Patients' access to ENHIS will increase their participation in disease prediction and prevention, and their role in diagnostic and treatment processes. Patients will also be able to add data from wearable electronic devices, remote monitoring, and self-assessment to the system. These roles will become increasingly important in the context of estimating risk for complex diseases and drug response rates. In addition, as all stages from reception to prescription are electronic, medical statistics are more comprehensive and easier to use. Thus, users can retrieve national statistics to identify systematic problems and measure health trends.

One of the most important tasks in establishing EHRs in the ENHIS was the establishment of an access rule for the usage of patients' health data. Discussions continued for more than 3 years and a special ethics committee was established. The issue was the need to find the right balance between personal rights and social expectations for delicate data usage. Aspects under discussion included the purpose of data usage, roles of the healthcare system's employees, patients' will, and specific situations (e.g., underage patients or those with restricted active legal capacity) [21]. When considering the incorporation of genetic data into the ENHIS, several issues will resurface and need to be discussed again, including the use of genetic data in health care, and research and development of new services and products with industrial partners. Implementation of the ENHIS has been one of the largest and most complex projects in the Estonian healthcare sector to date.

LINKING PHENOTYPE DATA WITH GENOTYPES AND SAMPLES

At baseline, medical personnel consisting of GPs and nurses filled out a thorough computer-assisted questionnaire covering health information, anthropometric measurements, personality traits among other information [12]. Project was open for all, not only for the patients visiting primary care providers or hospitals. As the usefulness of a population cohort for research purposes depends on the phenotypic data accompanying the samples analyses it is necessary to keep the phenotypic data regularly updated. The goal of maintaining the phenotypic data of a cohort of 52,000 donors regularly updated through recontacting participants seemed to be unrealistic for two reasons: it is time consuming and expensive, and the response rate is low for reasons we have not studied yet. Moreover, this rate decreased from 57.2% in the first wave of recontacting to 41.1% within a 4-year period [12]. There are preliminary discussions in order to increase the size of the biobank, however no decision has not made yet.

Although the recontacting of all gene donors remains the best option for some scenarios, we have had to use alternative methods for updating the database [14]. Based on participants' broad consent, which allowed the incorporation of health data from other medical databases into the EGCUT database, the biobank has negotiated with relevant institutions and obtained approval from the ethics committee and the Estonian Data Protection Inspectorate to do so. Since 2013, the EGCUT has retrieved health data from the hospital databases of Tartu University Hospital and the Northern Estonian Medical Center, two of the main hospitals in the country (Table 8.1). These data include diagnoses, hospital discharge records, laboratory data, and imaging data from as early as 1993. Overall, linkage with the two hospital databases has provided 320,000 unique confirmed diagnoses and up to 70,000 entries, including 5000 clinical biomarker measurements, to date. This step introduced a completely new dimension in terms of the amount and detail of health information on biobank participants available.

In addition, data regarding prescriptions given and used, specialty and other types of care provided, and billing are obtained from the Estonian Health Insurance Fund (HIF), which maintains the Digital Prescription Database. The HIF database contains information for all individuals covered by national health insurance, and the EGCUT has obtained more than 3.2 million treatment documents and 4.8 million diagnoses linked to biobank participants for the period of 2003–13 (Table 8.1). Linkage with the ENHIS was achieved in 2014: EHRs and digital prescriptions from all medical service providers have been retrieved from the ENHIS database, including 44,000 inpatient and 212,000 outpatient medical summaries for EGCUT participants [14].

DISEASE TRAJECTORIES

The baseline and subsequently added information enables the examination of disease prevalence and incident cases. With the addition of data from the Causes of Death Registry, full disease trajectories can be studied (Fig. 8.3). This ability has been extremely valuable for the development of risk prediction algorithms; prevalent cases were used to develop the tool and incident

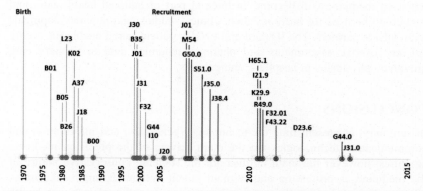

FIGURE 8.3   An example of a disease trajectory based on data from the Estonian Biobank. Health data coded according to the *International Classification of Diseases*, 10th Revision (ICD-10), collected at baseline (green) and obtained through linkage with national registries, hospital databases, and the ENHIS (red) are included. Such trajectories enable the analysis of prevalence at baseline and incident cases after recruitment. This trajectory is generated for a male born in 1970 who joined the biobank in 2007 at the age of 37 years.

cases were used to test it [22]. One could envision that with the incorporation of genetic data (variant—phenotype associations, likely drug response), the ENHIS could form a complete loop: Existing patient information (including genetic information) could be used, new information on disease treatment and outcomes would be incorporated into the system, and these data could be taken into account in future considerations. For example, this system could prompt consulting with extended families when an important genetic variant is identified. It may also require the implementation of new protocols or rules. In this way, the system will improve continuously based on the everyday work of physicians, nurses, and patients.

FUTURE ACTIVITY: THE PRECISION MEDICINE PILOT PROGRAM IN ESTONIA

The work described above has led to national precision medicine initiatives in Estonia: Exome sequencing was included into reimbursement program of the Estonian HIF, Dr. Eric Lander from the Broad Institute and his team visited the EGCUT and collaboration has started, Big Pharma companies have visited the genome center and are considering potential collaborative projects. Additionally, in 2014, the Estonian Government allocated funds to establish a plan for a shift toward precision medicine based on modern genetic technology. The Government in December 2014 proposed the pilot project, which will be conducted in 2016—18. A feasibility study was conducted in 2015 and the pilot project shall start in 2016. These activities would be important steps toward the precision medicine pilot project and hopefully in future precision medicine will become an integral part of national health care, and data could be used for basic research, clinical studies, and new developments with outside partners. The ID card and ENHIS patient portal enable the secure delivery of risk assessments and pharmacogenomics data to primary care providers, physicians in hospitals, and patients.

CONCLUSIONS

In summary, at present time the technology (genomics) and information and communication technologies are the main drivers of the precision medicine activities, however the main hurdles are the societal, conservative nature of the medicine as profession and reimbursement system. It might take up to 10 years to implement the precision medicine into the mainstream health care. But it is right time to start today.

ACKNOWLEDGMENTS

We would like to acknowledge all of the Estonian Biobank recruiters and participants, the team at the Estonian Genome Center, the University of Tartu, Ministry of Social Affairs, Ministry of Science and Education, Estonian Research Council, and the Archimedes Foundation for providing support throughout the years. In addition, we like to thank Dr. L. Milani, Mr. A. Novek, and Mrs. Mari-Liis Tammesoo for the help in preparing the manuscript.

REFERENCES

[1] An overview of the Human Genome Project [cited 2015 Jun 11]. Available from: http://www.genome.gov/12011238.
[2] Risch N, Merikangas K. The future of genetic studies of complex human diseases. Science 1996;273(5281):1516–17.
[3] Thorisson GA. The SNP Consortium website: past, present and future. Nucleic Acids Res 2003;31(1):124–7.
[4] International HapMap Project [cited 2015 Jun 11]. Available from: http://hapmap.ncbi.nlm.nih.gov/.
[5] Fodor SP, Rava RP, Huang XC, Pease AC, Holmes CP, Adams CL. Multiplexed biochemical assays with biological chips. Nature 1993;364(6437):555–6.
[6] Gunderson KL, Kruglyak S, Graige MS, Garcia F, Kermani BG, Zhao C, et al. Decoding randomly ordered DNA arrays. Genome Res 2004;14(5):870–7.
[7] Fears R, Poste G. Policy forum: health care delivery. Building populations genetics resources using the U.K. NHS. Science 1999;284(5412):267–8.
[8] Riboli E, Kaaks R. The EPIC Project: rationale and study design. European Prospective Investigation into Cancer and Nutrition. Int J Epidemiol 1997;26(Suppl. 1):S6–14.
[9] van Ommen G-JB, Törnwall O, Bréchot C, Dagher G, Galli J, Hveem K, et al. BBMRI-ERIC as a resource for pharmaceutical and life science industries: the development of biobank-based Expert Centres. Eur J Hum Genet 2014. Available from: http://www.ncbi.nlm.nih.gov/pubmed/25407005.

[10]   Riigikogu. Human Genes Research Act 2000 [cited 2014 Jun 27]. Available from: https://www.riigiteataja.ee/en/eli/531102013003/consolide.

[11]   Coalition agreement 2014 [cited 2014 Jul 3]. Available from: https://valitsus.ee/sites/default/files/content-editors/failid/2014_coalition_agreement_0.pdf.

[12]   Leitsalu L, Haller T, Esko T, Tammesoo M-L, Alavere H, Snieder H, et al. Cohort Profile: Estonian Biobank of the Estonian Genome Center, University of Tartu. Int J Epidemiol 2014;1−11.

[13]   WHO | International Classification of Diseases (ICD). World Health Organization [cited 2013 Jun 28]. Available from: http://www.who.int/classifications/icd/en/.

[14]   Leitsalu L, Alavere H, Tammesoo M-L, Leego E, Metspalu A. Linking a population bio-bank with national health registries—the Estonian experience. J Pers Med 2015. Available from: http://www.mdpi.com/journal/jpm/special_issues/biobanking.

[15]   Milani L, Leitsalu L, Metspalu A. An epidemiological perspective of personalized medi-cine: the Estonian experience. J Intern Med 2015. Available from: http://www.ncbi.nlm.nih.gov/pubmed/25339628.

[16]   Tiik M, Ross P. Patient opportunities in the Estonian Electronic Health Record System. Stud Health Technol Inform 2010;156:171−7.

[17]   Sepper R, Ross P, Tiik M. Nationwide Health Data Management System: a novel approach for integrating biomarker measurements with comprehensive health records in large populations studies. J Proteome Res 2011;10(1):97−100.

[18]   eHealth [cited 2013 Sep 17]. Available from: http://www.e-tervis.ee/index.php/en/2012-07-22-13-35-31/organization.

[19]   X-Road | e-Estonia [cited 2013 Jun 28]. Available from: http://e-estonia.com/components/x-road.

[20]   eHealth. Statistika [cited 2015 Jan 13]. Available from: http://www.e-tervis.ee/index.php/et/dokumentide-statistika.

[21]   Tiik M. Rules and access rights of the Estonian integrated e-Health system. Stud Health Technol Inform 2010;156:245−56.

[22]   Läll K, Mägi R, Morris A, Metspalu A. and Fischer K, Personalized risk prediction for type 2 diabetes: the potential of genetic risk scores. Genet Med. Available from: http://www.nature.com/gim/journal/vaop/ncurrent/full/gim2016103a.html.

Chapter 9

Electronic Health Records and Genomic Medicine

Daniel R. Masys
University of Washington, Seattle, WA, United States

Chapter Outline

INTRODUCTION: ELECTRONIC HEALTH RECORDS, AN ESSENTIAL INFRASTRUCTURE FOR GENOMIC AND PERSONALIZED MEDICINE

Just as advances in high-throughput laboratory technologies have made the acquisition of exomes and whole genomes and other forms of "omics" data feasible, so advances in information technology have made it possible to store, communicate, and interpret these data, and use them for guiding clinical care. This chapter addresses the desirable characteristics of clinical information systems that incorporate genomic data, discusses some of the limitations of currently available systems and provides recommendations for choosing among them, and describes a path forward for improving the state of the art.

The term "Electronic Medical Records" has been used as a synonym for computerized clinical systems and has its roots in the historical evolution of records of care delivery by health professionals from paper-based handwritten formats to electronic formats. As such, it carries an unfortunate connotation of systems that are primarily designed to be passive after-the-fact archives of the observations and actions initiated by clinicians on behalf of individual patients in the course of providing medical care. A more general

Genomic and Precision Medicine. DOI: http://dx.doi.org/10.1016/B978-0-12-800681-8.00009-8

and contemporary term is "Electronic Health Records" (EHRs), which expands the scope of systems to include data relevant to both health and disease and expands the sources of data beyond the records created by health professionals, to include patient observations about themselves, family history data generated by groups of related persons, and other novel electronic forms of data such as video and audio. But even this more general term misses what is arguably the most important functionality of contemporary systems, which is not their serving as an archive of records, but rather serving as *channels of communication and coordination* among providers, organizations, individuals, and families for the tasks of health maintenance, disease prevention, diagnosis, treatment selection and delivery, and long term follow-up. Thus, the incorporation into EHRs of messaging functions, such as secure e-mail, automated alerts, and reminders sent to providers in organizational settings and to individuals via personal devices such as smartphones, has recast the nature, scope, and reach of computerized clinical systems into a fast evolving array of information management and decision support tools. EHRs with these types of functionality have been shown to help contain healthcare costs and improve patient outcomes [1,2].

Because of its volume and innate complexity, data representing personal molecular variation is in many ways the "poster child" and driving case for new applications of information technology that will inevitably change the status quo in EHR design and functionality. For that reason, it is of value to consider the characteristics of an ideal genomic medicine system, in preference to focusing on the somewhat evanescent, limited functionality provided by current systems which are optimized for traditional forms of clinical data, such as history, physical exam, clinical chemistries, images, pharmacy data, procedure notes, etc.

CHARACTERISTICS OF GENOMIC DATA RELEVANT TO CLINICAL SYSTEMS

From a design perspective, there are several distinctive characteristics of high-throughput molecular data that need to be accommodated by EHR systems. These include its relatively large size (up to billions of individual observations at the level of nucleotides and amino acids), within which is a fraction of observations that potentially have relevance for an individual's experience of health or disease. Though it seems highly unlikely that the identity of every nucleotide will need to be accessible to EHR systems for clinical purposes, the fact that human genomes are currently only partially decoded and understood means that there is no rational basis for discarding molecular sequence data once the observations have been recorded. The EHR needs at a minimum to incorporate findings that are known to be clinically actionable, with the expectation that as understanding of personal molecular variation grows, the fraction of the data that becomes actionable will also grow. Hybrid models of genomic data storage, wherein the "raw" high-throughput data resides in a specialized laboratory information system archive and a subset of that data with its interpretation is transmitted to the EHR, will likely remain the norm for the foreseeable future.

Medical images from modalities such as digital radiography and computed tomography are often of a comparable size to molecular sequences (megabytes to gigabytes of data), but unlike images, where it is common to do data compression by methods that may average or discard individual pixel level detail, the details remain important in molecular genetics. Mendelian disorders resulting from a change of a single base pair, e.g., sickle cell disease, are reminder that whenever data reduction and compression strategies are applied to molecular data, they must be fully reversible so that a faithful copy of the original data can be reconstructed. The nucleotide data of each individual exceeds the bounds of unaided human cognition with respect to finding patterns predictive of health or disease and is essentially "naturally encrypted" in that it is impossible, by inspection, for a clinician to read it in its naturally occurring "raw" form. Additionally, there are no perfect laboratory methods for assessing whole genomes or proteomes at present; all sequencing technologies in use as of this writing generate data with blind spots and errors.

CHARACTERISTICS OF AN IDEAL EHR SYSTEM FOR GENOMIC MEDICINE

Given the features of molecular data described above, computerized data management, display, and decision support utilities for personalized medicine based on "omics" data need to have functionality that differs from the types of clinical systems prevalent today. This functionality will be described and then presented diagrammatically as a set of interconnected resources.

The US National Institutes of Health has sponsored several conferences on the topic of electronic clinical systems for these types of data. A 2011 workshop on incorporation of genetic testing results into EHR systems generated a consensus paper on the technical desiderata for management of genomic data, whose recommendations are summarized in Table 9.1 [3]. The recommendations are called "desiderata" because they represent desirable functional characteristics of systems, without specifying any particular technical approach to achieving that functionality.

The first of these desirable characteristics is the separation of the "raw" sequence data as determined by laboratory testing, from the clinical interpretations of those data and from the reference genomes and generalizable knowledge that supports those person-specific clinical interpretations. This separation is notably absent in most contemporary EHRs due to a workflow and reporting tradition similar to that used for surgical pathology specimens, where professional interpretation leads to creation of a human-readable

TABLE 9.1 Desiderata for Integration of Genomic Data into EHRs

1. Lossless data compression from primary observations to clinically relevant subsets
2. Since methods will change, molecular laboratory results should carry observation methods with them
3. Compact representation of clinically actionable subsets for optimal performance
4. Simultaneous support for human-viewable formats, with links to interpretation, and formats interpretable by decision support rules
5. Separation of primary sequence data, which remain true if accurate, from clinical interpretations of them, which will change with rapidly changing science
6. Anticipation of the boundless creativity of Nature: multiple somatic genomes, multiple germline genomes for each individual over their lifetime
7. Support for both individual care and discovery science

From Masys DR, Jarvik GP, Abernethy NF, Anderson NR, Papanicolaou GJ, Paltoo DN, et al. Technical desiderata for the integration of genomic data into Electronic Health Records. J Biomed Inform 2011, used with permission.

document that includes both the findings judged to be of interest and their clinical interpretation, both represented as narrative text. Since only a tiny subset of all possible observable changes in the human genome are currently judged to be clinically actionable, the document-based reporting method necessarily excludes the vast majority of the individual's sequence data in the case of whole-exome or whole-genome testing, and the remainder of the data is often either discarded or held inaccessible in a laboratory system outside of the EHR. The document produced and inserted into the EHR also represents a point-in-time interpretation that may need to be revised as genomic health science progresses, as described below. In addition, bundling of sequence variants with their interpretation in documents (whether paper or the electronic equivalent, the Portable Document Format—PDF) thwarts the creation of automated approaches to clinical decision support (CDS), since computer programs must first identify the findings of interest within documents that lack uniform structure. Thus, a related recommendation (#5 in Table 9.1) is that EHR systems simultaneously support both human-viewable and machine-readable formats specifically to facilitate implementation of patient-specific decision support logic. This functionality becomes useful only if clinical laboratories report results in both human-viewable and machine-readable formats, which is uncommon at present.

The second of the technical desiderata listed in Table 9.1—lossless data compression from primary observations to clinically manageable subsets—derives from both the volume of the high-throughput sequencing data compared to commonly available disk storage and network bandwidth in clinical systems, and the observation that the majority of the data is not currently of immediate relevance to clinical care. The terms "lossless" and "lossy" are used in information systems parlance to denote whether compressed data can be restored to their complete original form and size; as a general rule the more detail that can be sacrificed without losing the essential information contained in the data, the more dramatically the data can be compressed. As noted above, details count in molecular sequence data so lossless compression is required. A related recommendation (#4 in Table 9.1) regards the need for clinical systems to function with subsecond response times that preclude extensive examination or real-time analysis of molecular data. Thus, compact data formats in the form of codes that represent precomputed clinically relevant genomic states, such as an individual having a drug metabolizing pathway variant, should be supported by the EHR system.

The observation that no current laboratory method for genome sequencing is without errors suggests that EHR systems need to include a link between molecular results and the methods used to produce those results. This provides a plausible basis for automatically recognizing and annotating such errors as experience is gained with each of the technologies used to generate clinically relevant results. This linkage of results to methods is a component of widely used clinical laboratory data standards such as Logical Observation Identifiers Names and Codes (LOINC) [4], which is in turn, a component of a more general set of health data standards called HL7 that are a common interlingua of clinical systems [5]. Genomic data standards are in flux as of this writing, with candidate standards such as the Genome Variation Format proposed as a means of providing data comparability across different sequencing technologies [6,7]. LOINC has also been modified to encode DNA sequence variants (e.g., mutations, indels) using a variety of genetic data standards, such as gene symbols and identifiers from Human Gene Nomenclature Committee, variant nomenclature from the Human Genome Variation Society, variant identifiers from the Single Nucleotide Polymorphism Database, and reference sequences from National Center for Biotechnology Information and European Bioinformatics Institute [8].

The last two of the consensus desiderata listed in Table 9.1 represent a design philosophy more than a specific set of functional requirements. The first of these, to anticipate fundamental changes in the understanding of human molecular variation, suggests that there is no natural upper limit on the number of copies of an individual's full genomes, exomes, proteomes, etc., that may be found to have clinical value and thus justification for inclusion in EHR systems. Thus a firm estimate of the total amount of disc storage or communications bandwidth needed for clinical systems is simply unavailable, but expected to rise substantially over time, regardless of whether the primary molecular sequence data is stored inside of or external to the EHR. The recommendation that EHR systems "support both individual clinical care and discovery science" is both a radical departure from the traditional separation of research records from individual care records, and a recognition that each individual is a unique "experiment of Nature" that has the potential to extend our collective understanding of the relationship between molecular variation and personalized health care. In this context, EHR-derived phenotypes have already demonstrated substantial value for discovery science. These include the determination of genetic factors for complex traits, and the ability of EHR data to support "phenome-wide association studies," where a polymorphism associated with one disease may be found by use of clinical data to also be associated with other health conditions [9].

The original technical desiderata were subsequently extended by authors at the University of Utah, based on their finding that the original seven recommendations were necessary but not sufficient to guide implementation of automated CDS. Their additional desiderata are summarized in Table 9.2 [10].

A systems infrastructure for decision support has the potential, particularly in clinical domains where the volume and/or complexity of information exceeds the bounds of unaided human cognition, to reduce medical errors, improve clinician performance, and improve the quality of care [11]. These goals, and a technical means of achieving them, are highly relevant to genomic medicine, where the data representing personal molecular variation are both voluminous and complex. CDS systems use automated recognition logic to match individual patient characteristics as available within the EHR to predetermined scenarios where best practice guidance is available, and if a match is found, generate outputs of various types. These can include guided provider order entry, alerts and reminders that may be educational or have specific actionable recommendations, and/or computed results such as multivariate risk prediction programs [10]. As the desirable features of EHR systems in Table 9.1 derive from the characteristics of genomic data, so the features of an ideal genomic CDS (GCDS) system derive from the complexities of clinically actionable patterns in that data.

TABLE 9.2 Desiderata for Integration of Genomic Data with CDS

1. CDS knowledge must have the potential to incorporate multiple genes and clinical information
2. Keep CDS knowledge separate from variant classification
3. CDS knowledge must have the capacity to support multiple EHR platforms with various data representations with minimal modification
4. Support a large number of gene variants while simplifying the CDS knowledge to the extent possible
5. Leverage current and developing CDS and genomics standards
6. Support a CDS knowledge base deployed at and developed by multiple independent organizations
7. Access and transmit only the genomic information necessary for CDS

From Ullman-Cullere MH, Mathew JP. Emerging landscape of genomics in the electronic health record for personalized medicine. Hum Mutat 2011;32(5):512−516, used with permission.

Because many genetic disorders, particularly common, complex traits such as hypertension and diabetes, have measurable associations with up to hundreds of genes, as well as dependence on personal health factors such as age, weight, and diet, CDS systems must have the capacity to incorporate and reason with information from multiple genes as well as with an individual's "traditional" clinical information as represented in the EHR. In addition, since diseases such as cystic fibrosis and familial polyposis have demonstrated that many variants within a given gene may give rise to essentially identical physiologic effects, a related design goal for CDS systems is to support a large number of gene variants while simplifying the CDS knowledge to the extent possible.

Just as the laboratory methods for acquiring genomic data are in flux, so also are the methods that are used to define of clinically relevant variants. For this reason CDS systems need to keep the knowledge regarding variant interpretation and clinical actions to be taken in the setting of a particular variant separate from the rules or algorithms that are used to identify and classify the variant. Clinical actions recommended by such systems will also need to be kept up-to-date with best practice guidelines and regulatory guidance.

There are and will continue to be different technologies used to create EHR systems. In this pluralistic environment, three of the GCDS desiderata have particular relevance. First, it is clear that CDS knowledge must be in formats that have the capacity to support multiple EHR platforms with data representations that require minimal or no modification when shared among heterogeneous systems. To do this, it will be important that newly emerging systems leverage current and developing data standards for both general purposes CDS and genomic data. Since no single institution will have the resources to independently create a full complement of the decision support cases relevant to genomic medicine, it will be important for local CDS systems to support and be able to exchange information with a shared CDS knowledge base developed by multiple organizations. An open GCDS knowledge library for genomic medicine has been proposed [12], and pilot projects are underway to implement and test it and similar approaches [13].

Welch et al. [10] also propose the desiderata that clinical systems access and transmit only the genomic information necessary for CDS. This recommendation reflects an interest in parsimony and limiting of information in EHRs that may have no immediate relevance for clinical care. An alternative view supports the premise that being overly restrictive in the amount of information made available to clinicians does not well serve potential educational uses of the data for both clinicians and patients and may represent an unwarranted form of "genetic exceptionalism." An emerging set of requirements arise from federal incentives for adoption of EHRs, which include the ability of patients to download their EHR data, including laboratory results [14]. This download functionality has been termed "the Blue Button" within the user interface provided to patients by healthcare organizations, and it can

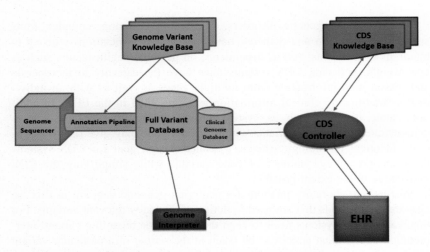

FIGURE 9.1 An architecture for whole-genome sequence enabled CDS. Source: *From Welch BM, Loya SR, Eilbeck K, Kawamoto K. A proposed clinical decision support architecture capable of supporting whole genome sequence information. J Pers Med 2014;4(2):176–199. doi:10.3390/jpm4020176, used with permission [15].*

reasonably be expected that the ability to download one's complete germline and somatic genomes will be a common feature of state-of-the-art EHR systems, as it exists already in direct-to-consumer genomic testing services.

A graphical depiction of the architectural components of an idealized genomic medicine CDS system is shown in Fig. 9.1.

The components of this modular approach include genome sequencing and annotation, reference genome databases, genome variant knowledge bases, a GCDS knowledge base, and the EHR system, which is the real-time recipient of decision support guidance sent by a computer program called the CDS controller. The CDS controller matches individual patient clinical characteristics contained in the EHR, genomic variants found for that patient, and knowledge about those variants, to create a patient-specific message or document for processing by the EHR. This approach limits the amount of data storage and analytical complexity that has to be incorporated directly into the EHR system and allows the infrastructure for acquiring and interpreting genomic data to evolve more or less independently of EHR technologies.

CHOOSING AN EHR SYSTEM FOR GENOMIC AND PERSONALIZED MEDICINE

The bulk of this chapter is forward-looking, enumerating the functional characteristics of ideal EHR systems and their associated CDS functionality.

One reason that these features are called "desirable" is that, by and large, they do not exist in current generation commercially available EHR systems, and exist only in subsets among even advanced academic medical centers that have their own "home grown" EHR with features the institution deems essential to high quality care delivery. The problem of clinical laboratories measuring many genomic loci by high-throughput methods and reporting only a small subset in a narrative document (PDF) format is currently endemic. When genomic data is reported in this fashion, every contemporary EHR can claim to be able to accept and display genomic data, since document-based information retrieval and display is a core functionality of all EHRs.

What then can a practitioner or an organization that is planning to implement clinical services involving exomes, genomes, and other forms of "omics" data reasonably do to choose an EHR system that will provide state-of-the-art functionality for the present as well as for a data-intensive future? The Office of the National Coordinator for Health Information Technology has developed certification criteria for EHR systems, and a certification process by which EHR systems are evaluated [16]. Use of certified systems is incentivized by payment rules of the Centers for Medicare and Medicaid Services, and eligible healthcare providers and organizations are required to demonstrate "Meaningful Use" of the EHR in clinical operations [14]. So the first step to choosing and implementing an EHR for genomic and personalized medicine is to focus on candidate systems that have met basic federal certification criteria. Within that certification there are a wide range of technologies, and a key part of the selection process is to map candidate EHRs' capabilities to the clinical mission and goals of one's healthcare organization.

Automated CDS capability is defined within Meaningful Use as "functionality that builds upon the foundation of an EHR to provide persons involved in care processes with general and person-specific information, intelligently filtered and organized, at appropriate times, to enhance health and health care" [17]. This broad definition enables many different information retrieval and display functions to qualify as CDS. As a result, it behooves a healthcare organization to "look under the hood" of a candidate EHR system to determine whether it will be suitable for genomic and personalized medicine. One plausible approach is to ask a vendor representative to examine the technical desiderata listed in Tables 9.1 and 9.2 and provide a point-by-point response as to which and how many of the desired functional characteristics their system(s) provide. The most important requirement for a potential buyer is to understand whether the EHR supports both human-readable and machine-interpretable genomic data, and how these two functions are synchronized within the system. Another important characteristic to investigate is how CDS knowledge can be imported, edited, or developed de novo within the EHR system. Authoring systems for CDS knowledge management vary widely among EHR vendors, and a user interface that is intuitively appealing and requires relatively little user training will contribute substantially to an organization's ability to develop and deploy a decision support infrastructure for genomic and personalized medicine [18].

THE PATH FORWARD

It is clear that a model of health care built on the capacity of autonomous individual practitioners reading and remembering an explosively growing professional literature will be inadequate for the era of genomic and personalized medicine [19]. A systems infrastructure for representing actionable clinical knowledge related to personal molecular variation, in formats that support automated CDS, will need to exist at a national or an international level. As noted above, the notion of developing a "Public Library" of interoperable CDS is a current topic of public discussion [20] and a focus of federally supported efforts such as the Clinical Decision Support Consortium [21], which seeks to develop both the technologies and the incentives for "learning healthcare organizations" [22] to share experience related to EHR-based CDS systems and develop a shared knowledge base at a national and an international level.

In this context, a unique opportunity exists for genomic medicine to advocate for a mechanism for harvesting the real-world experience of providers and patients, as represented in EHR systems across many independent healthcare provider organizations. Historically, CDS has been viewed as an information-providing process that ends when the information has been delivered to the intended user [23]. To the extent that rare genetic variants inform clinically meaningful differences among individuals, it will be important to aggregate data regarding clinical decisions made and the outcomes that follow those decisions, across very large numbers of individuals. Technology exists for a shared systems infrastructure of "closed-loop decision support" based on automated monitoring of events recorded in an EHR subsequent to a decision support event occurring. Correlating healthcare process and outcome measures with prior decision support events has the useful property that decision support logic and interventions can be refined whether or not practitioners follow the recommended guidance, and in fact benefits when healthcare providers are "smarter than the decision support guidance." The power of such a system would be proportional to the number of organizations that contribute to, and use knowledge from, a CDS Public Library on behalf of a learning healthcare system.

Effective CDS delivered in the context of EHR systems, which has been a useful accessory to clinical care in organizations that have adopted it to date, promises to become a much more mainstream foundation for evidence-based care in the 21st century. Genomic and personalized medicine will be a major driving force in achieving that future state.

REFERENCES

[1] Blumenthal D, Tavenner M. The "meaningful use" regulation for electronic health records. N Engl J Med 2010;363:501−4.

[2] e-Health. Saving lives, saving money: the imperative for computerized physician order entry in Massachusetts hospitals. The clinical baseline and financial impact study. Boston, MA: Massachusetts Technology Collaborative and New England Healthcare Institute; 2008. p. 2008.

[3] Masys DR, Jarvik GP, Abernethy NF, Anderson NR, Papanicolaou GJ, Paltoo DN, et al. Technical desiderata for the integration of genomic data into Electronic Health Records. J Biomed Inform. 2012;45(3):419−22.

[4] Logical Observation Identifiers Names and Codes. http://www.loinc.org [accessed 29.12.14].

[5] HL7 Clinical Genomics Working Group. http://www.hl7.org/Special/committees/clingenomics [accessed 29.12.14].

[6] Reese MG, Moore B, Batchelor C, Salas F, Cunningham F, Marth GT, et al. A standard variation file format for human genome sequences. Genome Biol 2010;11:R88.

[7] http://www.sequenceontology.org/resources/gvf.html [accessed 23.11.15].

[8] Ullman-Cullere MH, Mathew JP. Emerging landscape of genomics in the electronic health record for personalized medicine. Hum Mutat. 2011;32(5):512−16.

[9] Denny JC, Bastarache L, Ritchie MD, Carroll RJ, Zink R, Mosley JD, et al. Systematic comparison of phenome-wide association study of electronic medical record data and genome-wide association study data. Nat Biotechnol 2013;31(12):1102−10.

[10] Welch BM, Eilbeck K, Del Fiol G, Meyer LJ, Kawamoto K. Technical desiderata for the integration of genomic data with clinical decision support. J Biomed Inform. 2014;51:3−7.

[11] Jaspers MWM, Smeulers M, Vermeulen H, Peute LW. Effects of clinical decision-support systems on practitioner performance and patient outcomes: a synthesis of high-quality systematic review findings. J Am Med Inform Assoc 2011;18:327−34.

[12] Overby C, Heale B, Aronson S, Cherry JM, Dwight S, Milosavljevic A, et al. Providing access to genomic variant knowledge in a healthcare setting: a vision for the ClinGen Electronic Health Records Workgroup. Clin Pharmacol Ther. 2015;29. Available from: http://dx.doi.org/10.1002/cpt.270.

[13] Roundtable on Translating Genomic-Based Research for Health; Board on Health Sciences Policy; Institute of Medicine. Genomics-enabled learning health care systems: gathering and using genomic information to improve patient care and research: Workshop Summary. Washington (DC): National Academies Press (US); 2015 Jul 8. Available from: http://www.ncbi.nlm.nih.gov/books/NBK294179/.

[14] Centers for Medicare and Medicaid Services. EHR Meaningful Use requirements. Available from: http://www.cms.gov/Regulations-and-Guidance/Legislation/EHRIncentivePrograms/Stage_2.html [accessed 29.12.14].

[15] Welch BM, Loya SR, Eilbeck K, Kawamoto K. A proposed clinical decision support architecture capable of supporting whole genome sequence information. J Pers Med. 2014;4(2):176−99. Available from: http://dx.doi.org/10.3390/jpm4020176.

[16] ONC EHR Certification Process. Available from: http://www.healthit.gov/providers-professionals/certification-process-ehr-technologies [accessed 29.12.14].

[17] Clinical Decision Support Tipsheet. Available from: http://www.cms.gov/Regulations-and-Guidance/Legislation/EHRIncentivePrograms/Downloads/ClinicalDecisionSupport_Tipsheet-.pdf [accessed 29.12.14].

[18] Mardon R, Mercincavage L, Johnson M, et al. Findings and lessons from AHRQ's Clinical Decision Support Demonstration Projects (Prepared by Westat under Contract No. HHSA 290-2009-00023I). AHRQ Publication No. 14-0047-EF. Rockville, MD: Agency for Healthcare Research and Quality; June 2014.

[19] Masys DR. Effects of current and future information technologies on the health care workforce. Health Affairs 2002;21(5):33−41.

[20] Genomic Medicine Meeting VII: genomic clinical decision support—developing solutions for clinical and research implementation, https://www.genome.gov/27558904 [accessed 29.12.14].

[21] Clinical Decision Support Consortium. Available from: http://www.cdsconsortium.org/ [accessed 29.12.14].

[22] Olsen L, Aisner D, McGinnis JM, editors. The learning healthcare system. Washington, DC: National Academies Press; 2007.

[23] Kawamoto K, Houlihan CA, Balas EA, Lobach DF. Improving clinical practice using clinical decision support systems: a systematic review of trials to identify features critical to success. BMJ 2005;330(7494):765.

Chapter 10

Data Sharing and Privacy

Edward S. Dove[1,2], Graeme T. Laurie[1,2] and Bartha M. Knoppers[2,3]
[1]University of Edinburgh, Edinburgh, United Kingdom, [2]Global Alliance for Genomics and Health, Toronto, Canada, [3]McGill University, Montreal, Canada

Chapter Outline

INTRODUCTION

The world is more connected than ever as we craft far-reaching networks of communication. A tangible effect of our networked world in the genomic and personalized medicine context is the increased sharing of genomic and clinical data between professionals, laboratories, hospitals, organizations, patients, participants, and—increasingly—across national borders [1–3]. Sharing data facilitates both translational and precision medicine, not to mention research discovery [4,5].

Data sharing profoundly impacts both the pace and scope of international research collaboration and data-driven biomedical research discovery, and the sustainability of privacy [6]. Technological innovations and social movements are encouraging data sharing and accordingly cause many to continually evaluate the robustness of data sharing systems. Data sharing is becoming a norm in genomic research—so much so that many now speak of a "genome commons" [7]. Gradually, this might extend to clinical trials and clinical practice [8,9]. Data sharing is a social activity, and privacy is fundamentally a social consideration. Like any partnership, there are responsibilities on both sides.

One must also, of course, be attuned to the multiplicity of laws that are engaged by these activities. A central concern is that genomic and clinical data must be collected, used, and shared in ways that respect the fundamental rights and interests of individuals and society, as reflected in national and international laws, and most notably, privacy. Data sharing carries a spectrum of risks, particularly re-identification of individuals and related others, and thereby represents a prima facie threat to their privacy.

Notwithstanding, unrealistic or obsolete expectations of privacy protection, and irrational fixations on hypothetical but remote risks, can unduly impede the flow of genomic and clinical data and thereby harm valuable biomedical research and innovation [10,11]. This is the case with the "consent or anonymize" paradigm [12] and the "fetishization" of the consent form as a panacea for privacy concerns, both of which are unsustainable in the age of Big Data, social media, direct-to-consumer genetic testing, and patient-led data sharing initiatives [12,13]. This is compounded when individuals speak (often erroneously) of "owning" genomic information or having their genomic data "belong" to them. Such discussions of privacy and data sharing can run into potentially irreconcilable tension. This is not to suggest that concerns about privacy are unimportant; rather, we must be attuned to varying conceptions of privacy and autonomy and consider them *together* in particular contexts along with wider individual and public interests. To speak about responsible genomic and clinical data sharing is to speak about its proper contours. Privacy helps shape those contours and helps demonstrate the trustworthiness needed to allow biomedical research and innovation to flourish.

In this chapter, then, we discuss three themes: first, the practice and benefits of genomic and clinical data sharing; second, how privacy and data privacy regulation may challenge data sharing—or help us to remain vigilant of how data sharing is structured and governed; and finally, the ways in which one organization in particular, the Global Alliance for Genomics and Health (GA4GH), is addressing these challenges. We conclude with a message for sustaining privacy as a fundamental value in biomedical research and clinical practice, and with a clarion call for assuring that our conceptions of privacy, data protection, and oversight within data sharing systems are fit for purpose in 21st century genomic and personalized medicine.

TOWARD GLOBAL DATA SHARING

Data sharing has always existed in some form in science. In genomics, data sharing was propelled in the 1990s by the groundbreaking Human Genome Project data release policy, the "Bermuda Principles" [14]. These principles maintained that all genetic sequencing data generated by centers funded for large-scale human sequencing should be made freely and rapidly available in the public domain. Since then, these principles have been expanded and refined in further policies [15,16]. Large infrastructures such as the database of Genotypes and Phenotypes (dbGaP, http://www.ncbi.nlm.nih.gov/gap) and the European Genotyping Archive (https://www.ebi.ac.uk/ega/home) have further operationalized this approach by promoting the sharing of sequence data. Funders such as the Wellcome Trust, National Institutes of Health (NIH), and Genome Canada, and policy organizations such as the Organisation for Economic Co-operation and Development, have developed policies that encourage or require data sharing in science in general and in genomics in particular. Data sharing is now also an emerging norm in clinical research [9,17,18], and with good reason: making data accessible provides many benefits to individuals and society [19−21] (Box 10.1).

Several factors explain this shift to global data sharing. First, technological advances such as cloud computing, next-generation sequencing, social media, as well as long-term investments in infrastructure and international consortia, are fostering the creation of data linkable and sharable electronic medical record systems, large-scale disease cohort studies, population health data repositories, "citizen science", and biobanks [22]. Second, funders, journals, and institutions such as universities increasingly impose data sharing requirements, including the deposit of data in open or restricted access repositories [23]. Third, and not insignificantly, many patients and participants are requesting that the data they contribute to science and medicine be "opened up" for further, follow-on research use [24,25] (Box 10.2). Sharing data such as research results (whether aggregated or individualized) with those who contribute data to science and medicine operationalizes the ethical principles of reciprocity, solidarity, mutual responsibility, and respect. Fourth, the

BOX 10.1 The Multiple Benefits of Data Sharing

Among other benefits, data sharing can:
- reinforce open scientific inquiry;
- enable reproduction and verification of the results of past research;
- avoid unnecessary duplication;
- improve surveillance of drug safety and effectiveness;
- encourage diversity of analysis and opinion;
- allow for assurance of validity of the research or diagnosis and possible replication of research results;
- promote new research and interdisciplinary analyses;
- enable the testing of new or alternative methods, designs, and hypotheses;
- improve methods of data collection and measurement;
- help maintain data integrity and resource optimization;
- provide safeguards against misconduct related to data fabrication and falsification;
- encourage the recognition for discovery and innovation;
- facilitate teaching of new researchers and clinicians;
- permit the exploration of topics not envisioned by the initial study investigators;
- create new datasets through the combining and pooling of data from multiple sources;
- facilitate more effective use of funders' limited financial resources.

BOX 10.2 Examples of Patient-Centered Initiatives Encouraging Data Sharing

AllTrials (http://www.alltrials.net/)

The AllTrials campaign is an initiative of Bad Science, *BMJ*, Centre for Evidence-Based Medicine, Cochrane Collaboration, James Lind Initiative, *PLOS* and Sense About Science and is being led in the United States by Dartmouth's Geisel School of Medicine and the Dartmouth Institute for Health Policy & Clinical Practice. AllTrials was launched in January 2013 to call for all clinical trials to be registered and results reported.

Free the Data (http://www.free-the-data.org)

Free the Data is a consortium of organizations, managed by US-based Genetic Alliance and supported by Invitae, Private Access, the University of California, San Francisco, the International Collaboration for Clinical Genomics, Syapse, and Captricity. Free the Data enters *BRCA*1 or 2 mutations into a public database and gathers patients' health information so that associations between mutations and health outcomes can be discovered, and research can be advanced. Patients select their personal sharing preferences with researchers and disease advocacy organizations.

PatientsLikeMe (http://www.patientslikeme.com/)

A for-profit company and health information sharing website for patients cofounded in 2004 by three MIT engineers. The company has a health data sharing platform where patients can share and learn from real-world, outcome-based health data. PatientsLikeMe is aligned with patient and industry interests through data sharing partnerships; they work with nonprofit, research and industry partners who use their health data to improve products, services, and care for patients.

Personal Genome Project (http://www.personalgenomes.org/)

PersonalGenomes.org is a nonprofit organization working to generate, aggregate, and interpret human biological and trait data by making a wide spectrum of data about humans accessible to increase biological literacy and improve human health. They collaborate with individuals who are willing to share their data publicly with the understanding that re-identification is possible. The organization supports the Personal Genome Project (PGP) global network. The first PGP research study was founded at Harvard Medical School in 2005, and PGP sites now exist at institutions in four countries (the United States, Canada, Austria, and the United Kingdom).

Reg4All (https://www.reg4all.org/)

A partnership between the US-based Genetic Alliance and Private Access, Registries for All (Reg4All) seeks to enable patients to share their health information online with the medical and research community in order to find better treatment options for their health conditions. Patients have the option to share all, some, or none of their health information. Based on their preferences, patients may hear from interested researchers or disease advocacy organizations.

science demands it [26]. As genomic research has increased in complexity, resolution, and volume, being very much part of the "Big Data" movement, researchers need sufficient statistical power to determine the genes related to common and rare diseases. This can mean that tens or even hundreds of thousands—if not millions—of genes, transcripts, and/or proteins need to be studied simultaneously. Breakthroughs can more readily occur when researchers and clinicians are able to match patient genotype and phenotype data, wherever they may be in the world.

MODALITIES OF DATA SHARING

Data sharing occurs when individuals or groups intentionally make their data available to others for use in research, medicine, or other endeavors. Data sharing occurs in a variety of forms, which for heuristic purposes can be called the "how", "what", and "when." How genomic and clinical data are shared tends to fall into three modalities: (1) researchers or clinicians responding ad hoc to individual data requests; (2) researchers and clinicians establishing collaborative networks to pool their data to facilitate knowledge generation in areas of common interest; and (3) researchers and clinicians depositing their data in open/restricted access centralized or distributed databases.

Furthermore, genomic and clinical data tend to be made available through two main forms: open/restricted access and open/restricted data. The latter form speaks to actions performed on the data to make them more or less identifiable (e.g., key-coding or anonymization) and, depending on the actions performed, may make the data exempt from most parts of data privacy regulation (e.g., anonymized data are generally not considered "personal data" for the purposes of data privacy regulation, though pseudonymized/key-coded data usually are) (see Box 10.3 for definition of key data privacy terms). But, even with restricted data, *access* to them may remain relatively unrestricted. Conversely, restricted access may entail sharing unaltered identifiable personal data under strict access conditions and to which data privacy regulations will apply. Thus, while the data themselves might not be "safe", access requirements usually focus on "safe people" (such as accredited researchers) and/or "safe environments" (such as safe havens) [27].

Finally, there is a normative and temporal dimension to data sharing: at what point *should* data should be shared? Beyond the general agreement that genomic data (and arguably, anonymized clinical data) should be released in a "timely" manner, depending on policies, the data, and their quality, some data are shared immediately upon their collection and curation. Other data may be released simultaneously to, or shortly after, data producers have

BOX 10.3 Definition of Key Privacy Terms

Anonymized data

Data that were related to an identifiable individual when collected, but through a process of removing all direct identifiers, thereafter prevents the identity of an individual from being readily determined by a reasonably foreseeable method. Using state-of-the-art techniques, properly anonymized data helps prevent both direct and indirect identification of an individual.

Coded data

Data that are assigned one or more random codes. Direct identifiers are removed from the dataset(s) and held separately. The key(s) (i.e., a piece of data that an encryption algorithm uses to determine exactly how to unscramble the data) linking the code(s) back to direct identifiers are available only to a limited number of persons, e.g., a research team, or are held by a third party (such as the data holder or a trusted third party) and are unavailable to the researchers. Coded data are also known as "pseudonymized data" and "key-coded data."

De-identified data

A defensible, repeatable, and auditable process that consistently provides assurance, based on proven and repeatable statistical methodologies, including removing identifiers and other indices, such that there is a very small risk of re-identification of any data that are made accessible to researchers.

Identifiable data

Data that may reasonably be expected to identify an individual, alone or in combination with other data. One person's genotype data (i.e., the combination of sequence-related data) are, in theory, distinguishable (individuating) from all other people's genotypes, but they may not necessarily be identifiable.

Re-identification

The process of linking de-identified data to an individual.

Adapted from GA4GH Privacy and Security Policy.

completed data analysis, the production of publications derived from the data, and the securing of appropriate intellectual property, all within a reasonable time frame [26]. Longer term retention for likely future value—not currently realizable—brings its own challenges and privacy concerns.

Studies show that a majority of genomic researchers *perceive* value from sharing data and making their data available [20], though many fewer may *actually* share data. The gap between desire and actual practice is worrisome but somewhat understandable. While data sharing may be seen as an emerging scientific norm in genomics, many individual and institutional and policy barriers remain [3,4,20,28–32]. These include: difficulty finding relevant and usable quality data; lack of time and resources to make data accessible; concerns about intellectual property; perceived bureaucracy in getting authorization from a repository, organization, or individual to access data; difficulties in obtaining the required institutional approvals; difficulties in formatting data to ensure reliability and relevance and customizing data for specific applications; and the burdensome size (i.e., compute capacity) and time required to download and move data. Individual or cultural barriers also include concerns about the potential loss of future publications once unpublished data are shared; concerns that grant reviewers place little or no weight on data sharing plans in their reviews; and the lack of incentives to share data, as it is perceived as not contributing to publication records or professional promotion.

Clearly, work remains to be done in removing these barriers and demonstrating the value proposition of "intelligent openness" to maximize the public benefit. Incentives and rewards need to be built into health systems and organizations to create structures that facilitate data sharing for clinicians, scientists, and patient—participants alike. Governments can do their part. For example, Estonia has become a world leader in digital services; they have a national register that provides a single unique identifier for all citizens and residents, and identity cards that provide legally binding identity assurance and electronic signature [33]. The Estonian government has also begun developing an online system to making the genetic data of more than 52,000 Estonians in its national biobank freely available to participants and their physicians; citizens will be able to log on to a central online portal (the same online portal where Estonians file taxes, fill prescriptions, open a business, etc.) to see which research studies have used their sequenced DNA [34]. While the specific Estonian policies are not directly transposable to other countries due to Estonia's particular socio-legal context, an active and sustained government focus on achieving seamless sharing through digital platforms can and should be emulated across the globe.

Further, as the UK's Expert Advisory Group on Data Access has noted, future data sharing incentives should include research funders strengthening approaches for scrutinizing data management and sharing plans associated with their funded research, establishing data sharing as a formal criterion for assessing the track record and achievements of researchers during funding decisions, and ensuring key data repositories serving the data community have adequate funding to meet the long-term costs of data preservation [35]. Other examples include research institutions developing clear policies on data sharing and preservation and providing training and support for researchers to manage data effectively and recognizing data outputs in performance reviews. Journals can also establish clear policies on data sharing and processes [35,36].

Thus, changes from external factors like journals, funders, and universities can help significantly. Researchers in the United States have found, for instance, that NIH data sharing policies, and to a lesser extent, journal and other funder policies, have had a strong influence on increasing data sharing in the life sciences; they have also found, less fortunately, that there appear to be few sanctions or penalties for noncompliance [21]. Yet, there are other, more profound barriers to data sharing. Herein lies the challenge of privacy.

THE IMPACT OF PRIVACY ON DATA SHARING

Sharing genomic and clinical data requires oversight systems that protect privacy to the extent that is necessary and appropriate and relative to the overall objectives to be achieved. Privacy in its informational dimension is a state of affairs whereby information relating to a person is in a state of nonaccess. It is instrumentally valuable, as it enables people to flourish in developing personal relationships, and it is intrinsically valuable, as it is grounded in the universal human values of dignity, integrity, and autonomy. Genomic data are characteristically viewed as "sensitive" because they are both uniquely identifying and a source of familial information. Clinical data are also often perceived as sensitive. Respondents to a survey commissioned by the Wellcome Trust in 2013 strongly felt "that personal medical data are confidential, private and sensitive, and should not be shared outside secure, authorized bodies" [37].

Potential informational harms that could arise from privacy violations include psychological harms in the form of embarrassment, guilt, stress, anxiety, depression, and altered behavior; social harms in the form of stigmatization, embarrassment within one's social group, discrimination, and criminal prosecution; and economic harms in the form of loss of or change in employment. These harms can affect not just individual participants, but their families and communities as well. This said, hard evidence of harms arising from the misuse of data from health research and health care has been limited [38]. Yet, it bears emphasizing that individuals will likely only contribute their genomic and clinical data to science if they trust that their data are sufficiently protected. Robust privacy protection is needed, and it is the shared responsibility of governments, organizations, and individuals to effectively protect these data. The courts have long recognized that it is equally in the public interest that privacy interests be well protected.

Privacy impacts genomic and clinical data sharing especially because powerful cryptographic and statistical techniques allow the drawing of accurate inferences, possibly revealing the identity and characterization of individuals and connected others. Clinical data from trials concerning rare diseases or other small trials are difficult to anonymize effectively. Genomic data are arguably the most difficult type of data to de-identify. Research has shown that most standard genomic data anonymization techniques tend to be insufficient since even after anonymization, the data may still be capable of re-identifying an individual [39].

Some caution is in order, however. First, anonymizing genomic and clinical data may not be desired as it can significantly weaken the data's scientific utility and prevent ongoing communication with patients and participants. Second, the possibility of re-identification does not express the probability of the risk manifesting. Factors include: the expertise (or otherwise) of those working with data, the type of data, the time and costs involved, and the state of the data's identifiability. Moreover, we should be wary to perpetuate a worldview that absolute anonymity is ever achievable, let alone desirable: No data exist in splendid isolation from their social context.

Identifiability is thus best viewed as lying across a spectrum and depends on the ease of individuation (Table 10.1). Re-identification can be achieved in several ways: directly, by matching genomic data against a reference genotype; deductively or by linking them to nongenetic databases and matching them to genotype and other associated data; and inferentially, by profiling genomic data from DNA analysis [10]. Erlich and Narayanan also categorize data breaches in three forms: (1) identity tracing, i.e., exploiting quasi-identifiers (indirect identifiers) in DNA data to uncover an unknown genetic dataset's identity; (2) attribute disclosure, i.e., using known DNA data to link a person's identity with a sensitive phenotype; and (3) completion techniques, i.e., uncovering sensitive genomic areas that were masked [39].

The re-identification risk is real. A research team re-identified five people selected at random from a DNA database—without using a reference sample. Using their DNA, ages, and the US states where they lived, the team identified the five individuals, as well as some of their relatives, identifying in total nearly 50 people [40]. Other studies have also demonstrated the potential for re-identifying individuals through advanced techniques or by triangulating multiple publicly available data sources, such as census and genealogy data, obituaries, or voter registries [41,42]. One research group has found that relying on as few as 30−80 individual (statistically independent) single-nucleotide polymorphism (SNP) loci can enable unique individual identifiability [43]; another group found that an individual's SNP profile could potentially be identifiable even when it is aggregated with 1000 or more other samples [44].

Together, these findings have caused many organizations to revisit their data sharing systems and policies and institute more tiered access restrictions. Data repositories may establish data access committees (DACs) to govern third-party access to individual-level and potentially identifiable data [45]. It is common practice for individual-level genomic and clinical data to be shared (i.e., made accessible) only through controlled access portals, unless participants or patients have provided express consent for them to be freely shared [2,23]. Aggregate-level genomic and clinical data, on the other hand, may be publicly accessible or subject to many less access restrictions.

However, some criticize de-identification and data access mechanisms for not providing individuals enough "control" over their data, which is seen by some as a core element of data privacy [46]. Moreover, some DACs may be limited in their ability to detect violations of data access conditions or enforce restricted access restrictions once data have been downloaded by a

TABLE 10.1 The Spectrum of Identifiability

Identification Level	1 (Most Identifiable)	2	3	4 (Least Identifiable)
Definition	Data that are or can be fully identifiable to everyone	Data that are unidentifiable to most people, but remain re-identifiable to those with access to the key(s)	Data that are likely no longer identifiable to anyone	Data that never were identifiable to anyone
Synonymous terms	• Identified or identifiable • Personal • Nominative	• Coded • Key-coded • Pseudonymized • Reversibly de-identified • Linked anonymized • Masked • Encrypted	• Anonymized • De-identified • Irreversibly de-identified • Nonidentifiable • Unidentifiable • Unlinked anonymized	• Anonymous

Adapted from GA4GH Privacy and Security Policy.

user [47]. Even if instances of data misuse are rare, the ability to have fine-grained but flexible access controls that audit data uses is critical to maintain trust in the data sharing system.

Accountability is key here. Systems of monitoring and appropriate sanctioning can go a long way in rendering systems objectively trustworthy. Contrariwise, the requirement for separate access applications to individual data repositories each time data are needed can cause delays and additional costs and can also prove an obstacle to the long-term sustainability of data sharing [31]. Proportionality of regulation must be relative to the privacy risks and the likely public benefits [11]. Greater harmonization of access procedures through global, coordinated systems of access is needed [45].

Data privacy regulation also plays a big role in challenging genomic and clinical sharing. This type of regulation targets "personal data", which is often defined as data that relate to an identified or identifiable individual person. While some data privacy regulation seeks to promote responsible data sharing, often this is not the case. Regulations differ widely across locations and increasingly are outdated; they are geographically siloed and address point-to-point "data transfers" rather than borderless and dynamic data flows across global networks. Many create misleading notions of ownership or deep individual control over one's personal data. Inconsistency predominates. Some jurisdictions legally permit international sharing of data for research purposes, but many others still do not, making international data exchange between researchers difficult, if not impossible, to achieve. Other jurisdictions do not address cross-border data sharing at all. Thus, rather than enabling socially beneficial and efficient data flow, too many divergent regulations cause significant costs, delays, or inertia at the expense of research progress and human health. We are more optimistic, however, about the development of harmonized self-regulation and the turn toward principles-based, proportionate governance of responsible data sharing.

THE GLOBAL ALLIANCE FOR GENOMICS AND HEALTH

Even in the best of worlds, attempting to share genomic and clinical data while in a "culture of caution, confusion, uncertainty and inconsistency" [48] will aid neither individuals nor society. Some tools and initiatives are being created, however, that are making great strides in promoting the responsible sharing of genomic and clinical data and the expertise, resources, and innovative tools for the harmonization of biomedical research (see Box 10.4 for examples).

The GA4GH (www.genomicsandhealth.org) in particular is making great strides to harmonize data sharing tools and frameworks, as well as to both promote genomic and clinical data sharing and protect the legitimate privacy interests of participants. The Global Alliance is an international nonprofit organization formed in 2013 that develops and promulgates harmonized approaches (both technical and regulatory) for the effective and responsible

BOX 10.4 **Examples of Tools and Initiatives That Promote Responsible Sharing of Genomic and Clinical Data**

- Beacon Project (http://ga4gh.org/#/beacon)
- BRCA Exchange (http://brcaexchange.org/)
- Cafe Variome (http://cafevariome.org)
- ClinVar (http://www.ncbi.nlm.nih.gov/clinvar)
- DataSHIELD (http://datashield.org/)
- DNAdigest (http://dnadigest.org)
- Leiden Open Variation Database (LOVD, http://www.lovd.nl/3.0/home)
- Matchmaker Exchange (http://www.matchmakerexchange.org/)
- Open Researcher and Contributor ID (ORCID, http://orcid.org/)
- PhenomeCentral (https://phenomecentral.org)
- Public Population Project in Genomics and Society-International Policy interoperability and data Access Clearinghouse (P3G-IPAC, http://www.p3g.org/ipac)
- Research Data Alliance (https://rd-alliance.org)

sharing of genomic and clinical data across jurisdictions (see Fig. 10.1 for organizational structure). Comprised of a broad and diverse coalition of over 1200 organizations and individuals in health care, research, disease and patient advocacy, life science, and information technology in more than 70 countries, the Global Alliance's mission is to accelerate progress in science and medicine through global data sharing.

Within the Global Alliance are Working Groups that are tasked with developing tools and solutions in a specific issue area. The Regulatory and Ethics Working Group, which we help lead, has actively implemented a harmonized approach to improve data sharing by developing a *Framework for Responsible Sharing of Genomic and Health-Related Data*, which is freely available online (http://genomicsandhealth.org/framework-responsible-sharing-genomic-and-health-related-data).

The *Framework* contains foundational data sharing principles and core elements; uniquely, it is situated within a human rights framework, namely, the right of all citizens to benefit from advances in science and its applications, and the right for contributors to be recognized therefor [4]. Collaborative input in developing the *Framework* was provided from individuals as well as biomedical, patient advocacy, and ethical, policy and legal organizations, committees, and projects from all regions of the world. Translated into more than 12 languages and referenced in the Global Alliance's Constitution, the *Framework* initiates the first step in building a responsible global data sharing system that protects and promotes privacy in a proportionate manner.

In addition, to better address the specific challenges of privacy, the *Framework* is elaborated by a specific Privacy and Security Policy, which is also freely available online (http://genomicsandhealth.org/work-products-demonstration-projects/privacy-and-security-policy). This Policy is robust, proportionate,

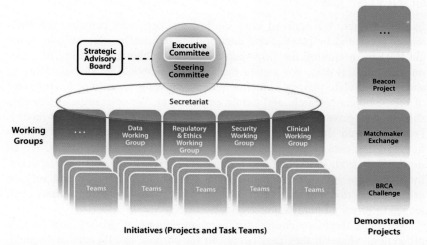

FIGURE 10.1 **Organizational structure of the Global Alliance for Genomics and Health.** The Global Alliance for Genomics and Health is a collaborative, global partnership led by a Steering Committee and an executive team. It also has a Strategic Advisory Board. Global Alliance members compose a plenary body and collaborate to advance its mission. These members participate in four Working Groups (Clinical, Security, Data, and Regulatory and Ethics). The Steering Committee and Working Groups are supported by professional staff, located at several host institutions. Working Groups develop initiatives, which include tasks and small-scale projects. A "Task Team" has a time-bound mission and delivers key tools and solutions. Small-scale projects are specific to a Working Group. Global Alliance "Demonstration Projects" use and improve on tools and solutions and demonstrate the value of data sharing through practical application of the organization's work. These, and other large-scale projects, have broad impact on data sharing stakeholders, reach a diverse community, and may require many resources.

and flexible in setting forth principled and practical guidance for sharing genomic and clinical data in a way that promotes and protects the privacy and security of data contributors. The Policy facilitates compliance with the obligations and norms set by international and national privacy laws, policies, and interoperable standards. It also identifies privacy best practices, tools, and benchmarks that support privacy and security in genomic and health-related data sharing. The Policy is intended to be interpreted in a proportionate manner that acknowledges different levels of risk and community cultural practices and, where appropriate, different contexts for data sharing and use. In contrast to data privacy law's notion of "data controllers," whose command-and-control remit is often to limit (if not prohibit) data uses, the Policy focuses on the notion of "data stewards," who take charge of the responsible and active *use* of data. In sum, if the Global Alliance's *Framework* and the Privacy and Security Policy are endorsed and implemented by data users, stewards and their organizations, as well as regulatory authorities across various jurisdictions, whether Global Alliance members or not, we believe it could make tremendous strides in building a sustainable global model for responsible genomic and clinical data sharing.

CONCLUSIONS

In this chapter, we outlined data sharing practices and benefits, the challenges presented by privacy and data privacy regulation, and the role of the GA4GH in promoting the *responsible* (privacy-enhancing) sharing of genomic and clinical data. We argued that while data should be shared, doing so should not violate privacy, which remains a pressing concern at the interface of biomedicine, research, and public engagement. We also stressed that data sharing and privacy are not zero-sum; they can and must work together. Promoting data sharing and privacy is a collective responsibility of regulators, funders, data producers, users, journals, institutions, and citizens. We all must ensure both individual and group-based privacy interests are respected, for it is certain that as biomedicine and technology advance and rely on greater collection and sharing of data, they will pose ever-more serious challenges to privacy interests.

We must therefore ensure that our conceptions of privacy and privacy protection, whatever they may be, are fit for purpose in 21st century biomedical research and practice. Point-to-point transfers have given way to dynamic flows of data and storage of databases in clouds with diminished attention given to national borders. This has led to improvements in human health; it has also led to increased risk of privacy-related harms. Yet, continually and solely relying on individual specific consent to minimize harms and to perform the task of ensuring individuals know what data are collected from them, and what they are being used for, is simply no longer feasible. Nor, for legal, ethical, technical, and scientific reasons, is anonymization always the suitable alternative. Instead, we must harness, if not forge, security, regulatory, and ethical frameworks that allow for responsible data sharing on a sound, risk-attuned, ethical, and principled basis. In other words, we must pivot toward good data governance.

Most patients and participants know that absolute guarantees of privacy in genomic research are impossible, yet this knowledge in itself likely will not deter them from engaging in research [49,50]. Therefore, with the help of the Global Alliance, we should work toward achieving several key goals in the near-term that will further promote responsible data sharing:

- engaging patients and participants to explain and learn about intended and actual uses of their data and to assure public resources arising from biomedical research benefit them;
- incentivizing research and clinician compliance with international privacy norms, professional codes of conduct, and flexible, practical, and principles-based policies that address reasonably foreseeable privacy risks;
- encouraging policymakers and regulators to regulate data use rather than technology or types of data;

- encouraging funders and institutions to develop sustainable financial mechanisms for data sharing;
- building capacity and infrastructure to facilitate data sharing in developing regions;
- developing robust governance and data access authorization mechanisms;
- holding data users accountable for their actions, which includes sanctioning them for deliberate and unauthorized attempts to identify (or re-identify) individuals.

Indeed, it is only through recognition and acceptance of a more contextualized approach to privacy, and through a global consensus on the values and principles of data sharing, that we can achieve a world of responsible genomic and clinical data sharing that respects the interests of individuals and society.

ABBREVIATIONS

DAC Data access committee
GA4GH Global Alliance for Genomics and Health
NIH National Institutes of Health (USA)
SNP Single-nucleotide polymorphism

REFERENCES

[1] Kosseim P, Dove ES, Baggaley C, Meslin EM, Cate FH, Kaye J, et al. Building a data sharing model for global genomic research. Genome Biol 2014;15:430.
[2] Joly Y, Dove ES, Knoppers BM, Bobrow M, Chalmers D. Data sharing in the post-genomic world: the experience of the International Cancer Genome Consortium (ICGC) Data Access Compliance Office (DACO). PLoS Comput Biol 2012;8:e1002549.
[3] Tenopir C, Dalton ED, Allard S, Frame M, Pjesivac I, Birch B, et al. Changes in data sharing and data reuse practices and perceptions among scientists worldwide. PLoS One 2015;10:e0134826.
[4] Knoppers BM, Harris JR, Budin-Ljøsne I, Dove ES. A human rights approach to an international code of conduct for genomic and clinical data sharing. Hum Genet 2014;133: 895–903.
[5] Council of Canadian Academies. Accessing health and health-related data in Canada. Available from: http://www.scienceadvice.ca/uploads/eng/assessments%20and%20publications%20and%20news%20releases/Health-data/HealthDataFullReportEn.pdf; 2015 [accessed 20.10.16].
[6] Kaye J. The tension between data sharing and the protection of privacy in genomics research. Annu Rev Genomics Hum Genet 2012;13:415–31.
[7] Contreras JL. Bermuda's legacy: policy, patents, and the design of the genome commons. Minn J Law Sci Technol 2011;12:61–125.
[8] Mello MM, Francer JK, Wilenzick M, Teden P, Bierer BE, Barnes M. Preparing for responsible sharing of clinical trial data. N Engl J Med 2013;369:1651–8.
[9] Institute of Medicine. Sharing clinical trial data: maximizing benefits, minimizing risk. Available from: http://www.nationalacademies.org/hmd/Reports/2015/Sharing-Clinical-Trial-Data.aspx; 2015 [accessed 20.10.16].

[10] Isasi R, Andrews PW, Baltz JM, Bredenoord AL, Burton P, Chiu IM, et al. Identifiability and privacy in pluripotent stem cell research. Cell Stem Cell 2014;14:427−30.

[11] Sethi N, Laurie GT. Delivering proportionate governance in the era of eHealth: making linkage and privacy work together. Med Law Int 2013;13:168−204.

[12] Laurie G, Postan E. Rhetoric or reality: what is the legal status of the consent form in health-related research? Med Law Rev 2013;21:371−414.

[13] Cate F. Protecting privacy in health research: the limits of individual choice. Calif Law Rev 2010;98:1765−803.

[14] Human Genome Organisation (HUGO). Principles agreed at the first international strategy meeting on human genome sequencing: 25−28 February 1996, Bermuda. Available from: http://web.ornl.gov/sci/techresources/Human_Genome/research/bermuda.shtml#1; 1996 [accessed 20.10.16].

[15] Wellcome Trust. Sharing data from large-scale biological research projects: a system of tripartite responsibility [Ft. Lauderdale Data Sharing Agreement]. Available from: http://www.genome.gov/pages/research/wellcomereport0303.pdf; 2003 [accessed 20.10.16].

[16] Toronto International Data Release Workshop Authors. Prepublication data sharing. Nature 2009;461:168−70.

[17] Strom BL, Buyse M, Hughes J, Knoppers BM. Data sharing, year 1—access to data from industry-sponsored clinical trials. N Engl J Med 2014;371:2052−4.

[18] Drazen JM. Sharing individual patient data from clinical trials. N Engl J Med 2015;372:201−2.

[19] Tenopir C, Allard S, Douglass K, Aydinoglu AU, Wu L, Read E, et al. Data sharing by scientists: practices and perceptions. PLoS One 2011;6:e21101.

[20] van Schaik TA, Kovalevskaya NV, Protopapas E, Wahid H, Nielsen FGG. The need to redefine genomic data sharing: a focus on data accessibility. Appl Transl Genomics 2014;3:100−4.

[21] Pham-Kanter G, Zinner DE, Campbell EG. Codifying collegiality: recent developments in data sharing policy in the life sciences. PLoS One 2014;9:e108451.

[22] Dove ES, Joly Y, Tassé AM, Public Population Project in Genomics and Society (P3G) International Steering Committee, International Cancer Genome Consortium (ICGC) Ethics and Policy Committee, Knoppers BM. Genomic cloud computing: legal and ethical points to consider. Eur J Hum Genet 2015;23:1271−8.

[23] Paltoo DN, Rodriguez LL, Feolo M, Gillanders E, Ramos EM, Rutter JL, et al. Data use under the NIH GWAS data sharing policy and future directions. Nat Genet 2014;46:934−8.

[24] McGuire AL, Oliver JM, Slashinski MJ, Graves JL, Wang T, Kelly PA, et al. To share or not to share: a randomized trial of consent for data sharing in genome research. Genet Med 2011;13:948−55.

[25] Oliver JM, Slashinski MJ, Wang T, Kelly PA, Hilsenbeck SG, McGuire AL. Balancing the risks and benefits of genomic data sharing: genome research participants' perspectives. Public Health Genomics 2012;15:106−14.

[26] Boddington P. Data sharing in genomics. In: Boddington P, editor. Ethical challenges in genomics research: a guide to understanding ethics in context. Heidelberg: Springer; 2012. p. 195−216.

[27] Burton PR, Murtagh MJ, Boyd A, Williams JB, Dove ES, Wallace SE, et al. Data Safe Havens in health research and healthcare. Bioinformatics 2015;31:3241−8.

[28] Blumenthal D, Campbell EG, Gokhale M, Yucel R, Clarridge B, Hilgartner S, et al. Data withholding in genetics and the other life sciences: prevalences and predictors. Acad Med 2006;81:137−45.

[29] van Panhuis WG, Paul P, Emerson C, Grefenstette J, Wilder R, Herbst AJ, et al. A systematic review of barriers to data sharing in public health. BMC Public Health 2014;14:1144.

[30] Simpson CL, Goldenberg AJ, Culverhouse R, Daley D, Igo RP, Jarvik GP, et al. Practical barriers and ethical challenges in genetic data sharing. Int J Environ Res Public Health 2014;11:8383−98.

[31] Kaye J, Hawkins N. Data sharing policy design for consortia: challenges for sustainability. Genome Med 2014;6:4.

[32] Budin-Ljøsne I, Isaeva J, Knoppers BM, Tassé AM, Shen HY, McCarthy MI, et al. Data sharing in large research consortia: experiences and recommendations from ENGAGE. Eur J Hum Genet 2014;22:317−21.

[33] Herlihy P. 'Government as a data model': what I learned in Estonia. Government Digital Service Blog, https://gds.blog.gov.uk/2013/10/31/government-as-a-data-model-what-i-learned-in-estonia/; 2013 [accessed 20.10.16].

[34] Schwab K. Estonia wants to collect the DNA of all its citizens. The Atlantic. Available from: http://www.theatlantic.com/health/archive/2015/10/is-a-biobank-system-the-future-of-personalized-medicine/409558/; 2015 [accessed 20.10.16].

[35] Expert Advisory Group on Data Access (EAGDA). Establishing incentives and changing cultures to support data access, https://wellcome.ac.uk/sites/default/files/establishing-incentives-and-changing-cultures-to-support-data-access-eagda-may14.pdf; 2014 [accessed 20.10.16].

[36] Expert Advisory Group on Data Access (EAGDA). Governance of data access, https://wellcome.ac.uk/sites/default/files/governance-of-data-access-eagda-jun15.pdf; 2015 [accessed 20.10.16].

[37] Wellcome Trust. Summary report of qualitative research into public attitudes to personal data and linking personal data, https://wellcome.ac.uk/sites/default/files/wtp053205_0.pdf; 2013 [accessed 20.10.16].

[38] Laurie G., Jones K.H., Stevens L., Dobbs C. A review of evidence relating to harm resulting from uses of health and biomedical data. Available from: http://nuffieldbioethics.org/wp-content/uploads/A-Review-of-Evidence-Relating-to-Harms-Resulting-from-Uses-of-Health-and-Biomedical-Data-FINAL.pdf [accessed 20.10.16].

[39] Erlich Y, Narayanan A. Routes for breaching and protecting genetic privacy. Nat Rev Genet 2014;15:409−21.

[40] Gymrek M, McGuire AL, Golan D, Halperin E, Erlich Y. Identifying personal genomes by surname inference. Science 2013;339:321−4.

[41] Im HK, Gamazon ER, Nicolae DL, Cox NJ. On sharing quantitative trait GWAS results in an era of multiple-omics data and the limits of genomic privacy. Am J Hum Genet 2012;90:591−8.

[42] Schadt EE, Woo S, Hao K. Bayesian method to predict individual SNP genotypes from gene expression data. Nat Genet 2012;44:603−8.

[43] Lin Z, Owen AB, Altman RB. Genomic research and human subjects privacy. Science 2004;305:183.

[44] Homer N, Szelinger S, Redman M, Duggan D, Tembe W, Muehling J, et al. Resolving individuals contributing trace amounts of DNA to highly complex mixtures using high-density SNP genotyping microarrays. PLoS Genetics 2008;4:e1000167.

[45] Shabani M, Knoppers BM, Borry P. From the principles of genomic data sharing to the practices of data access committees. EMBO Mol Med 2015;7:507−9.

[46] Erlich Y, Williams JB, Glazer D, Yocum K, Farahany N, Olson M, et al. Redefining genomic privacy: trust and empowerment. PLoS Biol 2014;12:e1001983.

[47] Ossorio PN. Bodies of data: genomic data and bioscience data sharing. Soc Res (New York) 2011;78:907−32.

[48] House of Lords, Science and Technology Committee. Genomic medicine: volume 1: report. Available from: http://www.publications.parliament.uk/pa/ld200809/ldselect/ldsctech/107/107i.pdf; 2009 [accessed 20.10.16].

[49] Rodriguez LL, Brooks LD, Greenberg JH, Green ED. The complexities of genomic identifiability. Science 2013;339:275−6.

[50] Nuffield Council on Bioethics. The collection, linking and use of data in biomedical research and health care: ethical issues. Available from: http://nuffieldbioethics.org/wp-content/uploads/Biological_and_health_data_web.pdf; 2015 [accessed 20.10.16].

Chapter 11

Designing Genetic- and Genomic-Based Clinical Trials

Benjamin French and Stephen E. Kimmel
University of Pennsylvania, Philadelphia, PA, United States

Chaptre Outline

INTRODUCTION

This chapter provides an overview of the randomized controlled trial (RCT) and its potential use in genomic medicine. We discuss strengths and limitations of RCTs in genomic medicine, the potential contribution of RCTs in demonstrating clinical utility of genomic interventions, and practical considerations for the design of genomic-based RCTs. We use the clinical example of genetic testing for determining the optimal initial dose of warfarin, a commonly used anti-coagulant, to illustrate several of the concepts introduced in this chapter.

The RCT is a uniquely unbiased study design for comparing therapeutic approaches. By randomly assigning treatments, preventive strategies, or other interventions, a properly designed RCT will balance the comparison groups on all factors other than the treatment assignments. Thus, the differences among the groups will reflect the effects of the treatment assignments. Nonrandomized study designs (e.g., observational studies) do not ensure equal balance between comparison groups and therefore are subject to confounding that can either mask true associations or create false associations.

Genomic and Precision Medicine. DOI: http://dx.doi.org/10.1016/B978-0-12-800681-8.00011-6

RCTs have additional advantages in genomic medicine. Unlike association studies that identify a relationship between a biomarker result (e.g., a genetic variant) and an outcome, an RCT can determine whether using that biomarker to alter clinical care can actually lead to improved outcomes. The RCT can therefore provide an unbiased, controlled comparison between, for example, a genomic-informed strategy versus a nongenomic-informed (i.e., one size fits all) strategy. In addition, designing and performing RCTs can provide an opportunity to address methodological and logistical challenges that might not be apparent without doing the trial, yet should be addressed prior to the use of genomics in clinical practice.

OVERVIEW OF CLINICAL UTILITY: IDENTIFYING AREAS OF STUDY WITH RANDOMIZED TRIALS

There is debate on if, and when, RCTs are needed prior to adopting genomics in practice. The purpose of this chapter is to discuss approaches to performing RCTs once a decision has been made that this is the appropriate approach. Nonetheless, it is worth briefly discussing when an RCT might be useful, noting that there is no objective criteria for when an RCT is needed versus other evidence (e.g., observational studies) prior to implementing genomics in practice. The debate typically centers on the need and methods to demonstrate "clinical utility" after a biomarker has proven to have adequate assay and clinical validity (discussed below).

The CDC's Office of Public Health Genomics established the ACCE Model Project as an analytical process for evaluating scientific data on emerging genetic tests [1]. Briefly, ACCE stands for Analytic validity, Clinical validity, Clinical utility, and Ethical, legal, and social issues. *Analytic validity* refers to how accurately and reliably the test measures the genotype of interest. *Clinical validity* refers to how consistently and accurately the test detects or predicts the intermediate or final outcomes of interest. Clinical validity is typically determined from observational, association studies. *Clinical utility* refers to how likely the test is to significantly improve patient outcomes. Clinical utility is the area in which the need for RCTs is most debated. For example, given the strong relationship between particular genetic variants and warfarin dose requirements (particularly in non-African Americans) in observational studies (clinical validity), one could question whether an RCT is needed to prove clinical utility (i.e., does using genetic information improve outcomes compared with not using genetic information). This is discussed further, below. *Ethical, legal, and social issues* are those that may arise in the context of using genetic information in practice.

The ACCE approach has been used to guide numerous entities in the United States and worldwide in their evaluation of genetic tests. One such entity, supported by the CDC, is the Evaluation of Genomic Applications in Practice and Prevention (EGAPP) initiative, which builds on the ACCE model. EGAPP provides "objective, timely, and credible information that is clearly linked to available scientific evidence. This information will allow health care providers and payers, consumers, policymakers, and others to distinguish genetic tests that are safe and useful" [2].

EGAPP prioritizes their reviews based on problems that have high-potential clinical utility [3]. These are the same areas that might identify high-priority areas for RCTs. Factors that determine the importance of clinical utility include the risk and cost of a disease and of testing, and the efficacy and effectiveness of the intervention. Greater risk can be conferred by highly prevalent conditions (thus affecting large numbers of people, such as hypertension), the severity of the disease and/or adverse effects of treating the disease (e.g., diseases with high mortality rates or treatments with narrow therapeutic indices), and the prevalence of the genetic factor affecting outcomes. Efficacy and effectiveness (discussed below) of an intervention are driven by the strength of the relationship between a test result (e.g., genetic marker) and the disease or treatment, the presence of an effective intervention, and the availability and appropriateness of using the test in clinical practice. In addition, the potential downstream consequences of using a new test in practice must be considered.

As an example, medications with a narrow therapeutic index, such as warfarin, are likely to benefit from using genetic information to better tailor therapy. The enzyme primarily responsible for the metabolism of warfarin to its inactive form is coded by the cytochrome P450 2C9 gene (*CYP2C9*), which has several variants. The target of warfarin is coded for by the vitamin K epoxide reductase complex 1 (*VKORC1*) gene, which also has variants that alter warfarin response. The ability to accurately detect these variants (analytic validity) has clearly been demonstrated; the FDA has approved several manufacturers' assays for clinical use. Numerous observational studies also have determined that variants in these genes are associated with variable warfarin dose requirements (clinical validity). However, until recently, there was little evidence of clinical utility. That is, does the use of genetic information to choose the initial dose of warfarin lead to improved anticoagulation control and fewer adverse events? Because of the many possible reasons other than genetics for differing warfarin dose requirement (e.g., adherence, interacting medications) and because dosing of warfarin can be adjusted on a frequent basis, it was not clear that using genetic information would improve outcomes. Until recently, the clinical utility of using genetics to help dose warfarin patients was only tested in small trials, many of which did not show benefit [4–7], while nonrandomized studies suggested benefits to pharmacogenetic-guided dosing [8,9]. Therefore, warfarin represented an example where large, randomized trials to determine clinical utility were needed [10].

DESIGN OF GENOMIC-BASED RCTs

Efficacy Versus Effectiveness

Clinical trials can be designed to test either efficacy or effectiveness of using genomic information. Efficacy trials (or, "explanatory" trials) test whether the use of genomic information, when used under optimal, ideal settings, will improve outcomes. For example, an efficacy trial of warfarin pharmacogenetic-based dosing would test whether the use of *CYP2C9* and *VKORC1* variants to determine initial warfarin dosing improves outcomes relative to not using these variants to determine dosing, when all other parameters (e.g., titration methods) are controlled and under blinded conditions (e.g., patients and clinicians do not know genotype or dose information). Effectiveness trials (or, "pragmatic" trials) test whether the use of genomic information improves outcomes under usual clinical practice settings. An effectiveness trial of warfarin pharmacogenetic-based dosing could, for example, randomize to providing versus not providing genetic information to clinicians and allowing them to use the information in any way they see fit. Typically, efficacy trials precede effectiveness trials because efficacy trials test whether the information gained from genomics has any effect on outcomes, all else being equal. However, effectiveness trials can be done in the absence of efficacy trials and can provide valuable information not gleaned from efficacy trials. For example, providing genetic information to clinicians and patients prior to warfarin therapy could alter monitoring vigilance, educational efforts by clinicians, patient adherence to therapy, patient adherence to diet, and the use of interacting medications. These effects could have a positive or negative influence on outcomes.

There is often a spectrum between efficacy and effectiveness trials. Clinical trials will often have characteristics of both designs but will tend to be closer to one or the other. A tool has been developed to assist with designing and characterizing trials, called "PRECIS" (pragmatic-explanatory continuum indicator summary) [11]. This tool incorporates 10 domains, each of which describes the pragmatic-explanatory balance of the trial. A PRECIS wheel can then be drawn by connecting points selected for each domain along a spoke of the wheel between the center hub (most explanatory) and the wheel rim (most pragmatic).

Here we use two published clinical trials of warfarin pharmacogenetics as an example [12,13]. It should be noted that there are no specific guidelines for where to place each point on the spoke of the wheel, so consensus among all researchers involved in a trial typically is required. Our example here is simply illustrative and not meant to represent a formal PRECIS analysis. The Clarification of Optimal Anticoagulation through Genetics (COAG) trial (Fig. 11.1) was more explanatory than the European Pharmacogenetics of Anticoagulant Therapy (EU-PACT) UK study (Fig. 11.2). This was primarily

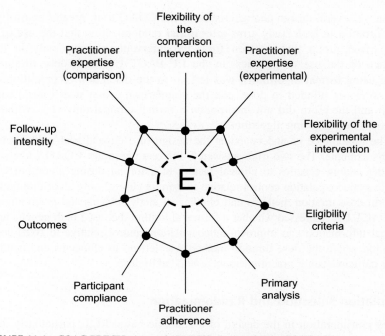

FIGURE 11.1 COAG PRECIS. A sample PRECIS diagram for the COAG trial.

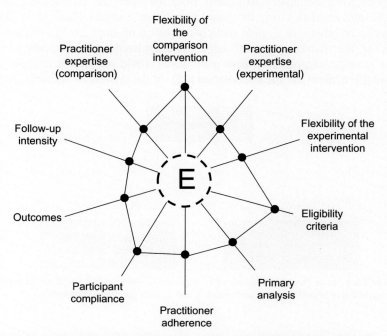

FIGURE 11.2 EU-PACT UK PRECIS A sample PRECIS diagram for the EU-PACT UK trial.

driven by several design characteristics of the COAG trial: greater control of dose titration in both study arms after initial randomization and the use of a comparison group that differed from the intervention group only in the absence of the use of genetics. In the EU-PACT UK trial, dose titration, while using formal algorithms, was left up to the discretion of practitioners who were not blinded to dose, and the comparison group was a usual care group and therefore did not incorporate a formal clinical-only (i.e., without genetics) initial dosing algorithm.

The importance of understanding the design of a clinical trial is illustrated by this example. The two trials answered different questions. COAG addressed whether adding genetics to clinical factors in a formal dosing algorithm can improve anticoagulation control relative to using a clinical-only algorithm under uniform dose titration methods and blinding of study subjects and practitioners. EU-PACT UK addressed whether a formal algorithm that used both genetic and clinical information can improve anticoagulation control relative to fixed-dose initiation where the dose titration method was left up to clinicians and neither study participants nor practitioners were blinded to dose.

Population Selection and Randomization

A key consideration in the design of a genomic-based RCT, particularly one to evaluate the efficacy of a genomic-based intervention, is whether to use genomic information in order to determine study eligibility [14−16]. For example, in an "enriched" design, study eligibility is restricted to a subset of the population anticipated to benefit from the intervention based on their genetic characteristics (or, the "marker-positive" subset) (Fig. 11.3). The exclusion of otherwise eligible participants based on their genetic characteristics (or, the "marker-negative" subset) is considered appropriate when evidence strongly suggests that these participants would not benefit from, or be potentially harmed by, the intervention [15]. In the context of warfarin, early

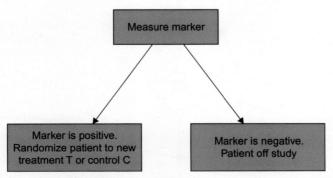

FIGURE 11.3 In an "enriched" design, study eligibility is restricted to a subset of the population anticipated to benefit from the intervention based on their genetic characteristics.

trials and observational studies indicated that patients with particular combinations of variants in *CYP2C9* and *VKORC1* might not benefit from pharmacogenetic-guided dosing [4]; therefore, these patients might be excluded from a trial evaluating the efficacy of pharmacogenetic-guided dosing. However, whether or not these subgroups would in fact differ in practice was not clear.

The exclusion of marker-negative patients in a targeted enrichment design can be dramatically more efficient compared to a traditional design that enrolls all potential participants, with respect to the required sample size. (The same number of patients would need to have genomic information measured and be screened, however, to determine eligibility.) For example, Simon and Maitournam [14] found that if the intervention is completely ineffective among the marker-negative subset, then the ratio of the number of participants required for the targeted design relative to the number required for the traditional design is approximately $1/\gamma^2$, for which γ denotes the proportion of marker-positive participants (i.e., if the marker-positive subset comprises 50% of the population, then the targeted design is four times more efficient than the traditional, untargeted design).

However, an untargeted design, while potentially requiring a larger sample size, facilitates the evaluation of assumptions regarding the potential benefit of genomic therapy. In addition, the results of an untargeted trial motivate consideration of the intervention's effectiveness in the population of all patients. Below, we discuss approaches to estimate the minimum detectable difference in untargeted trials that enroll marker-negative participants.

A potential alternative to either a targeted or an untargeted design is a marker-stratified design, for which randomization is stratified by discrete values of the marker (Fig. 11.4) [17]. In this case, stratified randomization based on a known baseline factor ensures covariate balance within strata,

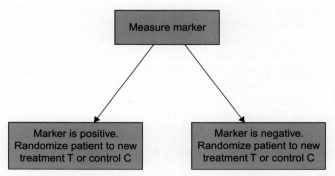

FIGURE 11.4   In the marker-stratified design, study eligibility is not restricted based on the marker, but randomization is stratified by discrete values of the marker.

which facilitates unbiased estimation of treatment effects overall and within marker strata. The corresponding analysis plan should specify these primary comparisons a priori.

Estimating the Minimum Detectable Difference

The design of an RCT requires specification of a sample size that provides adequate statistical power to detect a clinically relevant difference in the primary outcome between treatment groups, which typically depends on the minimum detectable difference, a measure of the outcome's variability, the significance level, and, often, the anticipated drop-out rate. In a genomic-based RCT that enrolls marker-negative participants, the minimum detectable difference can also depend on the size of the marker-negative and marker-positive subsets, as well as the within-subset treatment effects [18].

Let Y denote the primary outcome, which for presentation we assume to be a continuous variable (e.g., percentage of time in therapeutic range on warfarin), and let $\overline{Y}^+$ and $\overline{Y}^-$ denote the mean outcome among the marker-positive and marker-negative subsets, respectively. Denote by w^+ and w^- the proportion of participants in the marker-positive and marker-negative subsets, respectively, which are assumed to be balanced across treatment arms by randomization. The mean outcome among the treatment and control groups, denoted by $\overline{Y}^T$ and $\overline{Y}^C$, can then be viewed as a weighted average of the mean outcome and the corresponding treatment effect across the marker-positive and marker-negative subsets, in which the weights are determined by their relative size:

$$\overline{Y}^C = w^+\overline{Y}^+ + w^-\overline{Y}^- \tag{11.1}$$

$$\overline{Y}^T = w^+(\overline{Y}^+ \times \Delta^+) + w^-(\overline{Y}^- \times \Delta^-) \tag{11.2}$$

for which Δ^+ and Δ^- denote the relative treatment effects in the marker-positive and marker-negative subsets, respectively. Therefore, the absolute difference between arms depends on the relative differences within positive and negative subsets. Failure to account for differences in treatment effects across strata can therefore result in an overestimate of the minimum detectable difference, an underestimate of the sample size, and a resultant trial with insufficient power. In the COAG study, it was anticipated that the 40% of participants with only a single variant in either *CYP2C9* or *VKORC1* would not benefit from pharmacogenetic-guided warfarin dosing, and the sample size was inflated accordingly [18].

Significance Levels for Coprimary Analyses

Corrections for multiple comparisons are required when more than one primary comparison is made between treatment groups, so that the overall probability of a type-1 error is controlled at the nominal level, typically 0.05. Such corrections are usually calculated based on the number of tests (e.g., Bonferroni correction) and the observed p values (e.g., Benjamini–Hochberg procedure). Typically, this adjustment is made for different outcomes within a same trial or for multiple statistical tests of the trial as it progresses. However, another adjustment can be considered in genomic studies in which subsets of patients (e.g., those with a constellation of genetic variants) might be expected to have greater benefit than the population as a whole.

In previous sections, we discussed approaches to enroll subsets of participants who might respond differently to the intervention and corresponding adjustments to the anticipated minimum detectable difference. Here, we describe approaches to adjust the significance level when the intervention's efficacy or effectiveness is evaluated in all participants and in a subset of participants as coprimary analyses. Such approaches exploit the correlation between the analyses to inflate the significance level above that of a more conservative adjustment that ignores the correlation (e.g., Bonferroni correction). These "prospective alpha allocation" approaches are distinct from "alpha spending" approaches that adjust the significance level throughout the trial based on the number of interim analyses, which we do not discuss here.

Prospective alpha allocation facilitates formal evaluation of the treatment effect in the marker-positive subset as part of the primary analysis. By denoting the treatment effect in all participants and in the marker-positive subgroup by Δ and Δ^+, respectively, the null and alternative hypotheses are:

$$H : \Delta = 0 \text{ and } \Delta^+ = 0, \tag{11.3}$$

$$K : \Delta \neq 0 \text{ or } \Delta^+ \neq 0, \tag{11.4}$$

respectively. The statistical evaluation of the intervention among all participants is performed with a significance level of α, which is less than the overall significance level (0.05). Then, the statistical evaluation of the intervention among the marker-positive subset is performed with a significance level of $0.05 - \alpha = \alpha^+$. However, Joo and colleagues [19] have shown that this approach can be overly conservative when there is a positive correlation between the test performed among all participants (i.e., testing $\Delta = 0$) and that performed among the marker-positive subset (i.e., testing $\Delta^+ = 0$). For example, the treatment effect in all participants could be driven by the treatment effect in the marker-positive subset, resulting in correlated tests. Joo and colleagues provide simple formulas to calculate the significance levels based on the correlation between the test statistics, which for continuous outcomes depends on the standard deviations and the size of the marker-positive subset (w^+). Therefore, the significance level for the marker-positive subset is obtained with $\alpha^+ \geq 0.05 - \alpha$. In the COAG trial, the primary subgroup was comprised of participants who had an absolute difference of 1.0 mg or more in the predicted initial daily dose between the pharmacogenetic-guided dose-initiation algorithm and the clinically guided dose-initiation algorithm [20]. α was fixed at 0.04, and at the conclusion of the trial, α^+ for the subset was calculated as 0.016 (given that $w^+ = 0.41$).

Adaptive Designs

RCTs typically study a single intervention at a single point in time. Although useful for studying, for example, the efficacy of a new genomic therapy, this approach often does not inform the way existing therapies are used in clinical practice. Clinicians often do not just pick the same therapeutic strategy for everyone and continue it without regard to the patient's characteristics or response to therapy. In practice, clinical practitioners need to accommodate: heterogeneity in patient conditions, both prior to starting therapy and then during therapy; patient preferences; poor adherence; and interim outcomes and side effects when choosing or altering therapies. Therefore, many decisions in clinical medicine involve sequential decisions about when and how to apply and alter therapies in patients over time, based on patients' underlying characteristics and subsequent response to the initial therapies. The advent of genetic- and other biomarker-based testing represents perhaps the best example of the increased ability to identify patient heterogeneity. As more and more information is available about a patient's genotype and other markers of medication response, clinicians will need to know the best way to use this information for their patients' benefit.

Unfortunately, the traditional approach to RCTs does not formally test alterations in therapy after evaluating response to the initial strategic approach or based on new information obtained as the trial progresses. Another limitation is that, typically, one therapeutic intervention is tested in one trial, a different intervention in a different trial, and so on. There are two problems with this approach. First, this approach is very inefficient and costly. For example, a major cost in performing a clinical trial is recruiting patients, so performing multiple trials, each testing a different piece of a strategy, is highly inefficient. Second, making comparisons of different strategies that are tested across different trials may be confounded by intertrial differences in the study population, treating physicians, and calendar time.

A large body of statistical literature has focused on adaptive designs for RCTs. Brown and colleagues [21] conceptualized three types of adaptation: adaptive sequencing, which refers to the design of a new trial; adaptive designs, which refers to the conduct of an ongoing trial; and adaptive interventions, which refer to the intervention experience of a study participant. Here, we provide a brief overview on a particular adaptive intervention: the adaptive treatment strategy, tested via a sequential multiple assignment randomized trial (SMART) [22]. An adaptive treatment strategy is a sequence of individually tailored decision rules that specify whether, how, and when to alter the intensity, type, dosage, or delivery of treatment at critical decision points in the medical care process. To evaluate which of these strategies is best, the SMART approach randomizes patients to different therapeutic strategies at multiple time points in the trial and tailors the randomization to predictors of response at each randomization period. Thus, SMARTs allow testing of the tailoring variables and the intervention components in the same trial, and it allows clinicians to develop the best decision rules for their patients.

In the context of warfarin, Rich and colleagues [23] proposed a simulation-based approach to study personalized dosing strategies that take into account the therapeutic agent's pharmacokinetics and pharmacodynamics. In particular, they suggested that data simulations could be used at the initial design stage to reduce the complexity of the SMART and guide decisions regarding which adaptive strategies to investigate. Rich and colleagues also provide an illustration of the application of causal inference methods (e.g., G-estimation) and machine-learning techniques (i.e., Q-learning) to evaluate potential personalized dosing strategies.

Design of Pragmatic Trials

In previous sections, we contrasted "explanatory" trials (e.g., a trial to test whether the use of genomic information, when used under optimal, ideal settings, will improve patient outcomes) with "pragmatic" trials (e.g., a trial to test whether the use of genomic information improves outcomes under usual clinical practice settings) and discussed design challenges that arise primarily in explanatory trials. Here, we discuss approaches to the design of pragmatic trials.

Pragmatic trials have gained popularity as biomedical research has begun to focus on the generalizability and applicability of interventions to routine clinical care [24−26]. Pragmatic trials typically recruit from clinical practices that include patients who reflect the population affected by the condition the intervention aims to treat. Therefore, patients most likely to benefit from the intervention will be represented in the study, such that the observed effect represents what a typical patient could expect to experience. As noted earlier, a pragmatic trial of warfarin pharmacogenetics might randomize clinicians and/or patients to either receiving or not receiving their genetic information. All subsequent care would be based on clinical practice and not dictated by the trial.

In addition, the increasing availability of electronic health systems supports the implementation of a pragmatic trial across health care systems. In particular, these systems support the randomization of practices to intervention and control arms, in a typical "parallel" cluster-randomized design. For example, genetic information could randomly be provided to some clinics but not others. However, designs that randomize the time at which the clinic implements the intervention (i.e., a "stepped-wedge" cluster-randomized trial) can be more efficient than the typical cluster-randomized design and facilitate estimation of within-clinic effects, while adjusting for temporal trends in the outcome not due to the intervention. For example, randomization would occur within each clinic (e.g., the time at which a clinic begins to be provided with genetic information would be randomized), with data collected on patients before and after randomization. Hussey and Hughes [27] provided practical design considerations and power formulas for stepped-wedge designs.

The conduct of cluster-randomized pragmatic trials presents several design and analysis challenges. In particular, the Biostatistics and Study Design Core of the NIH Collaboratory has identified statistical challenges in the design of pragmatic trials [28]. These include the trade-off between sample size and potential contamination, the intraclass correlation that arises from clustering at different levels, varying cluster sizes due to differently sized practices, and the need for stratification or matching.

In addition, the use of electronic health records for data collection and follow-up presents the challenge of missing data and outcome-dependent data collection [29−31]. For example, individuals who are less healthy and have a greater burden of comorbidities are likely to have more health care visits. If an intervention were effective in improving health, then patients who received the intervention would be less likely to contribute follow-up data compared with those who did not receive the intervention. Ignoring the dependence between outcomes and the process of data collection can lead to biased treatment-effect estimates, although a growing body of statistical literature provides methods to correct for the bias [32].

CONCLUSIONS

The decision to perform an RCT in genomic medicine requires careful consideration of the need to demonstrate clinical utility. One must weigh the complexities and cost of a trial with the value of the RCT for demonstrating utility in a rigorous fashion and for the generally high level of acceptance that many practitioners and payors place on the RCT. Once the decision is made to proceed with an RCT, one must ensure that the proper population is studied, the correct trial design is chosen, the study is adequately powered, and that the analysis plan is sound. This approach requires a multidisciplinary team approach that includes clinical specialists (including as appropriate pharmacists), geneticists, clinical trial specialists, and biostatisticians.

ABBREVIATIONS

RCT randomized controlled trial
ACCE Analytic validity, Clinical validity, Clinical utility, and Ethical, legal, and social issues
EGAPP Evaluation of Genomic Applications in Practice and Prevention
PRECIS pragmatic-explanatory continuum indicator summary
COAG Clarification of Optimal Anticoagulation through Genetics
EU-PACT European Pharmacogenetics of Anticoagulant Therapy
SMART sequential multiple assignment randomized trial

REFERENCES

[1] Centers for Disease Control and Prevention (CDC) Genomic Testing. Available from: http://www.cdc.gov/genomics/gtesting/ACCE/index.htm [accessed 02.02.15].
[2] Centers for Disease Control and Prevention (CDC) Genomics Translation. Available from: http://www.cdc.gov/genomics/gtesting/EGAPP/index.htm [accessed 02.11.15].
[3] Evaluation of Genomic Applications in Practice and Prevention (EGAPP) Working Group: Topics. Available from: http://www.egappreviews.org/workingrp/topics.htm [accessed 02.11.15].
[4] Anderson JL, Horne BD, Stevens SM, et al. Randomized trial of genotype-guided versus standard warfarin dosing in patients initiating oral anticoagulation. Circulation 2007;116:2563−70.
[5] Caraco Y, Blotnick S, Muszkat M. CYP2C9 genotype-guided warfarin prescribing enhances the efficacy and safety of anticoagulation: a prospective randomized controlled study. Clin Pharmacol Ther 2008;83:460−70.
[6] Burmester JK, Berg RL, Yale SH, et al. A randomized controlled trial of genotype-based Coumadin initiation. Genet Med 2011;13(6):509−18.
[7] Wang M, Lang X, Cui S, et al. Clinical application of pharmacogenetic-based warfarin-dosing algorithm in patients of Han nationality after rheumatic valve replacement: a randomized and controlled trial. Int J Med Sci 2012;9(6):472−9.
[8] Epstein RS, Moyer TP, Aubert RE, et al. Warfarin genotyping reduces hospitalization rates results from the MM-WES (Medco-Mayo Warfarin Effectiveness study). J Am Coll Cardiol 2010;55(25):2804−12.
[9] Anderson JL, Horne BD, Stevens SM, et al. A randomized and clinical effectiveness trial comparing two pharmacogenetic algorithms and standard care for individualizing warfarin dosing (CoumaGen-II). Circulation 2012;125(16):1997−2005.
[10] Centers for Medicare and Medicaid Services (CMS) Medicare Coverage Database—Potential NCD Topics. Available from: http://www.cms.hhs.gov/mcd/ncpc_view_document.asp?id=19 [accessed 02.11.15].
[11] Thorp SR, Ayers CR, Nuevo R, Stoddard JA, Sorrell JT, Wetherell JL. Meta-analysis comparing different behavioral treatments for late-life anxiety. Am J Geriatr Psychiatry 2009;17(2):105−15.
[12] Kimmel SE, French B, Kasner SE, et al. A pharmacogenetic versus a clinical algorithm for warfarin dosing. N Engl J Med 2013;369(24):2283−93.

[13] Pirmohamed M, Burnside G, Eriksson N, et al. A randomized trial of genotype-guided dosing of warfarin. N Engl J Med 2013;369(24):2294–303.

[14] Maitournam A, Simon R. On the efficiency of targeted clinical trials. Stat Med 2005;24 (3):329–39.

[15] Simon R, Maitournam A. Evaluating the efficiency of targeted designs for randomized clinical trials. Clin Cancer Res 2004;10(20):6759–63.

[16] Simon R, Maitournam A. Evaluating the efficiency of targeted designs for randomized clinical trials: supplement and correction. Clin Cancer Res 2006;12:3229.

[17] Simon R. Clinical trial designs for evaluating the medical utility of prognostic and predictive biomarkers in oncology. Per Med 2010;7(1):33–47.

[18] French B, Joo J, Geller NL, et al. Statistical design of personalized medicine interventions: the Clarification of Optimal Anticoagulation through Genetics (COAG) trial. Trials 2010;11:108.

[19] Joo J, Geller NL, French B, Kimmel SE, Rosenberg Y, Ellenberg JH. Prospective alpha allocation in the Clarification of Optimal Anticoagulation through Genetics (COAG) trial. Clin Trials 2010;7(5):597–604.

[20] Kimmel SE, French B, Anderson JL, et al. Rationale and design of the Clarification of Optimal Anticoagulation through Genetics trial. Am Heart J 2013;166(3):435–41.

[21] Brown CH, Ten Have TR, Jo B, et al. Adaptive designs for randomized trials in public health. Annu Rev Public Health 2009;30:1–25.

[22] Lei H, Nahum-Shani I, Lynch K, Oslin D, Murphy SAA. "SMART" design for building individualized treatment sequences. Annu Rev Clin Psychol 2012;8:21–48.

[23] Rich B, Moodie EE, Stephens DA. Simulating sequential multiple assignment randomized trials to generate optimal personalized warfarin dosing strategies. Clin Trials 2014;11 (4):435–44.

[24] Tunis SR, Stryer DB, Clancy CM. Practical clinical trials: increasing the value of clinical research for decision making in clinical and health policy. JAMA 2003;290(12):1624–32.

[25] Weiss NS, Koepsell TD, Psaty BM. Generalizability of the results of randomized trials. Arch Intern Med 2008;168(2):133–5.

[26] Treweek S, Zwarenstein M. Making trials matter: pragmatic and explanatory trials and the problem of applicability. Trials 2009;10:10–37.

[27] Hussey MA, Hughes JP. Design and analysis of stepped wedge cluster randomized trials. Contemp Clin Trials 2007;28(2):182–91.

[28] NIH Collaboratory Health Care Systems Research Collaboratory. Rethinking Clinical Trials. Biostatistics and study design, https://www.nihcollaboratory.org/cores/Pages/Biostatistics.aspx [accessed 02.11.15].

[29] Burnum JF. The misinformation era: the fall of the medical record. Ann Intern Med 1989;110(6):482–4.

[30] van der Lei J. Use and abuse of computer-stored medical records. Methods Inf Med 1991;30(2):79–80.

[31] Li T, Hutfless S, Scharfstein DO, et al. Standards should be applied in the prevention and handling of missing data for patient-centered outcomes research: a systematic review and expert consensus. J Clin Epidemiol 2014;67(1):15–32.

[32] Tan KS, French B, Troxel AB. Regression modeling of longitudinal data with outcome-dependent observation times: extensions and comparative evaluation. Stat Med 2014;33 (27):4770–89.

Chapter 12

Developing the Evidence to Support Clinical Use of Genomics

Katrina Armstrong

Massachusetts General Hospital, Harvard Medical School, Boston, MA, United States

Chapter Outline

Sequencing human genomes has the potential to transform medical care. Much of this transformation will come from genome-driven advances in understanding human biology leading to the development of new therapeutic and preventive interventions. However, the clinical use of genomic tests is already a part of the daily practice of many areas in medicine and its impact will continue to grow as the pipeline of clinically applicable genomic tests continues to expand [1]. The clinical use of genomic tests falls into two major categories: diagnosis (in either symptomatic or asymptomatic individuals) and prediction (including risk of disease, outcomes of disease, or response to therapy). While the goal of genomic testing is to enable patients and providers to make better decisions, the value of such testing for clinical medicine has been controversial since the first introduction of sickle cell screening in the 1970s [2]. Such controversy is not uncommon for a new technology but is particularly pronounced for genomic testing, perhaps because of its potential ethical and social implications and the lack of an established and mandatory pathway for test approval at the federal level [3]. The need to develop a progressive and consistent approach to the translation of these new applications into clinical practice has been widely documented [4−6].

Genomic and Precision Medicine. DOI: http://dx.doi.org/10.1016/B978-0-12-800681-8.00012-8

Optimizing the translation of genomics into clinical practice requires research on the clinical effectiveness of genomic applications—a fundamental tenet of the paradigm of evidence-based medicine. Simplistically, this means that the decision about the use of a given genetic or genomic test should be based upon a "conscientious, explicit, and judicious use of current best evidence" and integrate "individual clinical expertise with the best available external clinical evidence from systematic research" (http://www.cochrane.org/about-us/evidence-based-health-care). Generation of evidence involves conducting studies that assess the benefits and harms of the proposed intervention with the quality of the evidence being judged according to the likelihood that the study results are valid [7]. However, perhaps not surprisingly, the application of this well-established paradigm to genomics has proven far from straightforward for several reasons. First, the evaluation of the benefits and harms of a diagnostic or predictive test is more complicated than the evaluation of a therapeutic or preventive intervention. The benefit of a diagnostic test does not come from a direct effect on clinical outcomes. Instead, the potential benefit is derived by the impact of the test results on subsequent health and health care decisions, thereby complicating any assessment process. Second, genetic testing can have implications beyond the prediction or diagnosis of the target condition in the tested individual. These include the phenomenon of pleiotropy where a single genetic variant can have multiple downstream phenotypic effects and the implications of the individual's germline DNA information for other members of his or her family. Third, the rapid pace of generation of new information in genomics has made it difficult for any evidence base to keep up, a challenge that is exacerbated by the ongoing innovation in testing platforms and approaches. Fourth, the initial focus of genomic discovery on populations of European ancestry limits the availability of information about genomic variation in minority populations, raising concerns about the generalizability of any evidence base [8] (Box 12.1). Given these challenges, it has become clear that the generation of evidence to guide the use of genomics in clinical care requires an expanded vision of evidence-based medicine. Such a vision

---

BOX 12.1 Challenges in the Evaluation of Genomic Testing

- Assessing the benefits and harms of a diagnostic or predictive test requires measuring the impact of test results on subsequent health and health care decisions.
- Genetic test results can have implications for multiple diseases and for other family members.
- Information about a genomic test often changes rapidly.
- Relatively few data are available about genomic variation in minority populations, raising concerns about generalizability.

must be based on a dynamic assessment process that considers and incorporates the incremental impact of new information, explicit integration with clinical expertise and decision support, and a recognition of the uncertainty involved in this new area of medicine.

The most commonly used framework for the assessment of genomic tests was developed by the CDC Office of Public Health Genomics and is referred to as the ACCE framework (http://www.cdc.gov/genomics/gtesting/ACCE/). It includes four components: Analytic validity, Clinical validity, Clinical utility, and Ethical, legal, and social implications [9] (Table 12.1). Together the four components create a chain of evidence that informs the decision about whether or not to recommend the use of a given test. In reality, the evidence generation process does not always proceed linearly through the steps from demonstrating analytic validity to the evaluation of ethical, legal, and social implications. For example, studies of clinical validity are often done without formal demonstration of analytic validity and studies of clinical utility may occur without definitive evidence of either analytic or clinical validity. In this setting, strong evidence in support of a step further down the evidence chain may preclude the need for further studies of earlier steps but a failure to demonstrate benefit at a later step does not determine where the chain is broken.

APPROACHES TO EVIDENCE GENERATION

A wide range of study designs are used to generate evidence about the clinical use of genomics [10]. Different designs provide different levels of evidence and are more or less appropriate for different components in the ACCE framework. In general, designs that involve randomized trials or meta-analyses of trial data are considered to provide the strongest level of evidence and case series to provide the weakest level of evidence (Table 12.2). Of course, within any type of study, the validity of the results is also influenced by specific design issues, including the selection of study subjects (particularly control subjects in observational studies), accuracy of the measurement of exposures and outcomes, loss to follow up, and analytic approach. Most of these issues are common to studies of any clinical intervention and are well described in several accepted methods for grading the quality of evidence including the GRADE criteria [11] (Table 12.2). However, some issues are specific or particularly relevant for studies of genomics and are discussed in the relevant sections below.

ANALYTIC VALIDITY

The analytic validity of any test is how well the test measures what it claims to measure. For tests that report DNA sequence variation, analytic validity is defined as the ability of the test to identify the presence or absence of the

TABLE 12.1 Components of Evidence Generation for Genomic Medicine

	Definition	Study Design(s)	Outcomes	Major Challenges
Analytic validity	How well does the test measure what it is supposed to?	Cross-sectional comparisons with another test, proficiency testing	Reproducibility, sensitivity, specificity	Lack of gold standard, rapidly changing technology
Clinical validity	How well does the test result predict the phenotype of interest?	Case-control and cohort studies, sometimes building on existing prospective studies including trials	Strength of association and measures of test performance including AUC	Multiple comparisons, need for external validation
Clinical utility	How does the use of the test affect health-related outcomes?	Randomized trials, quasi-experimental and observational studies of intervention effects, decision and simulation modeling	Health status including morbidity and mortality, health care, behavior, cost	Feasibility, generalizability to current practice, bias in observational studies
Social impact	What are the ethical, legal, and social implications of the use of the test?	Surveys, qualitative studies, case studies, policy analyses, stakeholder engagement	Attitudes and beliefs of different groups, policy impacts, consensus reports	Integration with clinical research, impact on relevant policy makers

TABLE 12.2 GRADE Approach to Quality of Evidence

Underlying Study Methodology	Quality Rating
Randomized trials; or double-upgraded observational studies	High
Downgraded randomized trials; or upgraded observational studies	Moderate
Double-downgraded randomized trials; or observational studies	Low
Triple-downgraded randomized trials; or downgraded observational studies; or case series/case reports	Very low

genetic alteration, such as a specific mutation/variant or set of mutations/variants. For tests of gene expression, analytic validity is the ability to measure the level(s) of the specific mRNA or the set of mRNAs [12].

The development of a genomic test involves establishing analytic validity including precision (reproducibility of results), robustness (performance in different settings), and, when possible, comparison of the results to a "gold" or reference standard. Comparison to a reference standard is necessary to assess the accuracy of the test including sensitivity (true positives/true positives + false negatives) and specificity (true negatives/true negatives + false positives). For tests of DNA sequence, the gold standard for the generation of evidence about analytic validity has generally been Sanger sequencing [13]. More recently, evaluations of the accuracy of next-generation sequencing methods have compared the accuracy of sequencing with fewer reads of each sequence (often referred to as the depth or coverage) with the "gold standard" of more reads of each sequence (i.e., deeper sequencing) [14]. While most efforts to determine the analytic validity of a sequencing test initially focused on the ability to accurately call single-nucleotide variants, the focus now includes measuring accuracy for the identification of small insertions and deletions (indels) and copy number variants, particularly for whole-genome or whole-exome sequencing [15]. Recognizing that all sequencing methods have an error rate, latent class analysis has been used to assess the accuracy of sequencing results in the absence of a clear gold standard [16].

For tests of gene expression, identifying a gold standard is more challenging and the development of a new test must often rely on the comparison to the results of a test that is both imperfect and potentially providing different information. Thus, assessment of analytic validity of gene expression tests has included the comparison of fluorescent in situ hybridization (FISH) to immunohistochemistry, array comparative genomic hybridization (aCGH) to FISH, and most recently RNAseq to aCGH, as well as other methods—each time raising uncertainty about the true gold standard and the possibility that different tests may actually be better for different questions [17,18]. Furthermore, analytic validity of a gene expression test often depends upon factors beyond the technical aspects of the test itself. For tissue-based tests, test results can be influenced by sample preparation issues including the use of fresh frozen tissue versus formalin-fixed, paraffin-embedded tissue and the selection of which cells are included in the test sample [19,20]. The validity of a gene expression profile is also dependent on the analytic approach including the calculation of a summary score and adjustment for "within-subject" variation in overall gene expression [21]. For any assessment of analytic validity, it is also important to ensure that the interpretation of the new test is blinded to the results of the reference standard and that the application of the new test is not guided by the results of the test being evaluated, either of which can lead to an overestimation of the new test's performance.

Because most genomic tests are laboratory-based tests, the assessment of analytic validity of a test in clinical practice falls under the requirements for proficiency testing as part of the Clinical Laboratory Improvement Amendments program of the Centers for Medicare and Medicaid Services [22]. In a proficiency testing program, a laboratory is sent a set of samples with known characteristics (e.g., genotypes) and the results of the laboratory are compared to the results from the reference laboratory. Differences may arise from the accuracy of the technology being used (e.g., microarray, next-generation sequencing), errors in sample preparation and processing, discordance in analytic approaches (particularly for gene expression profiles), and problems in data interpretation. Of particular relevance for genomic tests, the College of American Pathologists and the American College of Medical Genetics have collaborated to offer a proficiency testing program for genetic tests that provides an important resource for ensuring and improving individual laboratory performance and demonstrating the overall analytic validity of a test in clinical practice [23].

CLINICAL VALIDITY

The determination of clinical validity involves establishing an association between the results of the test (e.g., genotype, expression profile) and the condition or outcome of interest. Such evidence can be generated through cross-sectional studies of cases and controls, such as genome-wide association studies (GWAS) that link single-nucleotide polymorphisms to disease status, or longitudinal studies that link the test result to the condition of interest, such as disease progression. For some tests, studies may focus on the association between the test result and outcomes that are often considered intermediate steps in proving ultimate clinical impact, such as the level of a drug. Given that genomic tests can sometimes be done retrospectively on previously collected samples, the generation of evidence of clinical validity for many tests has benefited from harnessing existing studies, including randomized trials. For example, the initial evidence on the association between a commonly used 21 gene expression profile and breast cancer recurrence among node-negative, tamoxifen-treated patients was generated through additional analyses of the National Surgical Adjuvant Breast and Bowel Project clinical trial B-14 and the extension to node-positive patients was based upon a Southwestern Oncology Group Trial of adjuvant chemotherapy [24,25]. In studies of clinical validity, results are generally presented as measures of the strength of association, such as relative risks or odds ratios, or as measures of test performance, such as sensitivity and specificity and area under the Receiver Operating Curve. However, the gold standard is the presence or absence of the clinical condition of interest not the results of a different test as it is for the assessment of analytic validity. For many genomic risk prediction tests, the assessment of clinical validity includes the performance of the genomic test by itself and the incremental improvement in sensitivity, specificity, or Area Under the Curve (AUC) from adding genomic information to existing risk prediction models [26,27]. Although interest is growing in more clinically relevant measures of test performance such as the ability to reclassify individuals as above or below a clinically meaningful threshold of risk, such measures have made relatively little inroad in the evaluation of genomic tests to date [28].

Studies assessing the clinical validity of genomic tests are subject to several potential biases that are common to all studies of diagnostic tests [29]. First, test performance can be overestimated if the validation is not performed in a different set of samples/patients than the test development. Validation of the association in separate samples reduces the chance of a false-positive finding and generally leads to a reduction in the estimated strength of the effect. Cross-validation within the initial development sample is an alternative approach but its implementation is often problematic leading to potential bias [30]. Second, the selection of nonrepresentative patients for inclusion as either cases or controls can also lead to a misestimation of test performance. In general, this "spectrum bias" leads to an overestimation of test performance because study cases are often more advanced or pronounced than the patients in whom the test would be used [31]. However, it can also lead to an underestimation of performance if the study sample includes a greater proportion of patients with less clear-cut presentations than the patients in whom the test would be used, perhaps because the new test is only studied in patients who remain undiagnosed after previous tests. Ideally, the assessment of test performance should be conducted among patients who will be the primary target for testing in clinical practice. Third, inconsistent or inaccurate assessment of disease status or other outcome of interest can lead to biased results. Misclassification of the outcome that occurs because of random error generally leads to underestimation of the strength of the association, whereas misclassification that is influenced by knowledge of the test results may result in an overestimation.

Studies assessing the clinical validity of proposed genomic tests present several additional challenges [32]. Perhaps the most well described one of these challenges is the risk of false-positive associations generated by the ability to examine hundreds if not thousands of variants in a single study [33]. Although this problem continues to be a challenge for many large GWAS, the development and application of appropriate analytic approaches has greatly reduced the likelihood that a published genomic association will not be replicated in subsequent studies. In addition, the creation of major public resources for sharing data about genetic associations mitigates the concern about errors in the assessment of clinical validity arising from publication bias where "positive" findings of an association are more likely to be published than "negative" findings of the absence of an association [34]. Another challenge comes from a phenomenon referred to as population stratification where the correlation between the frequency of a genetic variant and an ancestral group can lead to a spurious association if the risk of the outcome of interest also varies by ancestry. Because ancestry can be difficult to measure phenotypically, studies seeking to demonstrate clinical validity can check for differences in genetic markers of ancestry between cases and controls and perform appropriate analyses if those differences are found [35]. The inclusion of populations from diverse ethnic backgrounds has been limited to date in genomic research and is needed to be able to establish clinical validity in these groups, a critically important issue for the eventual impact of precision medicine.

CLINICAL UTILITY

The clinical utility of a genomic test is based upon the ability of the test to improve outcomes and requires information about the benefits and harms of testing. The generation of evidence to assess clinical utility involves the same approaches as the assessment of the benefits and harms of any medical intervention. In general, the strongest evidence of clinical utility is derived by comparing outcomes between individuals who are exposed to the intervention and individuals who are not exposed using a randomized trial design. In this approach, individuals are randomized to receive either the new test or the current "standard of care" (that may or may not involve an existing diagnostic or predictive test) and followed forward to assess the outcomes of interest. Randomization balances potential confounders across the groups and reduces the likelihood of confounding and selection bias. Although most studies of clinical utility select a few outcomes of interest, multiple types of outcomes are relevant for decisions about the clinical use of genomic tests including health outcomes (morbidity, mortality, health status), health care outcomes (use of health care services including preventive, therapeutic, and other diagnostic/predictive interventions), behavioral/psychosocial outcomes (reassurance, anxiety, adherence), and economic outcomes [36,37]. Although the expense and time commitment of randomized clinical trials has made them relatively rare among studies of genomic tests, there is a rapidly growing body of randomized trial evidence about the clinical use of genomics and many more such trials are registered on clinicaltrials.gov [38].

When a randomized trial is not feasible or appropriate, quasi-experimental and observational studies can be used to assess the clinical utility of a genomic test [39,40]. Here, the exposure to the test is not determined by randomization but by variation in time, practice style, location, or other factors. Such studies are central to clinical effectiveness research and offer important advantages including generalizability to clinical practice and the relatively short time to results, a particularly important issue given the pace of change in genomics. However, they do not provide the level of protection against bias and confounding that occurs with a randomized controlled trial.

Specific issues of study design, including the selection of cases and controls, measurement of confounders, and the appropriate use of multivariate adjustment, become critical for assessing the validity of the study results. Confounding by indication, where a patient receives the test because of a greater or lower probability of the outcome of interest, is particularly problematic as the provider decision is often driven by factors that cannot easily be measured and thus cannot be included in adjustment models.

Although there is little debate that the direct assessment of clinical utility is the ideal evidence to inform the decision about the clinical use of a genomic test, the reality is that these decisions often have to be made before such evidence is available. The rapidity by which new information about clinical validity is generated and the cost and time involved to assess outcomes of a genomic test greatly limit the feasibility of clinical outcomes studies for determining clinical utility. In this setting, the assessment of clinical utility may involve linking several pieces of evidence often including the incremental predictive or diagnostic value of the test, the impact of the predictive or diagnostic information on treatment or other health-related decisions, and the impact of those decisions on health outcomes. Increasingly, such assessments are conducted using decision models that provide a structured and methodical approach to combining information about the decision alternatives and the probabilities of different outcomes given each alternative [41]. Such models have the advantage of being able to use information from different studies and assess the sensitivity of the results to assumptions and uncertainty. Although methodological advances have reduced concerns about the validity of decision models, their major limitation remains their dependence upon the quality and availability of the data that are available. The results of decision models are increasingly used to inform policy decisions when clinical utility data are not available and can help clinicians, patients, and payers understand the overall balance of benefits and harms associated with a genomic test and the factors that have the greatest influence on that balance.

ETHICAL, LEGAL, AND SOCIAL IMPLICATIONS

Health care exists within a social context and decisions about the clinical use of tests and treatments can have ramifications beyond the health care setting. Although these ramifications exist across health care decisions, the ethical, legal, and social implications of the use of genomic tests have received particular attention for several reasons [42]. Perhaps most importantly, the use of genomic information for purposes such as preimplantation analysis, prenatal testing, and carrier screening rises concerns about the societal implications of genetic engineering and the history of the eugenics movement. Even beyond the concern about genetic engineering, the potential for genetic information to be used to discriminate against vulnerable populations has long been recognized [43]. Most of the concern has focused on discrimination against racial and ethnic minority groups but similar possibilities exist for other populations, including the disabled. Genomic information also has potential implications for employment in some industries where exposures may have a differential effect by genotype and for the life and health insurance industries where asymmetric knowledge of genomic information could lead to adverse selection and a potential inability to cover claims.

Generating rigorous evidence about the expected societal impact of the use of genomic tests is challenging. Most of the evidence in this area is based upon adaptations of social science research methods including the use of attitudinal surveys, case studies, qualitative data collection, and even historical analysis. However, different and even multifaceted study designs may be appropriate for different questions. For example, an assessment of the impact of the *BRCA1/2* testing on the life insurance market included a survey of patients undergoing testing to assess their life insurance decisions, an assessment of the sensitivity of those decisions to test results, and actuarial modeling of the impact of test-driven purchasing decisions on the life insurance industry [44]. More recently, the implications of returning genomic results from research studies have been studied through the National Human Genome Research Institute funded Clinical Sequencing Exploratory Research consortium which specifically included Ethical Legal, and Social Implications projects in its approach [45]. Evidence about ethical, legal, and social implications is particularly important or policy decisions about genomic testing and the inclusion of relevant stakeholders in its development and dissemination is central to its impact.

FUTURE DIRECTIONS

The generation of evidence to inform the clinical use of genomics is a major priority for the next phase of genomics research. Although there is little doubt these activities will continue to be guided by the ACCE framework and to build upon the traditions of evidence-based medicine, they will also evolve to reflect developments in genomics, clinical research, and decision support.

The emergence of comparative effectiveness research (CER) and particularly patient-centered outcomes research (PCOR) has already had a major impact on the approach to evaluating medical interventions. These fields use the types of studies described under clinical utility but place additional emphasis on the need to examine the incremental benefit of a new intervention compared to existing strategies and to engage stakeholders in the research process [46]. CER is facilitated by the use of large data sets such as insurance claims data or electronic medical record (EMR) data repositories. Thus, while CER and PCOR have had relatively little overlap with genomics to date, this may change as more and more genomic applications enter practice, EMRs begin to include the results of genomic tests, and specific billing codes for specific genomic tests are moved into practice [37]. These changes will enable genomics to become part of the vision for learning health systems where the creation of new evidence is closely linked to the delivery of health care and the application of that evidence is an integral part of continuous quality improvement.

The paradigm of evidence-based medicine is founded in a probabilistic model of causality where the level, type, and amount of population-based evidence determine how likely an association is to be causal and thus what should be done for an individual patient. While biological plausibility plays into this model, Evidence-Based Medicine is very different than most models of personalized medicine, where the optimal decision is based upon understanding the biology underlying the question at hand, testing the patient for specific components of the pathway(s) of interest, and making a decision based upon those test results. This "personalized medicine" model is already emerging in oncology, where treatment decisions are driven by a biology-based prediction of response given the presence or absence of a certain molecular marker rather than a probabilistic prediction of response from population-based evidence [47]. Furthermore, it is already being extended to include the measurement of biological response in the individual patient and pathways that influence the relationship between measures of gene expression and actual protein function [48]. The ability to effectively integrate these two approaches into an effective and consistent approach to the use of genomics in clinical care will be central to the effective translation of these advances into population benefit.

As with many areas of medicine, the impact of genomic tests on economic outcomes at the societal, payer, or individual level remains largely unexplored. One recent review identified a total of 59 cost−utility analyses that address personalized medicine testing and many of these were evaluating the same test [49]. Overall, there is considerable uncertainty about how genomic tests will affect direct and indirect health care costs, largely because the greatest impact will come from the downstream impact on health care utilization rather than the cost of the test itself [50]. This "cost cascade" can make it particularly challenging to assess the full range of economic outcomes in a clinical trial or even a prospective observational study, so that economic analyses often rely upon simulation modeling.

Finally, the rapid dissemination of EMRs with the associated collection of clinical data has created major new opportunities for the implementation of clinical decision support. Those opportunities include the ability to create dynamic and modular algorithms that are more easily updated when new information becomes available than traditional guidelines and to tailor the support to individual patient characteristics. Eventually, these systems may even be able to generate new information that is specifically tailored to the question at hand by interrogating "big data" resources that link genomic and clinical variables. Such systems are already being built in oncology to support the use of tumor profiling, and there is little doubt that they will be necessary to support the use of whole-genome and whole-exome sequencing across clinical areas.

Genomic medicine has become a reality in the last decade. This reality has been supported by a rapid increase in the generation of empirical evidence to guide the appropriate use of genomic tests. However, there is little doubt that the need for such evidence will only continue to grow. Although much can be accomplished by building on the foundation of evidence-based medicine, any approach that requires the conduct of multiple randomized controlled trials that are summarized by meta-analyses and used for guideline development may have limited impact in a field that is moving so quickly and is driven by individual variation instead of population averages. New approaches that include advances in bioinformatics and modeling, the integration of biologic and probabilistic data, the collection and dissemination of "big data" from clinical and other platforms, and the continued education of patients and physicians about the strengths and limitations of genomic information will be central to the next decades of evidence generation and the eventual impact of genomic medicine on the health of our patients.

REFERENCES

[1] Scheuner MT, Sieverding P, Shekelle PG. Delivery of genomic medicine for common chronic adult diseases: a systematic review. JAMA 2008;299:1320—34.

[2] Shine I. Problems of sickle-cell screening. N Engl J Med 1973;288:971.

[3] Evans JP, Meslin EM, Marteau TM, Caulfield T. Genomics. Deflating the genomic bubble. Science 2011;331:861—2.

[4] Higashi MK, Veenstra DL. Managed care in the genomics era: assessing the cost effectiveness of genetic tests. Am J Manag Care 2003;9:493—500.

[5] Califf RM. Defining the balance of risk and benefit in the era of genomics and proteomics. Health Aff (Millwood) 2004;23:77—87.

[6] Berg AO. The CDC's EGAPP initiative: evaluating the clinical evidence for genetic tests. Am Fam Physician 2009;80:1218.

[7] Jaeschke R, Sackett DL. Research methods for obtaining primary evidence. Int J Technol Assess Health Care 1989;5:503—19.

[8] Armstrong K, Micco E, Carney A, Stopfer J, Putt M. Racial differences in the use of BRCA1/2 testing among women with a family history of breast or ovarian cancer. JAMA 2005;293:1729—36.

[9] Haddow JE, Palomaki GE. ACCE: a model process for evaluating data on emerging genetic tests. In: Khoury MJ, Little J, Burke W, editors. Human genome epidemiology: a scientific foundation for using genetic information to improve health and prevent disease. New York: Oxford University Press; 2003. p. 217—33.

[10] Evaluation of Genomic Applications in Practice and Prevention Working Group. The EGAPP initiative: lessons learned. Genet Med 2014;16:217—24.

[11] Schunemann HJ, Oxman AD, Brozek J, et al. Grading quality of evidence and strength of recommendations for diagnostic tests and strategies. BMJ 2008;336:1106—10.

[12] Teutsch SM, Bradley LA, Palomaki GE, et al. The Evaluation of Genomic Applications in Practice and Prevention (EGAPP) Initiative: methods of the EGAPP Working Group. Genet Med 2009;11:3—14.

[13] Sanger F, Nicklen S, Coulson AR. DNA sequencing with chain-terminating inhibitors. Proc Natl Acad Sci USA 1977;74:5463—7.

[14] Sims D, Sudbery I, Ilott NE, Heger A, Ponting CP. Sequencing depth and coverage: key considerations in genomic analyses. Nat Rev Genet 2014;15:121—32.

[15] Linderman MD, Brandt T, Edelmann L, et al. Analytical validation of whole exome and whole genome sequencing for clinical applications. BMC Med Genomics 2014;7:20.

[16] Walter SD, Riddell CA, Rabachini T, Villa LL, Franco EL. Accuracy of p53 codon 72 polymorphism status determined by multiple laboratory methods: a latent class model analysis. PLoS One 2013;8:e56430.

[17] Fumagalli D, Blanchet-Cohen A, Brown D, et al. Transfer of clinically relevant gene expression signatures in breast cancer: from Affymetrix microarray to Illumina RNA-Sequencing technology. BMC Genomics 2014;15:1008.

[18] Press MF, Slamon DJ, Flom KJ, Park J, Zhou JY, Bernstein L. Evaluation of HER-2/neu gene amplification and overexpression: comparison of frequently used assay methods in a molecularly characterized cohort of breast cancer specimens. J Clin Oncol 2002;20:3095—105.

[19] Meyerson M, Gabriel S, Getz G. Advances in understanding cancer genomes through second-generation sequencing. Nat Rev Genet 2010;11:685—96.

[20] Simon RM, Paik S, Hayes DF. Use of archived specimens in evaluation of prognostic and predictive biomarkers. J Natl Cancer Inst 2009;101:1446—52.

[21] Kim K, Zakharkin SO, Allison DB. Expectations, validity, and reality in gene expression profiling. J Clin Epidemiol 2010;63:950—9.

[22] Kalman LV, Lubin IM, Barker S, et al. Current landscape and new paradigms of proficiency testing and external quality assessment for molecular genetics. Arch Pathol Lab Med 2013;137:983—8.

[23] Feldman GL, Schrijver I, Lyon E, Palomaki GE, Biochemical CA, Molecular Genetics Resource Committee. Results of the College of American Pathology/American College of Medical Genetics and Genomics external proficiency testing from 2006 to 2013 for three conditions prevalent in the Ashkenazi Jewish population. Genet Med 2014;16:695—702.

[24] Paik S, Shak S, Tang G, et al. A multigene assay to predict recurrence of tamoxifen-treated, node-negative breast cancer. N Engl J Med 2004;351:2817—26.

[25] Albain KS, Barlow WE, Shak S, et al. Prognostic and predictive value of the 21-gene recurrence score assay in postmenopausal women with node-positive, oestrogen-receptor-positive breast cancer on chemotherapy: a retrospective analysis of a randomised trial. Lancet Oncol 2010;11:55—65.

[26] Wacholder S, Hartge P, Prentice R, et al. Performance of common genetic variants in breast-cancer risk models. N Engl J Med 2010;362:986—93.

[27] Pepe MS, Janes HE. Gauging the performance of SNPs, biomarkers, and clinical factors for predicting risk of breast cancer. J Natl Cancer Inst 2008;100:978—9.

[28] Cook NR, Ridker PM. Advances in measuring the effect of individual predictors of cardiovascular risk: the role of reclassification measures. Ann Intern Med 2009;150:795—802.

[29] Jaeschke R, Guyatt G, Sackett DL. Users' guides to the medical literature. III. How to use an article about a diagnostic test. A. Are the results of the study valid? Evidence-Based Medicine Working Group. JAMA 1994;271:389—91.

[30] Castaldi PJ, Dahabreh IJ, Ioannidis JP. An empirical assessment of validation practices for molecular classifiers. Brief Bioinform 2011;12:189—202.

[31]   Ransohoff DF, Feinstein AR. Problems of spectrum and bias in evaluating the efficacy of diagnostic tests. N Engl J Med 1978;299:926−30.

[32]   Attia J, Ioannidis JP, Thakkinstian A, et al. How to use an article about genetic association: B: Are the results of the study valid? JAMA 2009;301:191−7.

[33]   Colhoun HM, McKeigue PM, Davey Smith G. Problems of reporting genetic associations with complex outcomes. Lancet 2003;361:865−72.

[34]   Ioannidis JP, Khoury MJ. Improving validation practices in "omics" research. Science 2011;334:1230−2.

[35]   Barnholtz-Sloan JS, McEvoy B, Shriver MD, Rebbeck TR. Ancestry estimation and correction for population stratification in molecular epidemiologic association studies. Cancer Epidemiol Biomarkers Prev 2008;17:471−7.

[36]   Botkin JR, Teutsch SM, Kaye CI, et al. Outcomes of interest in evidence-based evaluations of genetic tests. Genet Med 2010;12:228−35.

[37]   Lin JS, Thompson M, Goddard KA, Piper MA, Heneghan C, Whitlock EP. Evaluating genomic tests from bench to bedside: a practical framework. BMC Med Inform Decis Mak 2012;12:117.

[38]   Kimmel SE, French B, Kasner SE, et al. A pharmacogenetic versus a clinical algorithm for warfarin dosing. N Engl J Med 2013;369:2283−93.

[39]   Concato J, Lawler EV, Lew RA, Gaziano JM, Aslan M, Huang GD. Observational methods in comparative effectiveness research. Am J Med 2010;123:e16−23.

[40]   Teutsch SM, Bradley LA, Palomaki GE, et al. The Evaluation of Genomic Applications in Practice and Prevention (EGAPP) Initiative: methods of the EGAPP Working Group. Genet Med 2009;11:3−14.

[41]   Veenstra DL, Piper M, Haddow JE, et al. Improving the efficiency and relevance of evidence-based recommendations in the era of whole-genome sequencing: an EGAPP methods update. Genet Med 2013;15:14−24.

[42]   Hudson KL. Genomics, health care, and society. N Engl J Med 2011;365:1033−41.

[43]   Holtzman NA, Rothstein MA. Eugenics and genetic discrimination. Am J Hum Genet 1992;50:457−9.

[44]   Armstrong K, Weber B, FitzGerald G, et al. Life insurance and breast cancer risk assessment: adverse selection, genetic testing decisions, and discrimination. Am J Med Genet A 2003;120:359−64.

[45]   Goddard KA, Whitlock EP, Berg JS, et al. Description and pilot results from a novel method for evaluating return of incidental findings from next-generation sequencing technologies. Genet Med 2013;15:721−8.

[46]   Institute of Medicine. Initial national priorities for comparative effectiveness research: report brief. Available from: http://www.iom.edu/ ~ /media/Files/ReportFiles/2009/ComparativeEffectivenessResearchPriorities/CERreportbrief08-13-09.ashx%3E; 2009 [accessed 07.11.09].

[47]   Buettner R, Wolf J, Thomas RK. Lessons learned from lung cancer genomics: the emerging concept of individualized diagnostics and treatment. J Clin Oncol 2013;31:1858−65.

[48]   Chen R, Mias GI, Li-Pook-Than J, et al. Personal omics profiling reveals dynamic molecular and medical phenotypes. Cell 2012;148:1293−307.

[49]   Phillips KA, Ann Sakowski J, Trosman J, Douglas MP, Liang SY, Neumann P. The economic value of personalized medicine tests: what we know and what we need to know. Genet Med 2014;16:251−7.

[50]   Armstrong K. Can genomics bend the cost curve? JAMA 2012;307:1031−2.

Chapter 13

Family Health History and Health Risk Assessment in Health Care

Lori A. Orlando and R. Ryanne Wu
Duke University, Durham, NC, United States

Chapter Outline

INTRODUCTION

In the continuum from health to disease there are several key transition periods. The first is from healthy to presymptomatic, where an individual still feels well and is asymptomatic but has developed a disease. An example of this health state is the beginning of diabetes when an individual's sugar is elevated but they are unaware of it. The second is from presymptomatic to disease diagnosis and the third is from diagnosis to disease status, which can be either well controlled or uncontrolled. Health risk assessments (HRAs) are an essential component of the healthy period. Their purpose is to estimate an individual's risk for developing common chronic diseases (see Table 13.1 for examples) allowing clinicians to tailor preventive care, screening, and testing to each individual's level of risk—with the goal of keeping healthy people healthy. Personalized care plans developed with the aid of HRAs balance effectiveness and harms with risk, in a way that maximizes benefit and minimizes harm not only for each individual, but when taken as a whole, for the population as well. Unfortunately, HRAs are not widely used in primary care, where they would be most effective, due to a

Genomic and Precision Medicine. DOI: http://dx.doi.org/10.1016/B978-0-12-800681-8.00013-X

TABLE 13.1 Examples of Conditions for Which Family Health History Based HRA Is Useful

	Risk Algorithm Based on Family Health History Only	Risk Algorithms Include Family Health History
Hereditary breast and ovarian cancer	x	
Hereditary non-polyposis colon cancer (Lynch syndrome)	x	
Alpha-1-antitrypsin deficiency	x	
Diabetes mellitus type 2		x
Abdominal aortic aneurysm		x
Coronary artery disease		x
Hemochromatosis	x	
Maturity onset diabetes of the young		x
Osteoporosis		x
Asthma		x
Melanoma		x
Prostate cancer		x
Age-related macular degeneration		x

number of constraints. This chapter discusses how HRAs were developed, their key aspects, and what needs to occur in order to integrate them into primary care settings.

IN THE BEGINNING

In 1948, Joseph Mountain, the Assistant Surgeon General, initiated the Framingham Heart Study, an innovative longitudinal study arising from the field of epidemiology. The goal, as devised by the director, Thomas Dawber, was to closely follow a group of individuals living in Framingham, MA, collecting as much data as possible over the course of many years in order to develop a risk prediction model for heart disease [1]. This was the first time the phrase "factor of risk," more commonly termed risk factor today, was introduced [2]. Despite initial skepticism among both the research and medical communities, the trial was successful beyond expectations and the field of HRA was born. In 2009, when Clay Christensen coined the term "Precision Medicine" he defined it as precisely predicting a medical outcome by combining a variety of data into rules [3]. By this definition, HRAs are simply the application of precision medicine to those who are healthy.

Today most HRAs include the following components: data collection (either through a Web-based or paper questionnaire), risk calculation, and report of risk results. This last component, the report, may or may not provide guidance about how to manage your risk. Some are exceptionally detailed and even indicate how much your risk can be lowered by initiating one or more recommended preventive actions, while others merely indicate that you are at increased risk for the specified condition. For the first component, data collection, the data collected varies depending upon which conditions are included in the risk assessment, but at a minimum they all include: demographics, lifestyle, personal health history, *family health history*, and biometrics (such as blood pressure, weight, cholesterol, etc.). Other types of data, such as genetic/genomic and individual preferences, are just now starting to be incorporated into some risk assessment models and have the potential to not only refine the accuracy of risk calculations, but to also improve shared decision making with medical providers [4,5].

WHY FAMILY HEALTH HISTORY IS CENTRAL TO HRAs

Family health history is an unassuming and often overlooked, but essential data element in HRAs. For many conditions, family health history is the strongest predictor of disease risk and for some, such as hereditary cancer syndromes, it is the only predictor (and thus the only component of the HRA) (see Table 13.1). An example of the impact of family health history on disease risk is Type II Diabetes, where a first-degree relative (parent or child) with the disease increases an individual's risk from an average of 3.2% to 14.3% [6]. In some cases, excluding a family health history can lead to missing those at highest risk for developing a condition. For example, many risk assessments for chronic obstructive pulmonary disease ask about environmental exposures (such as smoking and asbestos), but do not ask about family history; however, those with alpha-1-antitrypsin deficiency, a hereditary condition that runs in families, are at the highest risk of developing chronic obstructive pulmonary disease even without an environmental exposure [7]. Renal cell carcinoma, a tumor of the kidney, is another example. Almost all risk assessments include smoking, alcohol, and exercise, and some include family members with renal cell carcinoma, but most do not ask about a family history of other cancers even though renal cell carcinoma is part of the constellation of cancers that can occur in two hereditary cancer syndromes, Lynch and Von Hippel—Lindau [8]. While those with hereditary cancer syndromes or alpha-1-antitrypsin deficiency are only a small proportion of those developing these two conditions, they are the ones at the highest risk of developing disease.

In addition to being highly predictive, family health history also serves as the basis for a number of evidence-based guidelines that not only indicate the level of disease risk associated with a given combination of affected relatives, but also actions to take to manage risk. For example, the National Comprehensive Cancer Network's guidelines for breast and ovarian cancer recommend BRCA testing if an individuals' first-degree relative (parents or child) developed breast cancer at age 45 or younger [9]. Another example is abdominal aortic aneurysm screening. If an individual has a relative with the condition, then screening is recommended when they are aged 50 or older [10,11].

Thus family health history is the only data element in HRAs that is both highly predictive and actionable in combination with other data elements and by itself. Unfortunately family health history is often hard to obtain. Individuals often do not know much about their relatives' health and what they do know is often piecemeal or may be inaccurate [12]. This leads to the problem that one of the most informative data elements in HRAs is also one of the more difficult to collect.

AN IMPLEMENTATION CRISIS

Despite the acclaim surrounding the publication of the Framingham Heart Study results, there was little movement in the field of HRA until 1980 when the Center for Disease Control (CDC) released a publicly available HRA tool [13]. Incidentally, 1980 was also a time when employers and insurers were being faced with rapidly increasing healthcare costs. In their search for a way to manage these costs, they turned to HRAs [14]. To explore the impact of this resource, Prudential funded updates to the CDC's tool, which ultimately showed that use of an HRA tool in the workplace could lower company healthcare expenditures, as well as reduce absenteeism and increase productivity [15,16]. These results and Prudential's takeover of the program in 1986 led to rapid uptake among US employers and insurance companies; however, uptake continued to be anemic in the healthcare setting [13].

Explanations for why implementation in the healthcare system failed to take root include: the disconnect between public health and health care, increasing demands on primary care providers, and a perverse incentive system that rewarded interventions over maintaining health [13,17]. The combination of these factors encouraged the development of a healthcare system, incapable of responding to the needs of the healthy segment of the population, quickly leading to a negative feedback loop dominated by sick patients getting sicker, less time to manage risk among healthy patients, and ultimately healthy patients getting sick [18]. In this environment it is easy to see how adoption of HRA in clinical practice was slow.

Fortunately, recent studies have highlighted these findings and their unsustainable impact on the US healthcare system. In particular, the 5 Mirror, Mirror studies performed by the Commonwealth Fund to assess healthcare quality and cost in 11 international healthcare systems between 2004 and 2014, not only ranked the United States last in quality and highest in expenditures, but showed little improvement over the 10-year period [19]. In addition, the Affordable Care Act enacted in 2010 has emphasized the need for improvements in quality of care, maintaining health, and lowering costs. HRAs are neatly aligned with these objectives and are now viewed as a useful tool for redesigning healthcare systems. That being said, there are still a number of practical barriers to overcome before implementation in primary care can become widespread: ease of use for providers, ease of use for patients, quality of the data entered into the HRA (particularly family health history data), and its potential to improve quality of care in primary care populations. Each of these are describe in detail below.

WILL PROVIDERS USE IT?

Providers, primary care providers in particular, are frequently overloaded by the number of tasks to achieve within the constraints of the healthcare visit, and with face times shrinking to just over 9 min for most appointments many lower priority and/or complex tasks often lose out to higher acuity concerns [18]. Because HRA data collection, risk calculation, and evidence synthesis are complex and time consuming (particularly for family health history data), it often poses a significant challenge for integration into normal work flow [20]. In particular, algorithms are often complex requiring a computer to calculate, however, the calculators are typically scattered across the Internet and not integrated into electronic medical record (EMR) systems. In addition, the sheer magnitude of the literature available makes synthesizing an actionable risk management plan difficult and efforts to initiate provider education around these topics have fallen flat for many of the same reasons that implementing HRAs have [21].

One solution to these complex and interrelated barriers is to leverage the burgeoning field of health IT. Patient-facing Web- or computer-based HRA tools can eliminate the data collection component from the physician's office, moving it to the patient's home, and provide risk calculations and actionable risk management plans to the provider at the point-of-care. Similarly, mobile health technologies are beginning to demonstrate that they can be used to facilitate risk-related data collection (environmental, behavioral, psychological, and biological) and communication between patients and their healthcare system. To date a number of family health history based HRA tools have been built with just such capabilities [22]. See Table 13.2 for examples of currently available tools and their characteristics. Uptake of these tools has

TABLE 13.2 Examples of Patient-Facing Electronic Family Health History Based HRA Tools

FHH-Based CDS Software Programs for Adult	# Conditions Collected	Decision Support Diseases	Completed by	Validation of Data Accuracy	Who Receives Output	Available at Point-of-Care	Action Oriented Recommendations
MeTree[a]	100	27 (colon, breast, lung, and ovarian cancer; thrombosis, aortic aneurysm, heart disease, stroke, diabetes, hereditary cancer syndromes, hereditary cardiovascular syndromes, hereditary liver diseases)	Patient online or physician's office	Yes	Patient and physician	Yes	Yes
Schroy et al.[b]	1	1 (colon cancer)	Physician	?	Physician	Yes	No
GRACE[c]	1	1 (breast cancer)	Patient in physician's office	No	Patient, clinical nurse specialist, physician	Yes	No
Family Healthware[d]	6	6 (coronary heart disease, diabetes, stroke, colon, breast, and ovarian cancer)	Patient online	Yes	Patient	No	No

Family HealthLink[e]	38 (35 cancers)	Coronary heart disease, cancer	Patient online	?	Patient	No	No
Cancer Risk Intake system (CRIS)[f]	3	3 (breast, ovarian, and colon cancer)	Patient in physician's office	Yes	Patient and physician	Yes	No
Hughes Risk Apps[g]	~5 (cancers)	Breast cancer and hereditary cancer syndromes	Patient—clinician can revise online or physician's office	Yes	Patient and physician	Yes	No
Health Heritage[h]	87	15 (cancers, diabetes, neuromuscular diseases, and cardiovascular disease)	Patient online	No	Patient	No	No
Invitae[i]	6	Hereditary cancer and hereditary cardiac disease genetic testing	Physician online	No	Patient and Physician	?	No
MyFamily[j]	?	Cancer, cardiology, gastrointestinal (proprietary risk algorithms)	Patient	No	Physician	Yes	Yes
Myriad[k]	29 cancers	Hereditary cancer genetic testing	Patient	No	Patient	No	No

(Continued)

TABLE 13.2 (Continued)

FHH-Based CDS Software Programs for Adult	# Conditions Collected	Decision Support Diseases	Completed by	Validation of Data Accuracy	Who Receives Output	Available at Point-of-Care	Action Oriented Recommendations
Power Lineage[l]	Cancer	Six hereditary cancers (breast, colon, endometrial, ovarian, pancreatic, melanoma)	Patient and Physician	No	Physician	Yes	Yes
My Family Health Portrait[m]	~21	2 (diabetes and colon cancer)	Patient	Yes	Patient	No	Yes

[a]http://dukepersonalizedmedicine.org/disease-risk-and-diagnosis/risk-assesments/family-history/metree

[b]Schroy PC, 3rd, Glick JT, Geller AC, Jackson A, Heeren T, Prout M. A novel educational strategy to enhance internal medicine residents' familial colorectal cancer knowledge and risk assessment skills. Am J Gastroenterol 2005; 100(3): 677–684.

[c]Braithwaite D, Sutton S, Mackay J, Stein J, Emery J. Development of a risk assessment tool for women with a family history of breast cancer. Cancer Detect Prev 2005;29(5):433–439.

[d]Yoon PW, Scheuner MT, Jorgensen C, Khoury MJ. Developing Family Healthware, a family history screening tool to prevent common chronic diseases. Prev Chronic Dis 2009;6(1):A33.

[e]https://familyhealthlink.osumc.edu/Notice.aspx.

[f]Sugg Skinner C, Rawl SM, Moser BK, et al. Impact of the cancer risk intake system on patient-clinician discussions of tamoxifen, genetic counseling, and colonoscopy. J Gen Intern Med 2005;20(4):360–365. doi:10.1111/j.1525-1497.2005.40115.x.

[g]http://www.hughesriskapps.com/

[h]http://www.healthheritage.org/

[i]https://www.invitae.com/en/

[j]Doerr M, Edelman E, Gabitzsch E, Eng C, Teng K. Formative evaluation of clinician experience with integrating family history-based clinical decision support into clinical practice. J Pers Med 2014;4(2):115–136. doi:10.3390/jpm4020115.

[k]https://fht.myriad.com/app/#

[l]https://www.powerlineage.com/

[m]https://familyhistory.hhs.gov/FHH/html/index.html

been anecdotally promising. Currently, there is only one published study that reports physician experience with and uptake of an HRA tool. In this study, performed by the authors, primary care providers indicated that the HRA tool was easy to incorporate into workflow (100%), improved their quality of care (85%), made their practice easier (75%), and enhanced their understanding of the importance of family health history (62%) [23]. (See Fig. 13.1 for sample

MeTree© Personalized Risk Profile - Physician Summary

MeTree ID: A567
Questionnaire ID: 1234
Public, John
DOB: 12/13/1985 Age: 56
BMI: 26
MeTree© Assessment Tool recommendations are based on information supplied by patient. They may not represent a complete clinical assessment and are not intended to supplant physician discretion in risk management. Based on your needs, a genetic counselor may suggest additional screenings that are not included in this report.

Provider Risk Assessment, Prevention, and Screening Report

- If the patient has a condition prevention recommendations are not provided for that condition
- Cancer addresses breast, ovarian, colon, lung, and hereditary syndromes

Cancer

- **Recommend referral for genetic counseling for risk of hereditary cancer syndrome**[1,2,3,4,5] due to:
  - At least 3 relatives with HNPCC-related cancers (colorectal, uterine, gastric, ovarian, renal, small bowel, pancreatic, brain).
- **Recommend breast cancer surveillance via annual breast MRI and mammography**[7] due to:
  - Tyrer-Cuzick lifetime breast cancer risk = 20.1%, patient meets criteria to add breast MRI to annual mammography (>20%)[7,8].
  - Breast MRI NNT (number of women needed to screen to detect one breast cancer undetected by mammography) is 58 in increased-risk women.[9]
- **Colon cancer risk is increased, recommend early and more frequent colonoscopies (every 5 years beginning at age 40 or 10 yrs younger than the earliest diagnosis in the family, whichever comes first)**[4] due to:
  - 1 first-degree relative with colorectal cancer at age <60.

Cardiovascular Disease

- **Framingham**[10] **10 yr CHD risk score = %. Consider low-dose aspirin (75-81 mg) therapy for heart disease prevention** due to
  - Risk > 10%
  - Avoid if contradictions (such as aspirin intolerance, risk of bleeding and/or falls, poorly controlled hypertension) present.
- **Consider starting or switching to a high intensity statin due to**[13]:
  - Aged 40-60 without diabetes
  - Cardiovascular risk score >=7.5% 10 yr risk of atherosclerotic CVD events.
  - Number needed to treat (NNT) for one patient to avoid an MI = 60[11]
  - High intensity statins include Atorvastatin 40 & 80mg or Rosuvastatin 20 & 40mg are recommended.
  - If myalgias previously on statin(s) consider a lipophilic statin like Pravastatin, Atorvastatin, or Rosuvastatin or consider pharmacogenomic testing (see points to consider)

Point(s) to Consider

- Genetic testing results and genetic counselor consultation may help guide risk management decisions, including whether to recommend following NCCN guidelines for hereditary colon cancer syndromes[6].
- For most increased-risk women, MRI & mammography should begin at 30 & continue as long as the pt is in good health.
- Reynold's risk score has been shown in some studies to be better calibrated and have improved overall discrimination than Framingham, especially for women[10, 12].
- A trial of diet and exercise to lower LDL and recalculate lifetime risk is a reasonable first step in some individuals. Consider starting 7g psyllium fiber (decreases LDL by 20%), 2g plant sterols (decreases LDL by 15%), with decreased intake of animal proteins.

FIGURE 13.1   Sample MeTree provider report.

• For more information about available genetic tests, reasons to use them, and how to order see
www.ncbi.nlm.nih.gov/sites/GeneTests/clinic?db=GeneTests

Cancer Reference(s)

[1]U.S. Preventive Services Task Force. Genetic risk assessment and BRCA mutation testing for breast and ovarian cancer susceptibility: recommendation statement. Ann Intern Med. 2005;143:355-61. PubMed PMID: 16144894

[2]Hampel H et al. Referral for cancer genetics consultation: a review and compilation of risk assessment criteria. J Med Genet. 2004;41:81-91. PubMed PMID: 14757853

[3]Berliner JL et al. Risk assessment and genetic counseling for hereditary breast and ovarian cancer: recommendations of the National Society of Genetic Counselors. J Genet Counsel. 2007;16:241-60. PubMed PMID: 17508274

[4]Levin B et al. Screening and surveillance for the early detection of colorectal cancer and adenomatous polyps, 2008: a joint guideline from the American Cancer Society, the US Multi-Society Task Force on Colorectal Cancer, and the American College of Radiology. CA Cancer J Clin. 2008;58:130-60. PubMed PMID: 18322143

[5]Vasen HF et al. New clinical criteria for hereditary nonpolyposis colorectal cancer (HNPCC, Lynch syndrome) proposed by the International Collaborative group on HNPCC. Gastroenterology. 1999;116:1453-6. PubMed PMID: 10348829

[6]National Comprehensive Cancer Network. 2014. http://www.nccn.org/professionals/physician_gls

[7]Saslow D et al., American Cancer Society guidelines for breast screening with MRI as an adjunct to mammography. CA Cancer J Clin. 2007;57:75-89. PubMed PMID: 17392835

[8]Berry DA, et al., BRCAPRO validation, sensitivity of genetic testing of BRCA1/BRCA2, and prevalence of other breast cancer susceptibility genes. J Clin Oncol. 2002;20:2701-2712. PubMed PMID: 12039933

[9]Kuhl C et al. Prospective multicenter cohort study to refine management recommendations for women at elevated familial risk of breast cancer: the EVA trial. J Clin Oncol. 2010;28:1450-7. PubMed PMID: 20177029

Cardiac Reference(s)

[10]D'Agostino, RB, Vasan, RS, et al. General Cardiovascular Risk Profile for Use in Primary Care: The Framingham Heart Study. Circulation. 2008;117:743-753. PubMed PMID: 18212285

[11]Ray KK, Seshasai SR, Erqou S, Sever P, Jukema JW, Ford I, Sattar N. Statins and all-cause mortality in high-risk primary prevention: a meta-analysis of 11 randomized controlled trials involving 65,229 participants. Arch Intern Med. 2010 Jun 28;170(12):1024-31. Review. PubMed PMID: 20585067

[12]Cook, NR, Paynter, NP, et al. Comparison of the Framingham and Reynolds Risk scores for global cardiovascular risk prediction in the multiethnic Women's Health Initiative. Circulation. 2012;125:1748-56. PubMed PMID: 22399535

[13]Neil J. Stone, Jennifer Robinson, Alice H. Lichtenstein, et al. 2013 ACC/AHA Guideline on the Treatment of Blood Cholesterol to Reduce Atherosclerotic Cardiovascular Risk in Adults. Journal of the American College of Cardiology. 2013;10.1016/j.jacc.2013.11.002. PubMed PMID: 24239923

FIGURE 13.1 (Continued).

provider report.) These results suggest that with the right combination of features, electronic HRA tools can gain acceptance by busy primary care clinicians.

WILL THE PATIENTS USE IT?

With the expansion of Web access, the health systems' greater emphasis on patient engagement, and the increasing demands on medical providers' time, there has been a movement toward increased utilization of patient-derived data for HRAs. Some have raised concern that patients do not have the IT know-how to use these tools or the inclination to do so. But current trends would argue that that is not the case. Millions of patients use the Internet on a daily basis to better understand their health [24,25], and even among minorities and those with household incomes $25,000−$50,000 per year Internet use is over 66% [26]. In addition, individuals recognize the importance of family health history with over 90% of Americans surveyed reporting that they felt family health history was important to their personal health [27,28]. When HRAs are placed on health systems' patient portals significant uptake occurs. For example, when Health Heritage was launched on the Northshore University Health System's health portal in May 2014, 500 patients completed the assessment within the first 30 days (unpublished data). This was without any coordinated efforts by the health system to alert patients to the availability of the tool.

There are also concerns that such tools discriminate against those of lower socioeconomic status who may not have the IT literacy or access required to complete such tools. In fact, the reverse is true—HRA tools can reduce health disparities. Algorithms to assess risk and provide recommendations do not hold the inherent biases that providers may subconsciously bring to the patient encounter. Uptake has also not been shown to be limited by education level or ethnicity, two significant risk factors for healthcare disparity [23,29]. Patients using one HRA tool, MeTree, reported that the tool was easy to use (94%) and understand (97%) irrespective of education level or ethnicity [23].

When patients do use IT tools for HRA, reactions are overwhelming positive. In three studies, participants reported being generally satisfied with their experience (83–90%), that completing the tool did not cause persistent anxiety (96%), and that they would recommend it to others (93–99%) [23,29,30]. In addition to having a positive experience using the tool itself, patients incur the additional benefits of: raised awareness about their risk for disease and the actions they can take to mitigate their risk, and greater engagement in their health with the potential for an enhanced shared decision making experience with their providers [4,5]. (See Fig. 13.2 for HRA screenshot.)

WHAT ABOUT PATIENT-ENTERED DATA?

If we move HRA data collection out of the provider encounter, is the data entered by patients reliable? In particular, there are concerns about patients' ability to accurately collect and interpret their family's health history. Questions arise around differentiation of a primary cancer versus metastatic cancer, different types of cardiovascular disease such as coronary artery disease versus arrhythmias, and disease with names that sound similar such as cirrhosis and psoriasis. In a systematic review of family history questionnaires, four studies examined the agreement between patient-entered data and the presumed gold standard of genetic counselor acquired data with evidence of strong agreement between them (70–100%) [31]. In fact there is evidence that patient-entered data is significantly superior to what is collected in routine practice within primary care [20,29,31,32]. In our own experience, less than 4% of patients' medical records contained high-quality family history information documented for even one relative [33]. In

FIGURE 13.2   MeTree HRA screenshot.

comparison, with use of MeTree over 99% of pedigreess had at leat one relative with high-quality family history information and over 50% of pedigrees
had at least 50% of relatives with high-quality information [34].

Certainly the context within which family health history is collected will
affect its validity. When patients are offered education and the opportunity
to discuss their family health history with relatives, significant improvements
in accuracy can be seen. When patients are informed of the key components
of a quality family history and use that information to guide discussions with
family members, almost half will provide new or updated information and
16% will have a change in recommendations for disease risk management
following conversations with family members [35].

DOES IT MAKE A DIFFERENCE?

Knowing that providers and patients will use an HRA and that patient
reported data is reliable, does use of such tools make a difference in outcomes? The value of HRAs is that there are now a considerable number of
guidelines that tailor recommended risk management strategies to an individual's risk for disease. These higher risk strategies often have a greater sensitivity than strategies recommended for population-based screening but are

also typically associated with higher costs or adverse events rates that warrant limiting them to those that are most likely to benefit. This tailoring of risk strategies to risk level helps to balance benefits and harms at the individual level and when assessed at the population level is better than using "one size fits all" recommendations. To show that HRAs are in fact able to increase uptake of these guidelines and improve individual/population health they need to be able to: improve identification of individuals at increased disease risk, increase the match between risk level and risk management strategy, and improve patient acceptance of management recommendations above what is currently occurring in routine care.

Surveys of providers have shown that physicians frequently over- or underestimate risk and do not feel confident in their ability to assess risk [36–39]. Fortunately several studies have shown that family health history based HRAs can accurately estimate risk and identify a significant number of patients who were not identified during routine visits with their providers [32,40]. For example, The Family Healthware Trial found that 82% of participants were at strong or moderate risk for at least one of six conditions [41]; the Health Heritage trial found 42% to be at increased risk [29]; and the Genomedical Connection study found 44% to be at increased risk [42].

While evidence is clear that using HRAs increases identification of at-risk individuals, less has been done to evaluate the impact on provider and patient behavior as a result of this risk information. In the authors' experience with MeTree, use of the tool decreased overutilization of high-risk services among average risk individuals by 81% while increasing use among appropriate high-risk individuals [43]. In another study, the implementation of a breast cancer risk assessment tool, HughesRiskApps, resulted in significant increases in referrals to genetic counseling and genetic testing [44]. Other studies have also confirmed that understanding of family history increases appropriate utilization of cancer screening services [45,46]. Trials are ongoing to better understand how to engage patients and providers in shared decision making surrounding patients' behavioral and psychosocial risks [5]. One example is the NHGRI funded Implementing Genomics in Practice network which consists of six genomic implementation demonstration projects (http://www.genome.gov/27554264). Three of these projects address HRA: two are pilots exploring new HRA markers and one is a large-scale implementation of MeTree into primary care practices across five diverse national healthcare systems [47]. The MeTree study is also leveraging just-emerging health-related data transmission standards (SMART on FHIR) to evaluate the impact of bidirectional data flow between a patient-facing Web-based HRA application and EMR systems on patient care. Further work needs to be done in this area to better understand the impact of HRAs on patient and provider behavior as it relates to screening [48].

CONCLUSIONS

HRAs, a precision medicine tool intended to be used by well patients, hold promise for improving both individual health and population health by accurately estimating risk for disease and improving the match between risk level and risk management strategy above what is currently occurring in clinical care. These tools may also enhance patient experiences by increasing patient activation and shared decision making, thereby incorporating patient values into the patient–provider encounter and improving adherence to recommendations. Their accuracy and effectiveness may be further enhanced by combining genetic and genomic data about disease risk into risk algorithms. While uptake in the US healthcare system has been slow to start it is beginning to gain traction as the evidence builds around their benefit and innovations in health IT permit more seamless integration into clinical workflow. It is one of the few genomic and personalized medicine tools that is ready for immediate translation into clinical care (especially primary care) and as such can lead the way for translation of other genomic and personalized medicine in these fields by increasing clinicians' familiarity with these types of tools and their benefits.

ABBREVIATIONS

CDC Center for Disease Control
EMR electronic medical record
HRA health risk assessment

REFERENCES

[1] Dawber TR, Meadors GF, Moore Jr. FE. Epidemiological approaches to heart disease: the Framingham Study. Am J Public Health Nations Health 1951;41(3):279–81.
[2] Kannel WB, Dawber TR, Kagan A, Revotskie N, Stokes 3rd J. Factors of risk in the development of coronary heart disease—six year follow-up experience. The Framingham Study. Ann Intern Med 1961;55:33–50.
[3] Christensen CM, Grossman JH, Hwang J. The innovator's prescription: a disruptive solution for health care. New York: McGraw-Hill; 2009.
[4] Glasgow RE, Dickinson P, Fisher L, et al. Use of RE-AIM to develop a multi-media facilitation tool for the patient-centered medical home. Implement Sci 2011;6:118.
[5] Krist AH, Glenn BA, Glasgow RE, et al. Designing a valid randomized pragmatic primary care implementation trial: the my own health report (MOHR) project. Implement Sci 2013;8:73.
[6] Annis AM, Caulder MS, Cook ML, Duquette D. Family history, diabetes, and other demographic and risk factors among participants of the National Health and Nutrition Examination Survey 1999–2002. Prev Chronic Dis 2005;2(2):A19.
[7] Stockley RA. Alpha1-antitrypsin review. Clin Chest Med 2014;35(1):39–50.
[8] Chow WH, Dong LM, Devesa SS. Epidemiology and risk factors for kidney cancer. Nat Rev Urol 2010;7(5):245–57.

[9] Daly MB, Pilarski R, Axilbund JE, et al. Genetic/familial high-risk assessment: breast and ovarian, version 1.2014. J Natl Compr Cancer Netw 2014;12(9):1326–38.

[10] Kent KC, Zwolak RM, Jaff MR, et al. Screening for abdominal aortic aneurysm: a consensus statement. J Vasc Surg 2004;39(1):267–9.

[11] Joergensen TM, Houlind K, Green A, Lindholt JS. Abdominal aortic diameter is increased in males with a family history of abdominal aortic aneurysms: results from the Danish VIVA-trial. Eur J Vasc Endovasc Surg 2014;48(6):669–75.

[12] Quereshi N.W.B., Santaguida P., Carroll J., et al. Collection and use of cancer family history in primary care. AHRQ, Publication No. 08-E001. 2007. Available from: http://www.ahrq.gov/clinic/tp/famhisttp.htm.2015.

[13] Goetzel RZ, Staley P, Ogden L, et al. A framework for patient-centered health risk assessments—providing health promotion and disease prevention services to Medicare beneficiaries. Atlanta, GA: Department of Health and Human Services; Centers for Disease Control and Prevention; 2011.

[14] Breslow L, Fielding J, Herrman AA, Wilbur CS. Worksite health promotion: its evolution and the Johnson & Johnson experience. Prev Med 1990;19(1):13–21.

[15] DeFriese GH, Fielding JE. Health risk appraisal in the 1990s: opportunities, challenges, and expectations. Annu Rev Public Health 1990;11(1):401–18.

[16] Loeppke R, Nicholson S, Taitel M, Sweeney M, Haufle V, Kessler RC. The impact of an integrated population health enhancement and disease management program on employee health risk, health conditions, and productivity. Popul Health Manag 2008;11(6):287–96.

[17] McGinnis JM, Williams-Russo P, Knickman JR. The case for more active policy attention to health promotion. Health Aff 2002;21(2):78–93.

[18] Stange KC, Zyzanski SJ, Jaen CR, et al. Illuminating the 'black box'. A description of 4454 patient visits to 138 family physicians. J Fam Pract 1998;46(5):377–89.

[19] Davis K, Stremikis K, Schoen C, Squires D. Mirror, mirror on the wall, 2014 update: how the U.S. healthcare system compares internationally. The Commonwealth Fund; 2014.

[20] Sweet KM, Bradley TL, Westman JA. Identification and referral of families at high risk for cancer susceptibility. J Clin Oncol 2002;20(2):528–37.

[21] Schroy 3rd PC, Glick JT, Geller AC, Jackson A, Heeren T, Prout M. A novel educational strategy to enhance internal medicine residents' familial colorectal cancer knowledge and risk assessment skills. Am J Gastroenterol 2005;100(3):677–84.

[22] Orlando LA, Buchanan AH, Hahn SE, et al. Development and validation of a primary care-based family health history and decision support program (MeTree1). N C Med J 2013;74(4):287–96.

[23] Wu RR, Orlando LA, Himmel TL, et al. Patient and primary care provider experience using a family health history collection, risk stratification, and clinical decision support tool: a type 2 hybrid controlled implementation-effectiveness trial. BMC Fam Pract 2013;14(1):111.

[24] Fox S., Duggan M. Health online 2013: a project of the Pew Research Center. Available from: http://www.pewinternet.org/2013/01/15/health-online-2013/; 2013 [accessed 08.12.14].

[25] Bansil P, Keenan NL, Zlot AI, Gilliland JC. Health-related information on the Web: results from the HealthStyles Survey, 2002–2003. Prev Chronic Dis 2006;3(2):A36.

[26] File T., Ryan C. Computer and Internet use in the United States: 2013. Available from: http://www.census.gov/content/dam/Census/library/publications/2014/acs/acs-28.pdf; 2014 [accessed 23.12.14].

[27] Foland J, Burke B. Family health history data collection in Connecticut. Office CDoPHG. Hartford, CT: Connecticut Department of Public Health; 2014.

[28]   Centers for Disease Control and Prevention. Awareness of family health history as a risk factor for disease—United States, 2004. MMWR Morb Mortal Wkly Rep 2004;53(44):1044−7.

[29]   Cohn WF, Ropka ME, Pelletier SL, et al. Health Heritage© a web-based tool for the collection and assessment of family health history: initial user experience and analytic validity. Public Health Genomics 2010;13(7-8):477−91.

[30]   Qureshi N, Standen PJ, Hapgood R, Hayes J. A randomized controlled trial to assess the psychological impact of a family history screening questionnaire in general practice. Fam Pract 2001;18(1):78−83.

[31]   Reid GT, Walter FM, Brisbane JM, Emery JD. Family history questionnaires designed for clinical use: a systematic review. Public Health Genomics 2009;12(2):73−83.

[32]   Frezzo TM, Rubinstein WS, Dunham D, Ormond KE. The genetic family history as a risk assessment tool in internal medicine. Genet Med 2003;5(2):84−91.

[33]   Powell KP, Christianson CA, Hahn SE, et al. Collection of family health history for assessment of chronic disease risk in primary care. N C Med J 2013;74(4):279−86.

[34]   Wu RR, Himmel TL, Buchanan AH, et al. Quality of family history collection with use of a patient facing family history assessment tool. BMC Fam Pract 2014;15(1):31.

[35]   Beadles CA, Ryanne Wu R, Himmel T, et al. Providing patient education: impact on quantity and quality of family health history collection. Fam Cancer 2014;13(2):325−32.

[36]   Baldwin LM, Trivers KF, Andrilla CH, et al. Accuracy of ovarian and colon cancer risk assessments by U.S. physicians. J Gen Intern Med 2014;29(5):741−9.

[37]   Acton RT, Burst NM, Casebeer L, et al. Knowledge, attitudes, and behaviors of Alabama's primary care physicians regarding cancer genetics. Acad Med 2000;75(8):850−2.

[38]   Barrison AF, Smith C, Oviedo J, Heeren T, Schroy 3rd PC. Colorectal cancer screening and familial risk: a survey of internal medicine residents' knowledge and practice patterns. Am J Gastroenterol 2003;98(6):1410−16.

[39]   Gramling R, Nash J, Siren K, Eaton C, Culpepper L. Family physician self-efficacy with screening for inherited cancer risk. Ann Fam Med 2004;2(2):130−2.

[40]   Qureshi N, Armstrong S, Dhiman P, et al. Effect of adding systematic family history enquiry to cardiovascular disease risk assessment in primary care: a matched-pair, cluster randomized trial. Ann Intern Med 2012;156(4):253−62.

[41]   O'Neill SM, Rubinstein WS, Wang C, et al. Familial risk for common diseases in primary care: the Family Healthware Impact Trial. Am J Prev Med 2009;36(6):506−14.

[42]   Orlando LA, Wu RR, Beadles C, et al. Implementing family health history risk stratification in primary care: Impact of guideline criteria on populations and resource demand. Am J Med Genet C Semin Med Genet 2014;166C(1):24−33.

[43]   Orlando L, Wu RR, Myers RA, et al. Clinical utility of a web-enabled risk assessment and clinical decision support program. Genet Med 2016. Available from: http://dx.doi.org/10.1038/gim.2015.210.

[44]   Ozanne EM, Loberg A, Hughes S, et al. Identification and management of women at high risk for hereditary breast/ovarian cancer syndrome. Breast J 2009;15(2):155−62.

[45]   Murabito JM, Evans JC, Larson MG, et al. Family breast cancer history and mammography: Framingham Offspring Study. Am J Epidemiol 2001;154(10):916−23.

[46]   Codori AM, Petersen GM, Miglioretti DL, Boyd P. Health beliefs and endoscopic screening for colorectal cancer: potential for cancer prevention. Prev Med 2001;33(2 Pt 1):128−36.

[47]   Wu RR, Myers RA, McCarty CA, et al. Protocol for the "Implementation, adoption, and utility of family history in diverse care settings" study. Implement Sci 2015;10:163.

[48]   Qureshi N, Wilson B, Santaguida P, et al. Family history and improving health. Evid Rep Technol Assess (Full Rep) 2009;186:1−135.

Chapter 14

Genetics Aware Clinical Decision Support

Samuel J. Aronson[1] and Marc S. Williams[2]
[1]Partners HealthCare Personalized Medicine, Cambridge, MA, United States, [2]Genomic Medicine Institute, Geisinger Health System, Danville, PA, United States

Chapter Outline

INTRODUCTION

Genetics can only improve healthcare if clinicians and patients are able to identify when a genetic test may be useful, order the correct test for the given situation as well as manage and apply its results over time. The nature of genetic testing makes this challenging. Genetic tests often assess very specific conditions or pharmacogenomic effects. As with any intervention there are benefits and harms. The goal is to use interventions so that the benefits are likely to outweigh the harms. Ordered in the right context genetic tests can provide critical information that may improve care. However, ordered in the wrong context, the harms have the potential to overwhelm the benefits [1]. While this risk exists for any test, the combination of the complexity of indications for testing, the expense of the tests, and the interpretation of the results given the lack of genetics training for most clinicians is particularly daunting [2]. This suggests the need for expert guidance *at the point of test ordering* as evidenced by the high rate of inaccurate genetic test

Genomic and Precision Medicine. DOI: http://dx.doi.org/10.1016/B978-0-12-800681-8.00014-1

orders [2]. When genetic testing reports come back the results must be considered in the overall patient context and distilled into specific actions. Therefore, the results of these, at times complex and nuanced tests, must be made readily understandable to the clinician and the patient. Genetic testing results are one of a great many of pieces of information in the patient's healthcare record. In contrast to most laboratory results, the physical variants identified through a germline genetic test are not expected to change over time. Therefore, clinicians and patients must keep track of historical test results and determine when they should be considered in current decision making. In addition, new knowledge can emerge on previous genetic testing results. At times this knowledge has the potential to improve a patient's care but only if the clinician or patient learns that it exists. Under the right conditions, one patient's genetic test results may improve the care of other family members if properly communicated. For all these reasons, genetic test results not only compound the issues clinicians currently face managing patient health information, they also raise a new set of challenges that require new types of support.

Electronic clinical decision support (CDS) currently assists with these challenges in certain circumstances. It has the potential to do much more and, in fact, it must do much more to enable the use of genomics in healthcare to increase. Whole exome and genome sequencing combined with increasing understanding of the genetics of chronic diseases will increase both the power and complexity of clinical genetics. At the same time, the use of genetics has to move beyond highly specialized clinicians operating in the academic medical center environment. To benefit the largest number of patients it must expand to all care and community settings. CDS is also likely to play an important role in test result communication and interpretation [3].

WHAT IS GENETIC AWARE ELECTRONIC CDS?

Much of clinical Health Information Technology is focused on capturing information generated on a patient and storing it in an electronic health record (EHR). The EHR serves the function of the old manila folders that clinicians used to store health information except that it has the potential to improve care as a result of its functionality. CDS "...refers broadly to providing clinicians and/or patients with clinical knowledge and patient-related information, intelligently filtered, or presented at appropriate times, to enhance patient care" [4]. CDS can: (1) provide clinicians with additional information about the tests or therapies they are ordering and the results they have received; (2) alert clinicians if they take an action that is contraindicated by other information in the EHR; and/or (3) proactively notify clinicians if there is information or an action they should consider. CDS can either be built directly into EHRs or provided through ancillary systems.

TABLE 14.1 Challenges to Establishing Genetic Aware CDS

Challenge	
Structured electronic data not accessible	Genetic reports do not transfer from laboratories to clinicians with enough electronically structured data to use in CDS
	Clinical data that CDS rules depend upon may not be well structured or effectively collected
	Knowledge sources that genetic CDS depends upon for infobuttons or knowledge alerts are often not electronically available through structured interfaces
Rules must be established	In most cases each site must determine which genetic CDS rules to implement
CDS infrastructure not designed to handle rapid change	CDS engines generally require manual update and review of rules. Continuous learning infrastructure is uncommon but needed
Resources	Adding functionality to systems in the EHR ecosystem can be expensive. The funding models for this work must be determined

These ancillary systems may either be interfaced to EHRs, generally the preferred option, or operate entirely independently (Table 14.1).

Much of the CDS discussed in this chapter depends on the healthcare provider's ability to receive results in both human-readable and structured machine-readable form from the laboratories. Other than a few local examples, the interfaces required to enable this transmission do not exist. There are groups working to address this problem including the Institution of Medicine Genomic Roundtable's Displaying and Integrating Genetic Information Through the EHR (DIGITizE) Action Collaborative [5] focused on genetic enabling the EHR ecosystem, Logical Observation Identifiers Names and Codes, Health Level Seven, and ClinGen (Fig. 14.1).

TYPES OF CDS

CDS can be grouped into three general categories: passive, active, and asynchronous [6]. Passive CDS consists of nonmandatory resources available at the time of care. These support clinicians who have identified an information need. Electronic resources and infobuttons are examples of passive decision support. Asynchronous and active CDS have the potential to address the issue of a clinician not being aware of an important information need.

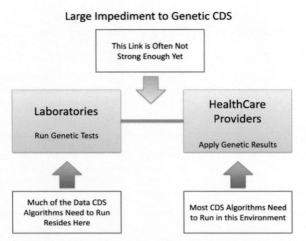

FIGURE 14.1 Robust information transfers between laboratories and providers are needed to enable genetic CDS.

PASSIVE DECISION SUPPORT: INFOBUTTONS

Infobuttons are a means of providing context-sensitive help to clinicians. These buttons are displayed within the EHR and link the clinician to internal or external resources. The infobutton can execute a query that takes into account patient data and clinical context to attempt to direct the clinician to the most pertinent resources. Infobuttons enhanced to target genetic specific references, including online databases maintained by the community, have been implemented in a few settings and have been shown to provide just-in-time education for clinicians using these tests [6–8]. The ClinGen resource has recently been configured to support infobutton requests through the OpenInfobutton standard with the hope that this will lower barriers to the accession of high-quality genetic information resources by patients and providers [9].

ACTIVE CDS IN SUPPORT OF PHARMACOGENOMICS

Pharmacogenomics, the process of using patient genetic results to guide drug selection or dosage, opens the need for several forms of CDS. Passively supplying information through infobuttons and calculators will be an important part of Pharmacogenomics CDS. There is also a need for two types of active CDS: testing ordering guidance and order validation. Test ordering guidance could be used to actively alert clinicians when they should order a genetic test for a patient. For example, Geisinger Hepatology service uses standard order sets for patients with chronic Hepatitis C that include genetic test orders which is a basic type of active decision support. Active CDS includes simple changes in computerized order entry, such as renaming an ambigu-ously named test [2], that improve clinician order accuracy. CDS can be used to prompt the order of a genetic test for a specific purpose [10]. To prevent severe adverse cutaneous events associated with the use of the anti-retroviral drug, Abacavir, it is recommended that patients be tested for the presence of HLA-B*57:01. A CDS "rule" could check whether an HLA result is available when abacavir is ordered. If no result is found, the clinician can be prompted to order the specific test. Ideally the alert would contain a link that would automatically generate an order for the HLA-B*57:01 test as well as a link (like an infobutton) to information explaining why the test should be ordered. A similar example could be a CDS rule that prompts for HLA-B*15:02 testing for individuals of Asian descent when carbamazepine is ordered. Such a rule could reduce the risk of Stevens—Johnson syndrome and toxic epidermal necrolysis [8].

Another form of pharmacogenomic CDS involves validating drug orders against *existing* patient genetic test results. The goal of this type of CDS is to alert the clinician if they order a drug or a dose that is contraindicated by a patient's genetic profile [11]. In 2006 Partners HealthCare launched a CDS rule that warned clinicians if they ordered gefitinib or erlotinib for a lung cancer patient who has an *EGFR* resistance variant that renders the drug inef-fective. This type of rule requires the ability to (1) trigger off of a drug order, (2) obtain previous test results, and (3) determine if any of them indicate resistance to the drug being ordered. Establishing this rule was possible at the time because the testing laboratory was within Partners HealthCare and the institution used a custom maintained EHR allowing use of its own data exchange formats in advance of a cross-organizational standard being devel-oped. However, standing up rules that uniformly work across institutions will require establishing standardized mechanisms for transferring results. There is also a need to monitor these rules over time. At the time of this writing Partners HealthCare is in the process of decommissioning this rule due to both evolution of the clinical thinking surrounding this scenario and the fact that it has been years since the rule triggered suggesting that clinical workflow has evolved. DIGITizE is currently working with the Clinical Pharmacogenetics Implementation Consortium (CPIC) and other groups to specify example CDS rules for validation for two drugs: abacvir and azathiopurine. The goal of specifying these rules is both to facilitate the construction of the rules themselves and also to provide an example that paves the way for more rapid development and dissemination of the CDS rules.

INFRASTRUCTURE NEEDED TO FACILITATE PHARMACOGENOMIC CDS DEVELOPMENT

The key challenge in delivering this kind of CDS lies in determining whether or not the pharmacogenomic effect has already been assessed. In order to do this the program must have unambiguous well-structured access to both the patients test history and the definition of the pharmacogenomic rules themselves. Modeling the latter is a problem referred to as knowledge management. From a computational standpoint the best way to do this would be to maintain, either locally or by referencing an external source:

- A list of all relevant pharmacogenomic genes and variants that should be assessed.
- A list of all drugs that could be affected by each gene/variant.
- All genetic tests run on each particular patient with a complete test definition, genes successfully tested, variants identified, and interpretations rendered based on this information.
- All additional or amended interpretations that have occurred on each genomic test result.

If all of this information were available, the CDS algorithm could take the drug being ordered and look up the genes that need to be tested for pharmacogenomic variants then gather all of the genes that have been tested in that patient together with the interpretations that have been done on those genes. Based on this information the clinician could be alerted if any of the required genes have not been tested in the patient sufficiently for the drug ordered. This logic should enable the clinicians to stay up-to-date on tests they need to order even as new pharmacogenomic drug gene interaction pairs are discovered.

There is an online collection of evidence-based pharmacogenetic guidelines generated by the CPIC [12], however, fully integrating this resource with the above functionality would require highly structured and complete transmissions between laboratories and providers. This in turn would require strong standardized data structures and ontologies for unambiguously identifying drugs, test definitions, pharmacogenomic variants, interpretations, and recommendations. Mechanisms are also needed for laboratories, providers, and knowledge resources to maintain all of this information over time. The technical and clinical knowledge exists to deliver each of these items. The challenge, as with many areas of CDS, revolves around achieving the cross-organizational and cross-industry consensus and action needed to put this into practice. The DIGITizE, CPIC, ClinGen, the National Human Genome Research Institute's genomic CDS workgroup, and other organizations are addressing these issues. The required infrastructure will need to be built and deployed incrementally with specific rules providing point solutions until the ideal state is reached.

ACTIVE CDS IN SUPPORT OF DIAGNOSIS

Beyond pharmacogenomics, CDS could also be useful to alert clinicians when a patient's risk for a disease or likelihood to respond to a particular treatment could be informed by a genetic test. For example, patients with numerous family members who have died of sudden cardiac arrest at a young age may benefit from hypertrophic cardiomyopathy testing. Sequencing patients with epileptic encephalopathy can provide a diagnosis that leads to a specific treatment such as folinic acid for those with a cerebral folate transport disorder due to mutations in *FOLR1*. As in pharmacogenomic CDS these forms of support also depend on obtaining structured clinical information in an unambiguous, machine-readable form. Much of the information within the medical record is in narrative form. While it is theoretically possible to extract needed information from notes through natural language processing, it is very difficult to do so with high sensitivity and specificity. There are many efforts underway to bring structured family history into the EHR [11,13]. If these efforts scale they could provide a rich basis for expanded CDS in this area.

ASYNCHRONOUS DECISION SUPPORT: KNOWLEDGE ALERTS

Germline genetic variants stay with a patient throughout their lifetime. Once a variant is confirmed in a patient, they do not need to be tested for that variant again, unless the reliability of the testing process later becomes suspect. This leads to "knowledge events" that occur when our understanding of a variant identified in a patient advances sufficiently to improve that patient's care. In most cases clinicians and patients will not be aware that a knowledge event has occurred unless it is brought to their attention. Many of these events occur when a laboratory reclassifies a variant based on new data that has emerged. The clinician and patient are unlikely to have direct access to the laboratory databases housing this information. Many laboratories contribute to ClinVar, a repository maintained by the National Institutes of Health. However, with over 100,000 variants residing in ClinVar there is no way for clinicians and patients to monitor all potentially relevant changes without support. At least one laboratory sends letters to clinicians informing them of variant interpretation changes that could affect care. However, these letters are difficult to track and manage in EHRs as the information in the letter is not represented in structured form meaning that searches and CDS cannot use the knowledge. This is also unlikely to be scalable as the amount of genomic information associated with a patient increases over time.

CDS can manage these updates. This is a situation where genetic aware CDS differs from other forms of CDS in that it must enable a new type of clinical process. Clinicians typically update patient care plans as a result of an event taken either by or for the patient: an office visit, the return of a test result the clinician explicitly ordered, or a decision to follow up with the patient after a predetermined period of time. However, a knowledge event is generated by an external source unrelated to any action taken by the patient or clinician. Even so it must be brought into the clinical flow. This can be done by proactively sending patient specific alerts to clinicians. These alerts can be e-mailed or preferably raised through their EHR's internal alerting mechanism.

E-mail alerts should either be sent via a HIPAA-compliant secure channel or the e-mails sent should contain no protected health information. Ideally these e-mails should provide a "deep" link that takes the clinician directly to patient-specific pages housed in a secure application after they authenticate.

Because of the number of variants involved and the rate at which variant changes occur, the generation of these knowledge alerts must be automated. This can be done by maintaining a knowledge base of variant interpretations that is updated as new genetic reports are generated. This process has been implemented through the GeneInsight system at Partners HealthCare and clinics that order from its Laboratory for Molecular Medicine (LMM) [14]. The LMM maintains a knowledge base within GeneInsight that it uses to automatically draft reports as variants are identified in patients. The LMM often reassesses previously reported variants when they are identified in a new case. If the variant classification changes, the knowledge base can be updated and used it to redraft the report. In this way the knowledge base is kept up-to-date through the clinical reporting flow itself. As changes to the knowledge base are made, alerts are fired to clinicians using the GeneInsight system to manage previously reported cases where the variant was identified. An analysis of historical activity found that 3.9% of covered cases per year received a medium or high alert meaning that the change in classification was significant enough to potentially prompt additional clinical action [15].

Concerns have been raised about the workflow and scalability of this process including: (1) what should happen when clinicians receive knowledge events for patient they are no longer seeing, (2) what if a clinician receives a large number of knowledge events all at once and cannot process them all, (3) what, if any, reimbursement is available for processing a knowledge event, and (4) should patients directly receive notices of these events and, if so, how can this be accomplished from a logistical perspective. Despite these questions, there is evidence that clinicians appreciate receiving these alerts [16]. The genetic tests we have studied were focused on patients suffering from or at risk for developing a life-threatening disease (hypertrophic cardiomyopathy). It is possible that clinicians would react differently to alerts related to a less serious illness and therefore a different process could be warranted in these circumstances. New knowledge about pharmacogenomic variants could be added to the knowledge management system without notifying the clinician as the knowledge is only relevant when a medication is prescribed, although an alert may be indicated if the patient is on a relevant medication. It is important to carefully consider which alerts are significant enough to push a proactive notification to a clinician. It is also worth evaluating the best way to push alerts to patients [17]. EHR patient portals offer one potential solution but they are institution specific. It is possible to establish directly accessible patient infrastructure through personal health record type infrastructure. However, the security, logical, and administrative challenges associated with doing so are significant (Fig. 14.2).

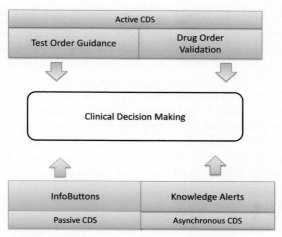

FIGURE 14.2 Types of genetic CDS.

IT IS NEVER PERFECT

Integrating genetic findings into patient care plans is a complex process. It is nearly impossible to proactively account for all potential scenarios when CDS functionality is initially built. Therefore, mechanisms for overriding CDS suggestions are important. Audit logs tracking when CDS rules are and are not overridden are also needed to understand clinician behavior [18]. Such audit logs can be used to assess rule effectiveness over time and can also reveal clinician knowledge that can be used to improve the utility of the CDS. Understanding when to present alerts in the clinician workflow has a major impact on clinician response to alerts. More work with clinicians to understand workflows impacted by genomic information is warranted.

Just like other tests CDS mechanisms have false-negative and false-positive rates. False-positives alerts consume clinician time without adding value or worse lead them in an inappropriate direction. If the false-positive rate is too high "alert fatigue" will set in and clinicians will begin ignoring all CDS alerts, one of the most significant problems impacting the effectiveness of CDS. For this reason it is important to control the alert false-positive rate from the outset. Doing so may increase the false-negative rate. Therefore, clinicians must understand the CDS will not be able to warn them in all situations.

The false-positive and false-negative rates will depend on the quality of the data and knowledge the CDS acts upon. Providers in conjunction with institutional content experts must validate the content that underlies their CDS rules—which represents a barrier to CDS adoption. Despite strong efforts being made to generate clinical grade public repositories, it will take time for these resources to grow, mature, and become interoperable with EHRs [6].

CONCLUSION: IT IS ALL ABOUT CONTINUOUS LEARNING

As genetic testing becomes increasingly useful in a growing number of clinical situations, we must make it widely accessible to clinicians and the patients they serve. One of the core barriers to doing so is the constant evolution of information surrounding tests and the genetic variants they can uncover. CDS can help clinicians stay on top of this evolution and thereby make it easier for them to incorporate genetics into their practices. Every member of the patient care ecosystem will learn information about different aspects of genetics at different rates. Our challenge as a society is to capture this learning, appropriately validate it, package it, and then distribute it as broadly as possible [19]. CDS has a critical role to play in doing so.

REFERENCES

[1]   Kohane IS, Masys DR, Altman RB. The incidentalome: a threat to genomic medicine. JAMA 2006;296(2):212–15.
[2]   Miller CE, Krautscheid P, Baldwin EE, Tvrdik T, Openshaw AS, Hart K, et al. Genetic counselor review of genetic test orders in a reference laboratory reduces unnecessary testing. Am J Med Genet A 2014;164A(5):1094–101.
[3]   Baron JM, Dighe AS, Arnaout R, Balis UJ, Black-Schaffer WS, Carter AB, et al. The 2013 symposium on pathology data integration and clinical decision support and the current state of field. J Pathol Inform 2014;5:2.
[4]   Osheroff JA, Teich JM, Middleton B, Steen EB, Wright A, Detmer DE. A roadmap for national action on clinical decision support. J Am Med Inform Assoc 2007;14(2):141–5.
[5]   http://iom.nationalacademies.org/Activities/Research/GenomicBasedResearch/Innovation-Collaboratives/EHR.aspx [accessed 22.10.15].
[6]   Williams MS. Information technology and primary care genetics. In: Medical genetics in pediatric practice. Chicago, IL: American Academy of Pediatrics; 2013.
[7]   Hoffman MA, Williams MS. Electronic medical records and personalized medicine. Hum Genet 2011;130(1):33–9.
[8]   Del Fiol G, Williams MS, Maram N, Rocha RA, Wood GM, Mitchell JA. Integrating genetic information resources with an EHR. AMIA Annu Symp Proc 2006;904.
[9]   https://www.clinicalgenome.org/tools/web-resources/ [accessed 30.10.15].
[10]   Bell GC, Crews KR, Wilkinson MR, Haidar CE, Hicks JK, Donald K, et al. Development and use of active clinical decision support for preemptive pharmacogenomics. J Am Med Inform Assoc 2014;21(e1):e93–9.
[11]   http://www.hughesriskapps.com [accessed 22.12.14].
[12]   http://www.pharmgkb.org/view/dosing-guidelines.do?source = CPIC# [accessed 24.12.14].
[13]   Orlando LA, Hauser ER, Christianson C, Powell KP, Buchanan AH, Chesnut B, et al. Protocol for implementation of family health history collection and decision support into primary care using a computerized family health history system. BMC Health Serv Res 2011;11:264.
[14]   Aronson SJ, Clark EH, Babb LJ, Baxter S, Farwell LM, Funke BH, et al. The GeneInsight Suite: a platform to support laboratory and provider use of DNA-based genetic testing. Hum Mutat 2011;32(5):532–6.

[15]   Aronson SJ, Clark EH, Varugheese M, Baxter S, Babb LJ, Rehm HL. Communicating new knowledge on previously reported genetic variants. Genet Med 2012.

[16]   Neri PM, Pollard SE, Volk LA, Newmark LP, Varugheese M, Baxter S, et al. Usability of a novel clinician interface for genetic results. J Biomed Inform 2012;45(5):950—7.

[17]   Stuckey H., Williams J.L., Fan A.L., Rahm A.K., Green J., Feldman L., et al. Enhancing genomic laboratory reports from the patients' view: a qualitative analysis (Manuscript in review).

[18]   McCoy AB, Thomas EJ, Krousel-Wood M, Sittig DF. Clinical decision support alert appropriateness: a review and proposal for improvement. Ochsner J 2014;14(2):195—202.

[19]   Aronson SJ, Rehm HL. Building the foundation for genomics in precision medicine. Nature 2015;526(7573):336—42.

Chapter 15

Implementation Science and Integration into Healthcare Systems

Marc S. Williams

Genomic Medicine Institute, Geisinger Health System, Danville, PA, United States

Chapter Outline

INTRODUCTION

Even the most compelling evidence is not sufficient to insure rapid translation into clinical care. Translation of evidence-based best practice into clinical practice is fraught with challenges that were extensively documented in two landmark publications from the Institute of Medicine, To Err is Human and Crossing the Quality Chasm [1,2]. Genomic interventions are no exception. While the information needed to achieve precision medicine may be more

Genomic and Precision Medicine. DOI: http://dx.doi.org/10.1016/B978-0-12-800681-8.00015-3

complex and voluminous compared to other types of health information, with few exceptions, the barriers to implementation are no different.

Recognition of this problem along with the hypothesis that there are generalizable principles and approaches that can be identified and used to facilitate implementation has led to the systematic study of implementation success and failure. The discipline referred to as "Implementation Science" is defined as the scientific study of methods to promote the systematic uptake of research findings and other evidence-based practices into routine practice, and, hence, to improve the quality and effectiveness of health services [3]. Implementation Science has three primary aims:

1. Develop reliable strategies for improving health-related processes and outcomes and facilitate widespread adoption of these strategies.
2. Produce insights and generalizable knowledge regarding implementation processes, barriers, facilitators, and strategies.
3. Develop, test, and refine implementation theories and hypotheses including methods and measures.

There are related activities that have created a confusing lexicon for this discipline. This includes terms such as clinical research, translational research (including the T1−T4 framework of Khoury [4]), dissemination research, and quality improvement research. It is important to understand the difference between quality improvement and implementation research. Quality Improvement focuses on the "here and now"—immediate, local improvement needs via rapid-cycle, iterative improvement, focused on a specific local quality problem. In contrast, Implementation Science attempts to develop, deploy, and rigorously evaluate a fixed implementation strategy across multiple sites, emphasizing theory, contextual factors, and mechanisms, addressing a common implementation gap. It intends to develop generalizable knowledge.

CLINICAL RESEARCH VERSUS IMPLEMENTATION SCIENCE

Until relatively recently the movement of knowledge from discovery in basic research, translation in clinical research to improvement in health outcomes was depicted as a simple linear progression as seen in Fig. 15.1A. This representation does not reflect the reality of translation. In 1999 the American Association of Medical Colleges published Clinical Research: A National Call to Action [5]. This landmark paper presented the results of a summit that critically reviewed the challenges associated with clinical research that encompassed the following areas [5]:

- There is not an agreed-upon definition of clinical research and its components.
- Clinical research is not adequately understood or valued by the public.

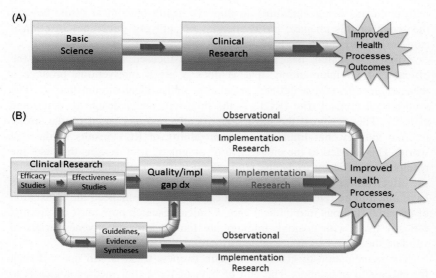

FIGURE 15.1   (A) Historical framework of translation from basic science discovery to implementation in clinical practice. (B) Contemporary conceptual framework of clinical and implementation research that reflects the range of methodologies needed to efficiently study efficacy, effectiveness, and implementation and improve health outcomes. Source: *Used with permission of Brian Mittman, PhD.*

- There is a lack of data on clinical research funding and productivity.
- There is insufficient funding for the conduct of some types of clinical research.
- There are insufficient numbers of clinical investigators.
- **There is insufficient emphasis on incorporating research findings into clinical practice**.
- There is inadequate coordination of clinical research among research entities and disciplines.
- The ability of academic health centers to conduct clinical research is at risk.
- There is a lack of a comprehensive, dynamic clinical research agenda.

The bolded item indicates that the concern about translation is long-standing. Subsequently the Institute of Medicine [6,7] and a group led by Sir David Cooksey in the United Kingdom [8] further delineated these challenges. These reports all concluded that without a fundamental change in approach the translational pipeline would continue to be inefficient due to barriers inherent in the current approach. The first approach was to synthesize clinical guidelines that transformed the evidence from clinical research into actionable information for use by clinicians. This resulted in some improvement, but a number of new issues were identified that interfered with clinicians using guidelines to improve health outcomes. While beyond the scope of this chapter, the challenges of applying guidelines in clinical care are explored in depth by Cabana [9].

Clinical research has historically been focused on determining the efficacy of an intervention, that is does a given intervention produce a measurable effect in a defined population under ideal circumstances? The randomized clinical trial (RCT) is the prototype study design to determine efficacy. This design is the most scientifically rigorous based on the strong internal validity, but limits the generalizability of RCT results in the real world. This has led to the emergence of pragmatic trial designs that take the results from efficacy trials and study them in the "real world" to determine the effectiveness of the intervention. The results of these effectiveness trials have much more relevance to clinical care and have enhanced the translation of research findings into clinical care. Clinical research now must ensure that efficacy trials with positive findings are moved into effectiveness trials to validate the intervention in real-world settings.

The combination of efficacy and effectiveness trials is still not sufficient for successful translation. The aim of both of these trial methodologies is to study the impact of a clinical intervention with the individual as the unit of study. The aim of implementation research is to study the impact of an implementation strategy. This requires interventions to change practices and systems with the clinician or organization as the unit of study. Table 15.1 compares the differences between clinical and implementation studies. Understanding the continuum of research from efficacy to effectiveness to implementation decreases the time and increases the fidelity of the translation of research into clinical care.

Within implementation research there is a continuum of methodologies that are akin to those in clinical research. One proposed model is that of

TABLE 15.1 Comparison of Clinical and Implementation Research

Study Type Study Feature	Clinical Research	Implementation Research
Aim: evaluate a/an...	Clinical intervention	Implementation strategy
Typical intervention	Drug, procedure, therapy	Clinician, organizational practice change
Typical outcomes	Symptoms, health outcomes, patient behavior	Adoption, adherence, fidelity
Typical unit of analysis, randomization	Patient	Clinician, team, facility

Table used with permission of Brian Mittman, PhD.

TABLE 15.2 Model of Phased Implementation

Phase	Study Type	Form of Evaluation
Pretrial	Program	Conceptual design of implementation program and design causal model from theory, prior empirical research
Phase 1	Pilot/ formative	Pilot test, assess feasibility, evaluation and formative refinement, develop intervention and evaluation protocols
Phase 2	Efficacy	Small-scale rigorous trial in controlled settings with ongoing intervention support; emphasis on internal validity
Phase 3	Effectiveness	Large-scale rigorous trial under routine conditions in diverse settings; emphasis on external validity
Phase 4	Monitoring	Ongoing monitoring and feedback

Table used with permission of Brian Mittman, PhD.

Phased Implementation (Table 15.2). The progression from pilot to effectiveness is similar to that in clinical research, however implementation research includes a monitoring and reevaluation phase that was adapted from quality improvement methods. This is a critical difference in that it allows for incorporation of new knowledge and evidence as well as changes in the delivery environment reducing the risk of reversion to prior practice. The phased implementation model is essential to the implementation of genomic medicine given the rapid pace of change of the knowledge and impact of the genome on health outcomes.

One might be concerned that moving an intervention linearly through the clinical and implementation research pipelines would require significant time and resources and would therefore not be possible for all interventions. Increasing pressures on economic and human resources in research and delivery necessitates the development of research methods that improve the efficiency of research without significantly compromising the determination of efficacy and effectiveness. Given these concerns, implementation scientists have recognized that well-constructed observational studies could accelerate implementation of effective interventions. The strengths of observational implementation studies include:

- The focus on naturally occurring (policy/practice-led) versus artificial (researcher-led) implementation processes.
- These studies maximize external validity and thus are more generalizable.
- The large sample sizes optimize the power to detect contextual influences.
- The ability to examine local adaptation processes and effects.

This study design has been systematically reviewed and there is little evidence for differences between well-designed observational studies and randomized controlled trials regardless of study design, heterogeneity, and the type of intervention studied [10].

The need to increase efficiency has led to development of a new trial design that combines aspects of clinical effectiveness studies and implementation research. These effectiveness—implementation hybrid study designs were first proposed in 2012 by Curran and colleagues [11]. The authors propose that these hybrid studies have several potential advantages that include more rapid translational gains, more effective implementation strategies, and more useful information for decision makers. They propose three different effectiveness—implementation hybrid typologies: (1) Hybrid type 1 testing effects of a clinical intervention on relevant outcomes while observing and gathering information on implementation; (2) Hybrid type 2 dual testing of clinical and implementation interventions/strategies; and (3) Hybrid type 3 testing of an implementation strategy while observing and gathering information on the clinical intervention's impact on relevant outcomes. The different components of clinical and implementation research presented in Table 15.1 are used to choose the appropriate hybrid design to use. Results of hybrid effectiveness trials are beginning to appear in the literature.

A synthesis of the information presented in this section radically alters the initial conceptual model. The resulting conceptual framework is represented in Fig. 15.1B. This framework allows consideration of the different options to study an intervention as it moves from research to implementation. The remainder of the chapter will use the example of the implementation of a genomic medicine intervention, tumor-based screening for the identification of patients with Lynch syndrome (LS), in a large integrated healthcare delivery system.

IMPLEMENTATION SCIENCE IN PRACTICE: TUMOR-BASED SCREENING FOR LS

The Clinical Problem—Identification of Patients With LS

Two to five percent of all colorectal cancer (CRC) is due to genetic mutations in one of four mismatch repair (MMR) genes collectively known as LS [12]. Carriers of one of these mutations are at significantly increased risk for development of CRC and other cancers. Identification of mutation carriers through sequencing of the genes can allow enhanced surveillance and prophylactic surgery that can reduce cancer risk. Identification of at-risk individuals has traditionally relied on family history criteria, or tumor characteristics, however this approach is insufficiently sensitive [13]. Based on a systematic review of the evidence, the Evaluation of Genomic Applications in Practice and Prevention Working Group (EGAPP) [13]

recommended testing colorectal tumors for molecular characteristics sugges-
tive of LS as an alternative to family history. It was determined that this was
worth exploring for possible implementation at Intermountain Healthcare
(IH). The process described below is represented graphically in Fig. 15.2.

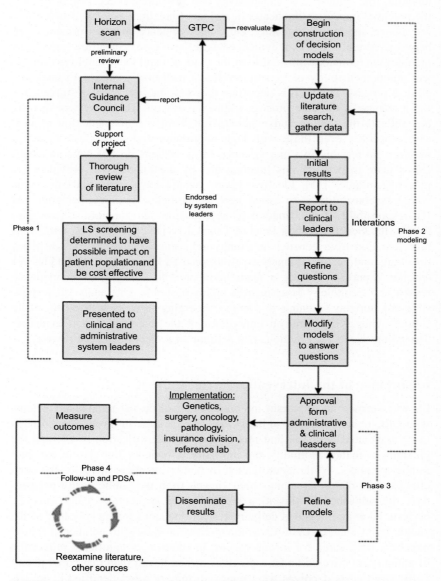

FIGURE 15.2   Schematic representation of review and implementation process for LS screening program. Source: *©Intermountain Healthcare Inc., 2010 all rights reserved.*

EVIDENCE REVIEW AND KNOWLEDGE SYNTHESIS

The first implementation task involves assessing the intervention in three general areas: importance of the intervention, scientific soundness, and feasibility [14]. The importance of the intervention has several subdomains including the relevance to the patients and providers in the health system; the potential impact on the health of patients (including aspects of incidence, prevalence, morbidity, mortality, etc.); and, evidence of a performance gap between current care and evidence-based best care. Scientific soundness includes two domains: the quality of the evidence and the reliability, validity, and comprehensibility of system-developed measures designed to track the clinical process. Finally it is important to assess the feasibility of implementing a change in the care process within the system. This is system-specific and reflects the culture and readiness for change. Implementation of new care processes, even those with high importance and impeccable evidence may not be feasible in some systems. Most systems do not have an established team to perform this function and by default defer to providers for new technologies [15]. Ideally, systems would have individuals or teams whose responsibility is to assess new technologies for implementation. At IH, responsibility for evaluation of new genetic and genomic technologies falls to the Genetic Testing Practice Council (GTPC in Fig. 15.2) which has representation from genetics, anatomic and clinical pathology, health plans, oncology, and healthcare analysts. Additional ad hoc members are added for specific expertise. Having a team member with background in evidence review and technology assessment is highly desirable. One analyst is responsible for horizon scanning to identify emerging tests as well as to create a brief synopsis focusing on importance of the intervention and evidence. The group selects interventions that warrant an in-depth evidence review.

Importance of the Intervention

LS was deemed an important intervention based on the number of patients with CRC treated annually in the system, the increase in morbidity and mortality associated with LS, and recognition that identification of LS had the potential to reduce morbidity and mortality in patients, but also in patients' family members. While identification of family members increases the cost-effectiveness of LS screening when viewed from the public health perspective, the relevance of this to a health system is not as clear. However, because IH is an integrated delivery system that cares for roughly half of the people in the state of Utah it was thought that a significant number of family members were cared for in our system. In addition, a significant proportion of patients are covered under a provider-owned health plan that could potentially avoid costs if polyps were removed, tumors were identified at an earlier stage, or through prophylactic surgery.

Analysis of Efficacy and Effectiveness

Literature relevant to LS screening was acquired, reviewed, and synthesized by the team analyst using a modification of the ACCE evidence framework [16]. The ACCE framework was chosen as it was specifically developed for the evaluation of genomic technologies and was used by EGAPP for their recommendation statement. It is important to note that while ACCE was developed for genomics, the comprehensive nature of its underlying conceptual framework as well as its explicit description of the process could allow it to be used for nongenetic technologies. However, any evidence synthesis approach can be used at this step. Local factors may determine the approach that is best suited for a given institution. Both clinical and economic data were analyzed. It was determined that the evidence was sufficient to recommend implementation of tumor-based screening for LS. During the course of the evaluation, newly published articles were identified and incorporated both into the evidence review and the modeling (see below) allowing improvement in an iterative fashion in real time. Accommodation for emerging evidence is critically important in a rapidly changing field such as genomics.

Decision analytic modeling customized to the setting was used to identify which of the many screening algorithms would be most effective and efficient at identification of patients as well as assessing the resource impact for the IH system. Outcomes of interest included cost-per-life-year saved, total costs of testing, number of cases detected, cost-per-case detected, and various budget impact metrics. The model incorporated not only published data, but unpublished data obtained from referral laboratories and researchers to refine the model assumptions. Outside experts provided consultation. The model was revised many times during the course of the evaluation as new data and testing strategies emerged [17].

Finally an assessment of the system's ability to capture measures relevant to the proposed screening process was performed. It was determined that the proposed outcomes were valid, reliable, and relevant and could be captured from clinical data sources with relatively minor modifications of existing systems.

Feasibility

Feasibility involves assessing the workflow of the process and the anticipated impact of proposed changes. The assessment team conducted preliminary inquiries regarding feasibility that included an assessment of the current pathology process to determine whether the samples could be appropriately handled and transferred to the referral laboratory, return of results, and impact on the patient care process. The conclusion was that the program appeared to be feasible in the system. An advisory group consisting of representatives from Oncology, Surgery, Pathology, and system medical, health plan, and business leaders was assembled and the proposed screening program was presented. The advisory group determined that the program met standards for patient impact and evidence and was feasible to implement in all system hospitals surgically treating CRC.

IMPLEMENTATION

Successful implementation depends on a number of factors including analysis of clinical workflow relevant to the current care process; assessment of how the new care process is likely to alter the workflow; working with representatives of the affected care processes to minimize workflow disruption; identification of early adopters to pilot the implementation; and revision of implementation plan based on pilot results. This provides the information needed to develop a program roll out that is likely to be successful.

Analysis of Clinical Workflow

Workflow ultimately occurs at the level of the clinician/care team, however depending on the care process it may involve several related systems and occasionally one or more organizations. The screening process involves analysis of the tumor necessitating interactions between the surgery and pathology departments across seven hospitals in the system each with different workflows. Following specimen accession and preparation it was necessary to transport the surgical specimen to the outside referral laboratory where analysis was performed in the immunohistochemistry and molecular pathology departments. The result is reported from the referral laboratory back to our organization. Normal results are stored in the electronic health record (EHR) and notification is given to the relevant clinical team for the patient. In the case of an abnormal result the result was also forwarded to the genetic oncology program. Evaluation of workflows at each of the seven hospitals revealed minor, but significant differences that had to be accounted for in the implementation plan.

While genetic testing and reporting of genetic results is integral to this particular implementation, it is important to note that these same processes could impact the implementation of nongenomic interventions. Detailed workflow analysis to understand the nature, context, and impact of an intervention is a universal requirement for successful implementation. Genomic medicine implementation may have differences in their nature and context, but this does not alter the approach to workflow analysis.

Assessment of Clinical Workflow Alterations

The biggest change in the workflow was the addition of genetic oncology to the workflow. This recommendation followed from the concern that abnormal results could be "lost" because the result would return to the surgeon of record after the patient would have concluded the episode of care. Care transitions are a major factor impacting a patient's care and safety [18]. A single point in the workflow was needed to ensure appropriate follow-up and coordination. The genetic counselor leading the oncology genetics program received all abnormal results and coordinated all patient (and family) follow-up. Minor alterations to the workflow in pathology were made to the processes already in place for tumor specimens requiring analysis at a referral laboratory for nongenomic indications. A new tracking system was implemented to ensure that all specimens were sent with appropriate orders and that results were received and were routed to the oncology genetic counselor when abnormal.

Pilot Implementation

No matter how much planning goes into a project, problems inevitably emerge with implementation. Pilot implementation using early adopters is a key to successful implementation. Pathologists from two smaller system hospitals volunteered to pilot the process. During this pilot difficulties with electronic lab ordering were identified, specifically a unique test code that included the reflex testing protocol was not available. Creation of a new test code required several months to implement. Pathology representatives indicated that a manual process would be unacceptable. The pilot created a work-around involving preprinted labels that could be placed on the lab order sheet. This simple process was used for the pilot. No other issues were identified and the tracking system was optimized.

Full Implementation

Following presentation of the successful pilot implementation other pathology departments committed to implementation once the electronic code was available. Implementation within each hospital has minor variation, so it is important to embed the implementation team with the system champions to customize the process to the specific system. This mass customization has been identified as one of the five keys to high-performance healthcare systems [19]. The work done in the pilot settings allowed the new process to be initiated in other sites at the same level of quality and consistency. All new specimens were immediately entered into the tracking system closing the care loop.

PROCESS MONITORING AND QUALITY IMPROVEMENT

Holding the Gains

System inertia can erode new processes as systems tend to revert to usual practice [20]. Active process management is needed to ensure optimal performance. Outcome measures were developed that would be regularly collected and reviewed.

Addressing Knowledge Gaps

With any new intervention there are educational issues for both patients and providers. Implementation of the LS screening program resulted in some patients being contacted by the genetic counselor so a brochure was developed that explained the screening process, the health implications of the screening, and the reason for the contact. A variety of methods were employed to educate providers about the new process including newsletter articles, Web-based information, CME presentations, departmental meetings, and tumor boards. Information sheets were developed and placed in an electronic resource accessible through the EHR. Addressing educational needs is essential for any implementation project.

Improving Patient Care

The reason for implementing an intervention is to improve patient care. Outcomes must be developed to assess the impact of the intervention. In the case of LS screening the purpose is to identify patients at increased risk of having LS. It is the responsibility of the genetic counselor to provide follow-up for all screen-positive patients. Outcomes for screen-positive patients include the number of patients lost to follow-up, number choosing to attend a counseling session, number pursuing confirmatory testing, and number attending a posttest counseling session. For those with negative confirmatory testing, a comprehensive assessment of other risk factors is done. For those that are confirmed to carry a mutation in one of the four MMR genes additional outcomes include communication of a cancer surveillance plan, adherence to the recommended surveillance, and identification of at-risk family members. Outcomes for at-risk family members would include the number of family members: contacted for counseling, attending counseling, undergoing family-specific mutation testing, etc. Some data were automatically abstracted from the EHR although some measures require other methods. The collection of information about family members constitutes a difference compared to most clinical processes and reflects the familial nature of genetic disease, as well as the recognition that there may be more benefit to family members from the screening program [13]. However, it is not unique to genomics in that conditions such as infectious diseases may also have a familial risk component that must be accounted for as part of implementation. Studying these outcomes may also lead to discoveries about patient or provider factors that assist or hinder adherence with recommendations promoting improvements to the process. It may also determine that significant numbers of patients and family members do not alter care based on the information limiting the utility of the program. Attention to this type of implementation research is essential if common principles that facilitate dissemination of best practices between systems are to be discovered.

Cost Outcomes

Cost of care is an outcome of importance to patients and systems. A significant variance in cost can influence decisions about ongoing investment in a given process. In a cost-constrained system, investment of resources in one area by necessity limits investment in other areas. Assessment of impact on patient outcomes and cost outcomes together can determine whether a given program should continue, or whether the resources should be redirected to another area with greater impact. Absent rigorous definition and measurement of outcomes, informed decisions within the system are not possible. This proved critical to the success of the program [17] as well as answering questions about modifying the program to utilize an age-cutoff [21] a change that was rejected based on the cost analysis.

CONCLUSION EMPOWERING THE LEARNING HEALTHCARE SYSTEM FOR GENOMIC MEDICINE

Following the two reports cited above, in 2007 the Institute of Medicine proposed a new paradigm in healthcare delivery called the Learning Healthcare System [22]. A Learning Healthcare System is designed to generate and apply the best evidence for the collaborative healthcare choices of each patient and provider; to drive the process of discovery as a natural outgrowth of patient care; and to ensure innovation, quality, safety, and value in health care. The key attributes of a learning healthcare system include:

- Adaptation to the pace of change
- Stronger synchrony of efforts
- New clinical research paradigm
- Clinical decision support systems
- Tools for database linkage, mining, and use
- Notion of clinical data as a public good
- Incentives aligned for practice-based evidence
- Public engagement
- Trusted scientific broker
- Leadership

In the United States, the movement toward a learning healthcare system environment appears inexorable. To achieve this requires the reengineering of clinical research as described above. If genomic medicine is to become an integral part of health care it is will be necessary to incorporate implementation science into the clinical research agenda. The author's current institution, Geisinger Health System (GHS), has identified genomic medicine as an institutional priority. To realize this vision, the GHS leadership has defined the value proposition [23] that guides the ongoing work to develop a robust, system-wide return of results program that utilizes sequencing-derived results to identify patients at risk for genetic conditions with effective medical interventions [24].

GLOSSARY

Care transitions The movement patients make between healthcare practitioners and settings as their condition and care needs change during the course of a chronic or acute illness.

Clinical workflow Clinical workflow is an established process describing:
- A series of tasks
- How they are accomplished
- By whom
- In what sequence
- At what priority

that accomplishes a defined step in an activity in the clinical care of patients.

Decision analysis (modeling) Decision analysis is a discipline comprising the philosophy, theory, methodology, and professional practice necessary to address important decisions in a formal manner. Decision models describe the relationship between all the elements of a decision—the known data (including results of predictive models), the decision, and the forecast results of the decision—in order to predict the results of decisions involving many variables. These models can be used in optimization, maximizing certain outcomes while minimizing others.

Gray (or Grey) data or literature The Grey Literature Network Service (see Web resources) defines grey literature as "information produced on all levels of government, academics, business and industry in electronic and print formats not controlled by commercial publishing i.e. where publishing is not the primary activity of the producing body."

Implementation science The scientific study of methods to promote the systematic uptake of research findings and other evidence-based practices into routine practice, and, hence, to improve the quality and effectiveness of health services and care.

Integrated delivery system An integrated delivery system (IDS) is a network of healthcare providers and organizations which provides or arranges to provide a coordinated continuum of services to a defined population and is willing to be held clinically and fiscally accountable for the clinical outcomes and health status of the population served. An IDS may own or could be closely aligned with an insurance product.

Learning healthcare system A healthcare system that has undergone a process of systematic organizational change that has incorporated and synthesized individual and system learning in order to improve and transform care delivery.

Mass customization Production of personalized or custom-tailored goods or services to meet consumers' diverse and changing needs at near mass production prices. Enabled by technologies such as computerization, Internet, product modularization, and lean production, it portends the ultimate stage in market segmentation where every customer can have exactly what he or she wants. In the context of health care this refers to individualization of the disease management or care process by customizing evidence-based best practice guidelines to meet the specific needs of a given patient including incorporation of patient preferences.

Process management Planning and administering the activities necessary to achieve a high level of performance in a process and identifying opportunities for improving quality, operational performance, and ultimately customer satisfaction.

Quality improvement Consists of systematic and continuous actions that lead to measurable improvement in healthcare services and the health status of targeted patient groups.

ACKNOWLEDGMENT

The author wishes to thank Dr. Brian Mittman for his insights and leadership in the field of implementation science. His work is foundational to the content provided in this chapter. Dr. Mittman also graciously gave permission for the use of Fig. 15.1A and B and Tables 15.1 and 15.2.

REFERENCES

[1] Kohn LT, Corrigan J, Donaldson MS. To err is human: building a safer health system. Washington, DC: National Academy Press; 2000.

[2] Institute of Medicine (U.S.). Crossing the quality chasm: a new health system for the 21st century. Washington, DC: National Academy Press; 2001.

[3] Eccles MP, Mittman BS. Welcome to implementation science. Implement Sci 2006;1:1–3.

[4] Khoury MJ, Gwinn M, Yoon PW, Dowling N, Moore CA, Bradley L. The continuum of translation research in genomic medicine: how can we accelerate the appropriate integration of human genome discoveries into health care and disease prevention? Genet Med 2007;9:665–74.

[5] https://www.umassmed.edu/PageFiles/64044/NationalCalltoAction.pdf [last accessed 02.11.15].

[6] Sung NS, Crowley Jr WF, Genel M, Salber P, Sandy L, Sherwood LM, et al. Central challenges facing the national clinical research enterprise. JAMA 2003;289:1278–87.

[7] Crowley Jr WF, Sherwood L, Salber P, Scheinberg D, Slavkin H, Tilson H, et al. Clinical research in the United States at a crossroads: proposal for a novel public-private partnership to establish a national clinical research enterprise. JAMA 2004;291:1120–6.

[8] https://www.gov.uk/government/uploads/system/uploads/attachment_data/file/228984/0118404881.pdf [last accessed 02.11.15].

[9] Cabana MD, Rand CS, Powe NR, Wu AW, Wilson MH, Abboud PA, et al. Why don't physicians follow clinical practice guidelines? A framework for improvement. JAMA 1999;282:1458–65.

[10] Anglemyer A, Horvath HT, Bero L. Healthcare outcomes assessed with observational study designs compared with those assessed in randomized trials. Cochrane Database Syst Rev 2014;4:MR000034.

[11] Curran GM, Bauer M, Mittman B, Pyne JM, Stetler C. Effectiveness-implementation hybrid designs: combining elements of clinical effectiveness and implementation research to enhance public health impact. Med Care 2012;50:217—26.

[12] Hampel H, Frankel WL, Martin E, Arnold M, Khanduja K, Kuebler P, et al. Screening for the Lynch syndrome (hereditary nonpolyposis colorectal cancer). N Engl J Med 2005;352:1851—60.

[13] EGAPP. Recommendations from the EGAPP Working Group: genetic testing strategies in newly diagnosed individuals with colorectal cancer aimed at reducing morbidity and mortality from Lynch syndrome in relatives. Genet Med 2009;2009(11):35—41.

[14] AHRQ. Selecting quality and resource use measures: a decision guide for community quality collaboratives. Available from: http://www.ahrq.gov/professionals/quality-patient-safety/quality-resources/tools/perfmeasguide/index.html [last accessed 02.11.15].

[15] Davis K. Slowing the growth of healthcare costs—learning from international experience. N Engl J Med 2008;359:1751—5.

[16] Gudgeon JM, McClain MR, Palomaki GE, Williams MS. Rapid-ACCE: experience with a rapid and structured approach for evaluating gene-based testing. Genet Med 2007;9:473—8.

[17] Gudgeon JM, Williams JL, Burt RW, Samowitz WS, Snow GL, Williams MS. Lynch syndrome screening implementation: business analysis by a healthcare system. Am J Manag Care 2011;17:e288—300.

[18] Hughes RG, Clancy CM. Improving the complex nature of care transitions. J Nurs Care Qual 2007;22:289—92.

[19] Schoenbaum SC. Keys to a high-performance health system for the United States. Healthc Financ Manage 2006;60:60—4.

[20] Giovino JM. Holding the gains in quality improvement. Fam Pract Manag 1999;6:29—32.

[21] Gudgeon JM, Belnap TW, Williams JL, Williams MS. Impact of age cut-offs on a Lynch syndrome screening program. J Oncol Pract 2013;9:175—9.

[22] Institute of Medicine. In: Olsen L, Aisner D, McGinnis JM, editors. The learning healthcare system: workshop summary. Washington, DC: National Academies Press; 2007.

[23] Wade JE, Ledbetter DH, Williams MS. Implementation of genomic medicine in a health care delivery system: a value proposition? Am J Med Genet C Semin Med Genet 2014;166C:112—16.

[24] Murray MF. Your DNA is not your diagnosis: getting diagnoses right following secondary genomic findings. Genet Med 2015. Available from: http://dx.doi.org/10.1038/gim.2015.134.

Chapter 16

Pharmacogenetics and Pharmacogenomics

Deepak Voora
Duke University, Durham, NC, United States

Chapter Outline

INTRODUCTION

Most medications are prescribed based on prior evidence collected from clinical trials, observational data, physician intuition, or anecdotal experience. In general, the guiding principal driving physician behavior is a "one size fits all" model where the effect of a drug in a population is extrapolated to the expected effect in an individual. Identifying the most effective drug for a patient is often empiric (i.e., trial and error) with little guidance beyond age, kidney/liver function, and the physicians own experience. However, for centuries, it has been known that there are individual differences in drug response that cannot be explained by known factors. Pythagoras noted that ingestion of fava beans resulted in an adverse reaction—hemolytic anemia—in some but not all individuals [1]. More recently in 1957 Motulsky posited that inherited defects of metabolism may explain individual differences in drug response [2]. Soon after, in 1959, Vogel coined the term "pharmacogenetics," which is now known as the study of genetic variation, primarily single-nucleotide polymorphisms (SNPs), to identify subgroups of patients that may respond differently to a different medication type or dose [3]. Pharmacogenetics has expanded from SNPs to include other forms of genetic variation such as insertions, deletions, and copy number variation (CNV). "Pharmacogenomics," a broader term often used interchangeably with

Genomic and Precision Medicine. DOI: http://dx.doi.org/10.1016/B978-0-12-800681-8.00016-5
233

"pharmacogenetics." In contrast to pharmacogenetics, which focuses on the inherited genome, pharmacogenomics instead focuses on the "expressed genome" and includes techniques such as gene expression, proteomics, or metabolomics to identify factors that impact drug response. In this chapter, we will use the term pharmacogenomics to encompass the field of study where the genome (inherited or expressed) is considered in describing drug responses. The ultimate goal of pharmacogenomics is to prescribe the right dose of the right drug to the right individual at the right time. By moving away from the "one size fits all" approach to one that takes into account individual differences the goal of pharmacogenomics is to maximize efficacy and minimize adverse effects.

PRINCIPLES OF PHARMACOGENOMICS

The response to a drug can be measured in at least two major ways. The first is a surrogate, (usually) laboratory-based or physiology-based measurement of drug response and can include a drug metabolite concentration, activity of a drug's intended target, or function of a pathway, cell type, or organ targeted by a drug. Such measures are often easily measured and are very close to the intended target of given medication. Examples include clopidogrel active metabolite concentration, heart rate in response to metoprolol, or blood pressure in response to diuretics. Alternatively, drug response can be defined as the clinical outcome that treated by the drug of interest. Such clinical responses are often several steps downstream from the target of the drug, are multifactorial and complex in nature with multiple determinants of outcome. Though harder to measure, these clinical responses are often more meaningful to the patient and provider. Examples include stent thrombosis for clopidogrel, relief of chest pain symptoms for metoprolol, and stroke in the setting of diuretics. In considering the application of pharmacogenomic findings into clinical practice a critical first step is to consider the types of responses used in the discovery process.

Pharmacogenomics requires the presence of heterogeneity in drug response. In other words, if a uniform effect is seen among all individuals, it is unlikely that any intervention centered on personalized or individualized therapy will improve health outcomes. For example, aspirin is a well-known inhibitor of platelet COX-1 and with typical doses (81–325 mg) a single dose is sufficient to inhibit platelet COX-1 activity and true "aspirin resistance" is rare [1]. Therefore, strategies to individualize aspirin therapy on the basis of COX-1 activity are unlikely to improve clinical outcomes. If there is variation in a drug response, that variation must be clinically significant in order for interventions that tailor drug therapy to make a meaningful impact on health outcomes. With the class of cholesterol-lowering agents known as "statins," there is well-known variation in the magnitude of cholesterol lowering between individuals [2]. However, the magnitude of the variation in cholesterol-lowering response is small compared to the average drug effect and not large enough to translate to differences in preventing heart disease. When considering a pharmacogenomics strategy, it is also important to understand that not all variation in drug response are due to genomic causes, as they can be a result of patient nonadherence as well as drug-to-drug interactions. For example, drug resistance is impossible to distinguish from noncompliance if measures of adherence are not taken into account. Strategies to improve outcomes to drug therapy should first and foremost focus on improving adherence to existing therapies followed by the use of genomics to fine-tune and optimize outcomes.

Once a clinically relevant variation in drug response is identified, there are typically four major classes that describe pharmacogenomics profiles that correlate with drug response: (1) pharmacokinetic, (2) pharmacodynamic, (3) underlying disease pathway, and (4) off-target (Table 16.1). Variation in pharmacokinetics or "what the body does to the drug" reflects variation in concentration of a drug at the site of the drug's target. The pharmacokinetic profile of a drug is dependent upon four main factors, known as ADME: (1) *absorption* of the drug from the site of its administration, (2) *distribution* of the drug into different tissue compartments, (3) *metabolism* of the drug within the body, and (4) *excretion* of the drug from the body. The majority of this variability occurs in drug-metabolizing enzymes and transporters that are responsible for clearing drugs from the circulation. Many variants are found in the cytochrome P450 (CYP) system of the liver, which metabolizes the majority of medications in use today. For instance, in the case of clopidogrel, carriers of reduced-function alleles in *CYP2C19*, when treated with the drug, have lower levels of the active metabolite [4].

Pharmacodynamic variation is a complement to pharmacokinetics and is described as "what the drug does to the body." More specific, pharmacodynamic variation reflects a drug's ability to influence its "on-target" effects, typically the intended receptor or enzyme. Examples include variants that alter the expression levels of a drug target such that the amount of target varies from one individual to the next such as the case with *VKORC1* genetic variants and warfarin dosing [3]. Higher target levels may translate to a higher dose of a drug required to produce a certain level of inhibition. In other examples, the target itself may be subject to genetic variation as in the case with variants in *ADRB1* and metoprolol [5]. Finally, in a unique example of how alternative splicing in a drug target can lead to variation in drug response, genetic variation in *HMGCR* produces an alternatively spliced *HMGCR* transcript that produces a version of HMGCR that is less sensitive to simvastatin inhibition [6]. Specific examples of pharmacodynamic variation that lead to clinically actionable results are presented later in this chapter.

In most cases, drugs are intended to not only affect a specific enzyme or receptor but also to modulate the activity of a physiologic pathway. Therefore, variation in the activity of that underlying pathway forms another source of variability in drug response. Profiles that belong to this class of variability are often downstream or independent of the drug target. Due to

TABLE 16.1 Major Classes of Pharmacogenomic Variants Underlying Drug Responses

Class	Pharmacokinetic	Pharmacodynamic	Pathway	Off-Target
Description	Genes related to ADME	Genes that encode the intended target of a drug	Genes that encode proteins in the downstream pathway of the drug target	Genes that encode proteins that are not in the intended drug pathway
Example gene products	CYP enzymes and drug transporters	Cell surface receptors and intracellular enzymes	Signaling molecules, transcription factors, metabolic pathways	Immune recognition proteins
Primary effect	Alter the concentration of active drug metabolite at the site of drug action	Alter the levels of the drug target or alter the ability of a drug to bind to its drug target	Alter the activity of the pathway targeted by drug therapy	Generate an immune response to the drug or its metabolite
Prototype	*CYP2D6* and codeine	*VKORC1* and warfarin	*LDLR* and statins	*HLA-B* and abacavir
Potential mitigating approaches	Change drug dose; choose alternate drug with different pharmacology	Change drug dose; choose alternate drug with different target	Change drug dose; choose alternate drug with different drug pathway	Drug avoidance or alternate drug with different chemical structure

the complex and multifactorial nature of cardiovascular disease, there has been limited success in applying this concept to pharmacogenetics research. However as the genetic architecture of complex disease unfolds, it is likely that these variants will represent important determinants of drug response. A related concept is that once a drug is placed within in an intact human system, there are a series of on-target and off-target events that are complex and cannot be predicted by the known biology of a drug. The field of systems biology attempts to quantitatively describe the complex interactions of genes, proteins, pathways, networks, cells, and organs as intact systems. Coupled with traditional pharmacology concepts of pharmacokinetics and pharmacodynamics, a new field termed "systems pharmacology" is emerging that is going beyond traditional views of genetics and pharmacology to better describe drug responses. Although nascent, this field is likely to deliver important discoveries that underlie variation in drug response.

Last, there are a series of off-target effects of a drug that underlie variation in drug response. Unlike genomic variation in pharmacokinetics, pharmacodynamics, and the underlying disease pathways, this class of profiles is often unpredictable and idiosyncratic. Examples include hypersensitivity reactions to carbamazepine, which are driven by specific alleles in the HLA system [7] that create specific epitopes in the presence of carbamazepine (or its metabolites) that trigger autoimmune reactions. Although important for determining responses to carbamazepine, these variants typically do not generalize to adverse reactions to other agents with different chemical structures.

With these perspectives in mind, the remainder of this chapter will focus on current examples of pharmacogenomics that are currently being implemented or can be expected to be implemented in the near (i.e., 5 years) future. The principles described above will be referred to in each example. The examples described in this chapter are not meant to be exhaustive, but instead are meant to be illustrative of particular concepts surrounding the translation of pharmacogenomics into clinical medicine. Because other volumes in this textbook are dedicated to cardiovascular disease and oncology, we have chosen examples outside of these fields in this chapter. An additional source of well-curated, pharmacogenetic information can be found at the Pharmacogenetics Knowledge Base (www.pharmgkb.org) which helps to curate publicly available pharmacogenetic data. The Cytochrome P450 Nomenclature Committee at http://www.cypalleles.ki.se maintains a comprehensive catalog of genetic variants and haplotypes for CYP enzymes. In addition the Food and Drug Administration (FDA) has approved the addition of clinically relevant pharmacogenetic information in drug label information for over 150 drug labels. The Pharmacogenomics Research Network Clinical Pharmacogenomics Implementation Consortium (CPIC) has issued a series of guidelines for drug—gene combinations (i.e., warfarin and *CYP2C9/ VKORC1*). The goal of the CPIC guidelines is not on whether or not a pharmacogenetic test should be ordered, but instead focuses on how to interpret a pharmacogenetic test result in the context of patient care. Table 16.2 lists

TABLE 16.2 Medications with Pharmacogenetic Guidance on Drug Dosing and Selection

Drug	Gene	CPIC	FDA Label
Abacavir	*HLA-B*	X	X
Amitriptyline	*CYP2D6*	X	X
Azathioprine	*TPMT*	X	X
Capecitabine	*DPYD*	X	X
Carbamazepine	*HLA-B*	X	X
Clopidogrel	*CYP2C19*	X	X
Codeine	*CYP2D6*	X	X
Desipramine	*CYP2D6*	X	X
Doxepin	*CYP2D6*	X	X
Fluorouracil	*DPYD*	X	X
Imipramine	*CYP2D6*	X	X
Ivacaftor	*CFTR*	X	X
Mercaptopurine	*TPMT*	X	X
Nortriptyline	*CYP2D6*	X	X
Phenytoin	*HLA-B*	X	X
Rasburicase	*G6PD*	X	X
Thioguanine	*TPMT*	X	X
Trimipramine	*CYP2D6*	X	X
Warfarin	*CYP2C9, VKORC1*	X	X
Citalopram	*CYP2C19*	X	X
Peginterferon alfa-2b	*IFNL3*	X	X
Telaprevir	*IFNL3*	X	X
Clomipramine	*CYP2D6*	X	X
Dapsone	*G6PD*		X
Dextromethorphan and Quinidine	*CYP2D6*		X
Divalproex	*POLG*		X
Eltrombopag	*F5, SERPINC1*		X
Lenalidomide	*del (5q)*		X

(Continued)

TABLE 16.2 (Continued)

Drug	Gene	CPIC	FDA Label
Mafenide	G6PD		X
Mipomersen	LDLR		X
Mycophenolic acid	HPRT1		X
Nitrofurantoin	G6PD		X
Pegloticase	G6PD		X
Primaquine	G6PD		X
Quinine sulfate	G6PD		X
Sodium nitrite	G6PD		X
Thioridazine	CYP2D6		X
Tolterodine	CYP2D6		X
Valproic acid	POLG, NAGS, CPS1, ASS1, OTC, ASL, ABL2		X

commonly used medications and where to find pharmacogenomic information regarding dosing or drug selection.

PHARMACOKINETIC EXAMPLE—CODEINE

The largest source of clinically significant findings to have emerged from genome-wide association studies (GWAS) of drug responses is genes that encode drug-metabolizing enzymes and transporters. Medications are often prescribed in their active form or in inactive prodrugs that must be converted to an active metabolite in vivo. Codeine is a commonly used pain medication in the pediatric population, particularly in the postoperative setting. Codeine is a prodrug that is bioactivated by CYP2D6 in the liver into its active metabolites morphine and morphine-6-glucuronide [8]. Because codeine has an approximately 200-fold lower affinity for the opioid receptor, the analgesic properties of codeine are attributed to the production of its more potent metabolites. *CYP2D6* is a well-known "pharmacogene" with over 30 different alleles that impact the function of CYP2D6 enzyme. In addition, *CYP2D6* is also subject to CNV such that certain individuals carry multiple, functional, copies of *CYP2D6*. The combination of *CYP2D6* genotypes and copy numbers can be used to predict CYP2D6 function through the use of an activity score that ranges from 0 (nonfunctional CYP2D6) to 1 (normal function). For each additional copy of *CYP2D6* the score for each allele is multiplied by the number of copies of *CYP2D6* that are detected. The total score is the sum of scores assigned to each allele. Total scores of 0 are referred to "poor metabolizers," 0.5 as "intermediate metabolizers," 1−2 as "extensive metabolizers," and more than 2 as "ultrarapid metabolizers" [9]. Including additional information on concomitant drugs that inhibit CYP2D6 can be used to modify activity scores and thus further refine predicted CYP2D6 activity [10]. The degree of analgesia produced by codeine is linked to genetic variation in *CYP2D6*. Poor metabolizers produce lower concentrations of morphine and an experience of reduced analgesia in response to codeine in controlled studies of healthy volunteers [11]. Adverse effects of codeine (i.e., constipation, nausea, and dry month) do not appear to differ by *CYP2D6* genotype, however [11]. Although larger studies are needed to definitively link *CYP2D6* status to the side effect profile of codeine, there is a clearer link with toxicity. Several case reports have established that severe or life-threatening respiratory depression using standard doses of codeine in ultrarapid metabolizers [4,12,13]. Recent changes in the FDA label for codeine strongly warn against using codeine in all children undergoing tonsillectomy; however, codeine remains widely used in adult and pediatric populations. Therefore, based on availability of alternative analgesics that do not depend on *CYP2D6* such as morphine and non-opioid analgesics, one approach would be to avoid codeine in *CYP2D6* poor and ultrarapid metabolizers due to lack of efficacy and toxicity, respectively. Alternative approaches include modification of the starting dose or increasing the frequency of monitoring. Many commercial laboratories provide *CYP2D6* genetic testing. Certain centers have adopted *CYP2D6* testing into routine clinical care [14] coupled with clinical decision support systems embedded within the electronic medical record to assist providers with interpreting genetic test results and subsequent medication management [15,16]. However, it is not yet clear if incorporating genetic data has improved patient outcomes. In addition, while other opioid analgesics (tramadol, oxycodone) also utilize the CYP2D6 pathway for bioactivation, it is not yet clear if genetic variation affects the response to these alternate opioids to the same degree as codeine. Therefore, the example of codeine and *CYP2D6* represents a clear example of how genetic variation in a key drug-metabolizing enzyme explains variation in response to a commonly used medication that forms a paradigm for future drug−gene pairs to emulate.

PHARMACODYNAMIC EXAMPLE—WARFARIN

Warfarin is a commonly used oral anticoagulant that, for decades, has been the standard for prevention of thrombosis in atrial fibrillation, venous thromboembolism, and mechanical heart valves. Despite its well-known efficacy, warfarin is characterized by large interindividual variability in dose requirements such that the daily dose required to produce a therapeutic level of anticoagulation can vary 10-fold between individuals [17]. Warfarin has a narrow therapeutic index such that excessive levels of anticoagulation can lead to life-threatening hemorrhage and subtherapeutic levels leave patients at risk for thrombosis. As a consequence, ensuring adequate dosing to achieve therapeutic levels of anticoagulation is critical to successful warfarin therapy. After identifying the gene encoding for warfarin's target, vitamin K epoxide reductase complex, subunit 1 (*VKORC1*) [18,19] and subsequent resequencing of *VKORC1* [3] a key genetic variant in the promoter region was identified, rs9923231 (also referred to as $-1639\,G > A$ and the 3673 SNP). Variation at this position is associated with variable levels of *VKORC1* gene expression [20], therefore carriers of the minor allele require a 30% per allele reduction in warfarin dose requirements [21]. In an effort to identify additional variants that may underlie warfarin dose requirements, a GWAS identified variants in *CYP4F2* which reduce levels of hepatic CYP4F2, lead to higher levels of hepatic vitamin K (the substrate of VKORC1) and thus lead to higher warfarin dose requirements [22–26]. To meet the translational requirements of converting a *VKORC1*, *CYP4F2*, and other genotypes into an clinically actionable result, an online dosing calculator (www.warfarindosing.org) has been developed based on validated models [27]. The inputs into these calculators require not only genotype data but also known determinants of warfarin dose: age, sex, body size, race, smoking, and concomitant medications. The result is a predicted warfarin dose based on a patient's individual clinical and genetic profile. To test the hypothesis that addition of genetic data during warfarin initiation improves clinical outcomes, three recently completed randomized clinical trials (RCTs) compared warfarin initiation using genetic + clinical data to either usual care or clinical dosing alone [28–30]. These trials produce mixed results with some benefit observed when genetic information was compared to usual care, but when added to a clinical dosing model did not improve laboratory outcomes. As a result, it is unlikely that pharmacogenetics-guided warfarin therapy will become standard of care at this time. Instead, the use of genetic data may be useful in selected cases where the risk of bleeding is high (or would be catastrophic) or in selection of warfarin versus novel oral anticoagulants (e.g., rivaroxaban, apixaban, dabigatran, and endoxaban). Ongoing RCTs of genotype-guided warfarin therapy that are focused on clinical outcomes versus laboratory outcomes may provide different results; however, in the meantime the optimal role for genetic testing during warfarin initiation has not been settled.

OFF-TARGET EFFECTS EXAMPLE—ABACAVIR

Abacavir is a nucleoside reverse transcriptase inhibitor indicated for the treatment of HIV infection as part of a combination of other antiretroviral medications. Compared to tenofovir-based regimens, including abacavir leads to shorter time to virologic failure and also a shorter time to first adverse event in patients with baseline viral loads more than 100,000 copies/mL [31]. Although efficacious and, in general, well tolerated approximately 5−8% of patients experience a hypersensitivity reaction characterized by fever, rash, gastrointestinal symptoms (e.g., nausea, vomiting, abdominal pain), fatigue, cough, and dyspnea during the first 6 weeks of treatment that requires immediate drug discontinuation. Delayed recognition or rechallenge can lead to clinical worsening and may be fatal. Therefore, several efforts to identify genetic variants associated with abacavir hypersensitivity reactions led to the identification of allele HLA-B*5701 [32−34]. Compared to typical pharmacogenetics variants which confer an odds ratio of approximately 2−3, the HLA-B*5701 was associated with an odds ratio in excess of 100. Based on the unusual strength of association, several subsequent prospective studies (single-arm and RCT) that incorporated HLA-B*5701 testing prior to abacavir initiation and limiting abacavir use to noncarriers of this allele were able to eliminate the incidence of abacavir hypersensitivity [35,36]. In response to these landmark studies, the FDA changed the label for abacavir to include information regarding the risk of hypersensitivity reactions and HLA-B*57:01 and also that all patients be screened before being treated with abacavir and that abacavir not be initiated in carriers of HLA-B*57:01 because such a strategy eliminates hypersensitivity reactions to this drug. Many commercial laboratories provide testing for this allele and interpretation is a straightforward, binary, interpretation as the presence or absence of the allele. In combination, the case of abacavir and HLA-B*57:01 remains the exemplar case in pharmacogenomics and the standard to which all other drugs/genes should be held to. This is not to imply that all pharmacogenomic markers must meet the same standards established by abacavir before clinical implementation. There are many scenarios where either weaker levels of evidence or retrospective data of efficacy may be sufficient to alter management. Each drug, gene, and clinical scenario needs to be evaluated on a case-by-case basis in order to decide to what extent testing and tailored therapy based on test results is warranted.

FUTURE OPPORTUNITIES AND OBSTACLES

One area that holds major promise for human health, disease, and drug response is the microbiome. The community of organisms that lives on us and within us has been finely tuned over thousands of generations to exist in a mutualistic relationship with us. The microbiome can rapidly adapt to changes in our diet [37] and in response to a wide array of medications. In the instance of digoxin, which has long been known to undergo metabolism by the enteric microbiome, recent findings demonstrate that digoxin exposure can induce the proliferation of a specific bacterial species capable of inactivating this drug [38]. Through the ability to use polymerase chain reaction to identify specific bacterial sequences in fecal samples, we can envision a novel, fecal biomarker, for an individual's ability to metabolize digoxin. As a consequence, starting doses of this drug with a narrow therapeutic index could be individualized. Similarly, interactions between the enteric microbiome and statins have been suggested [39]. Therefore, we are only beginning to realize the powerful role of the microbiome and its impact on drug responses. Ultimately, demonstrating that the use of our knowledge about the microbiome impacts health outcomes will be required before widespread adoption.

Although the vast majority of pharmacogenomic variation is embedded in the "inherited genome," there will likely emerge novel profiles based on the "expressed genome." As our ability to synthesize the products of the genome (i.e., transcripts, proteins, and metabolites) as well as the complex interaction between the genome, diet, and environment, we can anticipate pharmacology to shift to a "systems pharmacology" approach to emerge. Under this new approach, the tools of systems biology (e.g., biologic networks, cells, tissues, organs, and organisms) are studied for the simultaneous actions of a drug in an intact system. Although a daunting task, by being able to not only handle the complexity of such systems but also make predictions of drug response based on these data, we can anticipate more precise markers of drug response to enable personalized and precision drug selection.

Last, the evidentiary framework for evaluating novel pharmacogenomic markers requires a reevaluation. For decades the RCT has been the gold standard for demonstrating the benefit of a new drug, test, or strategy over the existing standard of care. Although laudable and a powerful approach, it is impossible to demand an RCT for each new drug/gene pair. Furthermore, it may not be required to perform an RCT for certain clinical decisions. For example, when decided between a range of approved doses of a given drug, perhaps a study based on pharmacokinetics/pharmacodynamics is sufficient. When deciding between a range of FDA-approved drugs for the same clinical indication, perhaps retrospective, observational data is enough to demonstrate benefits of genetic testing. However, for decisions that require adding new medications or withholding evidence-based ones, the RCT will remain a gold standard for tailored drug therapy decisions. In summary, the field of pharmacogenomics/pharmacogenetics will not realize its full potential of improving health outcomes unless we allow different levels of evidence, outside of the RCT such as registries, prespecified secondary end points in clinical trials, or pragmatic/cluster randomized comparative effective studies, to guide clinical practice.

CONCLUSIONS

In conclusion, the field of pharmacogenomics is an established field with several proof-of-concept examples where use of genomic information can improve health outcomes. The major classes of pharmacogenomic variation are pharmacokinetics, pharmacodynamics, underlying pathways, and off-target effects. With respect to the path from discovery to clinical implementation, each drug is unique and will require its own path toward clinical implementation based on alternative treatments and the clinical context. For each drug/gene/clinical scenario, different levels of evidence will likely be required before clinical adoption. The challenge for pharmacogenomics is to ensure that predictions regarding drug response are sufficiently precise to justify a change in drug therapy. This will likely require a much more detailed mechanistic understanding of how drugs work (or do not) such that novel biomarkers can be designed to monitor and/or adjust drug therapy. Providers will also require enhanced tools to be able to interpret expanding arrays of drug response markers, ideally integrated into the electronic medical record and assisted by decision support tools. Therefore, the future challenges for pharmacogenomics mirror those of genomic medicine as a whole.

ABBREVIATIONS

CNV copy number variation
CPIC Clinical Pharmacogenomics Implementation Consortium
CYP cytochrome P450
FDA Food and Drug Administration
GWAS genome-wide association study
RCT randomized clinical trial
SNP single-nucleotide polymorphism

REFERENCES

[1] Grosser T, Fries S, Lawson JA, Kapoor SC, Grant GR, FitzGerald GA. Drug resistance and pseudoresistance: an unintended consequence of enteric coating aspirin. Circulation 2012.

[2] Simon JA, Lin F, Hulley SB, et al. Phenotypic predictors of response to simvastatin therapy among African-Americans and Caucasians: the Cholesterol and Pharmacogenetics (CAP) Study. Am J Cardiol 2006;97(6):843−50.

[3] Rieder MJ, Reiner AP, Gage BF, et al. Effect of VKORC1 haplotypes on transcriptional regulation and warfarin dose. N Engl J Med 2005;352(22):2285−93.

[4] Dalen P, Frengell C, Dahl ML, Sjoqvist F. Quick onset of severe abdominal pain after codeine in an ultrarapid metabolizer of debrisoquine. Ther Drug Monit 1997;19(5):543−4.

[5] Perez JM, Rathz DA, Petrashevskaya NN, et al. Beta(1)-adrenergic receptor polymorphisms confer differential function and predisposition to heart failure. Nat Med 2003;9 (10):1300−5.

[6] Medina MW, Gao F, Ruan W, Rotter JI, Krauss RM. Alternative splicing of 3-hydroxy-3-methylglutaryl coenzyme A reductase is associated with plasma low-density lipoprotein cholesterol response to simvastatin. Circulation 2008;118(4):355−62.

[7] Chen P, Lin J-J, Lu C-S, et al. Carbamazepine-induced toxic effects and HLA-B*1502 screening in Taiwan. N Engl J Med 2011;364(12):1126−33.

[8] Mignat C, Wille U, Ziegler A. Affinity profiles of morphine, codeine, dihydrocodeine and their glucuronides at opioid receptor subtypes. Life Sci 1995;56(10):793−9.

[9]   Gaedigk A, Simon SD, Pearce RE, Bradford LD, Kennedy MJ, Leeder JS. The CYP2D6 activity score: translating genotype information into a qualitative measure of phenotype. Clin Pharmacol Ther 2008;83(2):234—42.

[10]   Borges S, Desta Z, Jin Y, et al. Composite functional genetic and comedication CYP2D6 activity score in predicting tamoxifen drug exposure among breast cancer patients. J Clin Pharmacol 2010;50(4):450—8.

[11]   Eckhardt K, Li S, Ammon S, Schanzle G, Mikus G, Eichelbaum M. Same incidence of adverse drug events after codeine administration irrespective of the genetically determined differences in morphine formation. Pain 1998;76(1-2):27—33.

[12]   Gasche Y, Daali Y, Fathi M, et al. Codeine intoxication associated with ultrarapid CYP2D6 metabolism. N Engl J Med 2004;351(27):2827—31.

[13]   Ciszkowski C, Madadi P, Phillips MS, Lauwers AE, Koren G. Codeine, ultrarapid-metabolism genotype, and postoperative death. N Engl J Med 2009;361(8):827—8.

[14]   Hoffman JM, Haidar CE, Wilkinson MR, et al. PG4KDS: a model for the clinical implementation of pre-emptive pharmacogenetics. Am J Med Genet C Semin Med Genet 2014;166C(1):45—55.

[15]   Bell GC, Crews KR, Wilkinson MR, et al. Development and use of active clinical decision support for preemptive pharmacogenomics. J Am Med Inform Assoc 2014;21(e1): e93—9.

[16]   Hicks JK, Crews KR, Hoffman JM, et al. A clinician-driven automated system for integration of pharmacogenetic interpretations into an electronic medical record. Clin Pharmacol Ther 2012;92(5):563—6.

[17]   Gage BF, Eby C, Johnson JA, et al. Use of pharmacogenetic and clinical factors to predict the therapeutic dose of warfarin. Clin Pharmacol Ther 2008;84(3):326—31.

[18]   Rost S, Fregin A, Ivaskevicius V, et al. Mutations in VKORC1 cause warfarin resistance and multiple coagulation factor deficiency type 2. Nature 2004;427(6974):537—41.

[19]   Li T, Chang CY, Jin DY, Lin PJ, Khvorova A, Stafford DW. Identification of the gene for vitamin K epoxide reductase. Nature 2004;427(6974):541—4.

[20]   Wang D, Chen H, Momary KM, Cavallari LH, Johnson JA, Sadee W. Regulatory polymorphism in vitamin K epoxide reductase complex subunit 1 (VKORC1) affects gene expression and warfarin dose requirement. Blood 2008;112(4):1013—21.

[21]   The International Warfarin Pharmacogenetics Consortium. Estimation of the warfarin dose with clinical and pharmacogenetic data. N Engl J Med 2009;360(8):753—64.

[22]   Singh O, Sandanaraj E, Subramanian K, Lee LH, Chowbay B. The influence of CYP4F2 rs2108622 (V433M) on warfarin dose requirement in Asian patients. Drug Metab Pharmacokinet 2010.

[23]   Caldwell MD, Awad T, Johnson JA, et al. CYP4F2 genetic variant alters required warfarin dose. Blood 2008;111(8):4106—12.

[24]   Takeuchi F, McGinnis R, Bourgeois S, et al. A genome-wide association study confirms VKORC1, CYP2C9, and CYP4F2 as principal genetic determinants of warfarin dose. PLoS Genet 2009;5(3):e1000433.

[25]   Cooper GM, Johnson JA, Langaee TY, et al. A genome-wide scan for common genetic variants with a large influence on warfarin maintenance dose. Blood 2008;112(4):1022—7.

[26]   McDonald MG, Rieder MJ, Nakano M, Hsia CK, Rettie AE. CYP4F2 is a vitamin K1 oxidase: an explanation for altered warfarin dose in carriers of the V433M variant. Mol Pharmacol 2009;75(6):1337—46.

[27]   Klein TE, Altman RB, Eriksson N, et al. Estimation of the warfarin dose with clinical and pharmacogenetic data. N Engl J Med 2009;360(8):753—64.

[28] Kimmel SE, French B, Kasner SE, et al. A pharmacogenetic versus a clinical algorithm for warfarin dosing. N Engl J Med 2013;369(24):2283−93.

[29] Pirmohamed M, Burnside G, Eriksson N, et al. A randomized trial of genotype-guided dosing of warfarin. N Engl J Med 2013;369(24):2294−303.

[30] Verhoef TI, Ragia G, de Boer A, et al. A randomized trial of genotype-guided dosing of acenocoumarol and phenprocoumon. N Engl J Med 2013;369(24):2304−12.

[31] Sax PE, Tierney C, Collier AC, et al. Abacavir-lamivudine versus tenofovir-emtricitabine for initial HIV-1 therapy. N Engl J Med 2009;361(23):2230−40.

[32] Mallal S, Nolan D, Witt C, et al. Association between presence of HLA-B*5701, HLA-DR7, and HLA-DQ3 and hypersensitivity to HIV-1 reverse-transcriptase inhibitor abacavir. Lancet 2002;359(9308):727−32.

[33] Hetherington S, Hughes AR, Mosteller M, et al. Genetic variations in HLA-B region and hypersensitivity reactions to abacavir. Lancet 2002;359(9312):1121−2.

[34] Saag M, Balu R, Phillips E, et al. High sensitivity of human leukocyte antigen-b*5701 as a marker for immunologically confirmed abacavir hypersensitivity in white and black patients. Clin Infect Dis 2008;46(7):1111−18.

[35] Rauch A, Nolan D, Martin A, McKinnon E, Almeida C, Mallal S. Prospective genetic screening decreases the incidence of abacavir hypersensitivity reactions in the Western Australian HIV cohort study. Clin Infect Dis 2006;43(1):99−102.

[36] Mallal S, Phillips E, Carosi G, et al. HLA-B*5701 screening for hypersensitivity to abacavir. N Engl J Med 2008;358(6):568−79.

[37] David LA, Maurice CF, Carmody RN, et al. Diet rapidly and reproducibly alters the human gut microbiome. Nature 2014;505(7484):559−63.

[38] Haiser HJ, Gootenberg DB, Chatman K, Sirasani G, Balskus EP, Turnbaugh PJ. Predicting and manipulating cardiac drug inactivation by the human gut bacterium *Eggerthella lenta*. Science 2013;341(6143):295−8.

[39] Kaddurah-Daouk R, Baillie RA, Zhu H, et al. Enteric microbiome metabolites correlate with response to simvastatin treatment. PLoS One 2011;6(10):e25482.

Chapter 17

Clinical Genomic Testing

Ken J. Hampel[1], Debra G.B. Leonard[1,2] and Nikoletta Sidiropoulos[1,2]
[1]University of Vermont Medical Center, Burlington, VT, United States, [2]University of Vermont College of Medicine, Burlington, VT, United States

Chapter Outline

INTRODUCTION

Optimism regarding the potential impact of genomics on healthcare began in the years preceding the completion of the Human Genome Project [1−3]. Since the project was officially declared complete in 2003, anticipation of a transformation of the current clinical care model to one that applies genomic science at the bedside has been growing. Full realization of genomic medicine however has not been as swift as predicted. This perceived slowness should not be surprising because new scientific discoveries take upward of 17 years for full-scale clinical adoption [4].

Interrogation of the genome is broadening from a focus on an individual or a small subset of Mendelian variants of large effect to simultaneous analysis of large swaths of the genome. Technical advances and rapidly decreasing costs of genomic technologies are increasing clinical access to genomic information resulting in an expanding knowledge base of genomic alterations. However, "knowing" genomic information is not equivalent to "understanding" it.

Genomic and Precision Medicine. DOI: http://dx.doi.org/10.1016/B978-0-12-800681-8.00017-7
© 2017 Elsevier Inc. All rights reserved.

Deciphering the human genome for medical use requires much effort. Many harmless, or silent, differences between the genomes of individuals are observed in the protein-coding regions that must carefully be delineated from harmful variation. Genome variants proposed to be harmful must be robustly proven to be causative or significantly and reproducibly associated with human disease and this scientific exercise can take many years. Only when a proposed harmful variant has gone through this level of scrutiny and confirmed as disease related should it be considered for clinical use. More recently, the emerging understanding of the functionality of non-protein-coding DNA is also adding tremendous complexity to the effort achieve full understanding of the human genome in general and the specific genome sequence of individual patients.

Challenges to implementation of genomic testing exist within the clinical laboratory and its interface with clinical specialists. Clinical care pathways, laboratory workflows, and supportive infrastructure must be developed to facilitate the delivery and comprehension of genomic data at the point of care in a busy clinical setting. Thoughtful consideration of patient autonomy during the design process should ensure opportunities for genetic counseling, informed consent, and financial counseling.

In any attempt to develop clinical genomics the question will arise, "who will pay to test for this information?" While reimbursement is a challenge and will likely remain so over the next decade, the rapidly decreasing costs of broadly interrogating the genome for clinical purposes are approaching those of single-gene testing [5,6]. An expanding knowledge base and vast technological improvements are minimizing logistical complexities; therefore, this environment is rapidly enabling wide-spread feasibility and scalability of genomic testing for clinical care. The efforts of early adopters of clinical genomic medicine have proven that some universal implementation challenges must be addressed [7]. This chapter describes state-of-the-art genomic technologies and outlines common challenges that arise in the implementation of a clinical genomic medicine program.

GENOMIC TECHNOLOGIES

The first DNA sequencing method was developed by Allan Maxam and Walter Gilbert in 1976, which was largely replaced by a more streamlined chain termination DNA sequencing method developed by Fredrick Sanger in 1977 and earned him a share of his second Nobel Prize for Chemistry in 1980 [8]. Sanger sequencing relies on the enzyme chemistry of nucleotide addition based on a template DNA strand. In 2003, the human genome project was completed as it started, with a Sanger sequencing run [9–11]. Notwithstanding this monumental achievement, Sanger sequencing is not well suited to common usage in sequencing panels of 10 or more genes, much less exome or genome sequencing. Robust Sanger sequencing instruments can simultaneously sequence 384 pieces of DNA, 800 bp each, and generate 3×10^5 bp of DNA sequence. By comparison, the lower sequence throughput of next-generation sequencing (NGS) instruments is 6 million reads with a read length 150 bp, which generates 9×10^8 bp of DNA sequence, in approximately equivalent instrument time. This section of the chapter will review state-of-the-art technology applied in the clinical laboratory with a significant focus on NGS, also referred to as massively parallel sequencing.

Next-Generation Sequencing

NGS utilizes methods immobilize distinct DNA fragments on solid supports, commonly referred to as "chips" or "flow-cells." *Parallel* sequencing of millions of individual DNA fragments is carried out with each DNA fragment occupying a unique location on the solid support. Parallel sequencing of the random fragments means individual bases are sequenced many times [12]. The number of individual reads of a base is referred to as the *depth of coverage (DOC)*. Increasing the DOC improves the reliability of calling sequence variants relative to a reference sequence and is critical when variant allele frequency is low. High DOC is, for example, essential in tumor sequencing since cellular heterogeneity reduces the variant frequency to levels as low as 10% [13,14].

Several NGS technologies are currently used in the clinical setting (Table 17.1), each of which has unique applications [15]. For instance, clinical genome or exome sequencing requires instruments that can provide 30X to 100X DOC. An exome, or protein-coding regions of the genome, represents approximately 1% of the human genome. The diploid genome contains 6 billion bases; therefore, 30X coverage of the genome requires instruments that can generate 180 Gb of sequence data. The standard range of coverage that is sought for human exome sequencing, conservatively 100X to 200X, requires a system that can deliver 6−12 Gb of sequence, although the needed DOC is higher for cancer testing (500X to 1000X) and lower for inherited disorder testing (30X to 100X).

An emerging innovation in sequencing technology is *single-molecule methodology*. This technology mitigates introduction of base variants during polymerase-driven DNA fragment amplification [16]. At present, the leading technology in the field is the Pacific Biosciences (Menlo Park, CA) RS system (Table 17.1), which offers the possibility of single DNA fragment read lengths of 20,000 bp at high levels of accuracy [17] (Table 17.1).

Gene Panel, Exome and Genome Sequencing Applications

One of the most important challenges of genomic medicine today is matching the scale of sequencing to the clinical application [13,18]. Genome and exome sequencing are relevant to a host of clinical applications, but the cost

TABLE 17.1 Current Clinical Sequencing Instrument Platforms and Performance Characteristics

Company	Sequencing Instrument	Sequencing Technology	Base Pair Yield/Run, Gbp	Mean Sequence Length, bp	$ Cost/Mbp
Illumina	MiSeq	Sequence by Synthesis (SbS) using reversible dye termination	15	150–250 PE*	0.13
	HiSeq 2500		600	50–100 PE	0.06
	NextSeq		120	150 PE	0.05
Life Technologies	Ion Torrent 318 v2 Chip	SbS using proton detection	2	200–400	0.71
	Ion Torrent Proton		10	200	0.15
	ABI SOLiD 5500 xl	Ligation base sequencing	200	75 PE	0.5
Roche	454 GS FlX	SbS pyrosequencing	0.7	600	8.5
	454 GS Junior		0.035	450	20
Pacific BioSciences	RS	Single-molecule SbS	0.04	1000–20,000 PE	3.5

*PE, capable of paired-end sequencing.

of sequencing and the time and resources needed to attain a clear description of the data are challenging. It is informative to compare and contrast genome and exome sequencing at this point. Both methods are best suited to applications where the genotype of a specific disorder or set of symptoms is not well characterized. The main attribute of genome sequencing is that the method is nearly comprehensive of genomic regions of potential clinical importance. Exome sequencing covers all protein-coding sequences, which currently are the genome regions understood to be responsible for specific disease states, but few if any noncoding regions that also are sites of disease-linked variants.

From an analytical view, the principal weakness of genome sequencing is the constraint on DOC in an attempt to contain costs below $5000. This is not as much an issue with germline testing for inherited disorders, but is problematic for cancer sequencing where lower frequency alleles may not be identified. In addition, the bioinformatics challenges are significant. Both of these drawbacks are very likely to fade in the near future as sequencing technology and bioinformatics tools become more robust and less expensive. One significant drawback that is not likely to be reversed anytime soon, however, is the difficulty with interpreting and annotating sequence variants of unknown biological or clinical significance.

Exome sequencing also is limited in identification of low frequency alleles with high confidence, but compared to genome sequencing, exome methods can achieve greater sequencing depth at lower overall cost. In addition, the pipelines for bioinformatics analysis of exome sequencing data are better developed and include commercialized software packages from vendors of exome capture kits. Finally, although variants that lie outside of exome-captured regions will be missed by this method, exome sequencing has been shown to have a diagnostic yield of as great as 25% in cases of unidentified inherited disorder testing [19–21].

It is important to note that many clinical applications of NGS cannot be effectively performed with sequencing methods that do not permit deep coverage. For example, genomic testing of cancer requires coverage as high as 1000X to provide confidence that very low frequency alleles ($\sim 5\%$), which are frequently encountered due to tissue and tumor heterogeneity in cancer specimens, are correctly identified. So it should not be surprising that exome or genome sequencing has not been adopted routinely in clinical oncology. Instead, targeted panels composed of genes with evidence-based clinical relevance have become common place in clinical cancer testing.

Sequencing targeted gene panels offer significant advantages over genome or exome sequencing. First, even panels composed of 50–100 genes can be covered at a depth greater than 1000X on a single run from sequencing instruments such as the Illumina (San Diego, CA) MiSeq and the Life Technologies (Carlsbad, CA) Ion Torrent PGM [13,14]. These sequencers are much less expensive than the instruments required for genome and exome sequencing, such as the Illumina HiSeq. High DOC, greater than 1000X, means that variants that are clinically relevant can be identified with 95% confidence limits in samples where the variant allele is present at a frequency of as little as 5%, such as cancer testing. A second important benefit of targeted sequencing is the reduced complexity of bioinformatic analysis and gene annotation relative to that which is required for exome and genome sequencing. In many instances, targeted panels are composed of genes deemed clinically relevant because they are associated with supportive medical literature, clinical trial activity, and/or FDA-approved therapies. A supportive knowledge base for genes in targeted panels has the potential to facilitate clinical interpretation of genetic variation and as the knowledge base grows, it has the potential to decrease variants of unknown clinical significance. ClinGen and ClinVar are two notable examples of public databases that are available for the support of clinical genomics [22,23].

An alternative to gene panel sequencing is amplicon sequencing. In this method genomic locations of interest are targeted for polymerase chain reaction (PCR) amplification on microarray platforms or by digital PCR. The amplified products, *amplicons*, are pooled and subjected to NGS. These amplicon capture methods have recently been reviewed and tested [24,25]. Amplicon methods limit results to targeted hotspots yet typically limit variants of unknown significance and the bioinformatics workload.

Steps in Clinical NGS Testing

NGS workflow within the clinical laboratory begins with proper specimen collection. The sample type should be appropriate for the application (Table 17.2) and appropriately qualified by staff or a pathologist, where applicable. Proper tissue block selection by a surgical pathologist is critical to ensure viable tumor with a high tumor to nontumor cell ratio. Also, tissue processing must be reviewed to ensure that processes consistent with DNA testing were used (see below).

DNA quality is determined by three factors. First, extracted DNA needs to be at or above the minimal fragment length for the library preparation that will be used, typically in the range of 100−300 bp. Second, because library preparation is an enzymatic process, samples must not contain inhibitors of enzymes needed to process the DNA. Fortunately most inhibitors are excluded by proper preparation of samples and DNA purification kits. Third, DNA damage can result from anatomic pathology preparatory methods used for morphologic diagnosis. Formalin-fixed, paraffin-embedded (FFPE) tissue commonly contains DNA that is cross-linked to cellular proteins. Fortunately, cross-linking can be reversed by heat treatment, which is now a common step in commercial FFPE DNA extraction kits [26]. DNA isolated from FFPE tissue is often extensively, and irreversibly, fragmented as a result of formalin fixation. This is a significant cause of quality control

TABLE 17.2 Applications of Genomic Medicine and Typical Sample Types

	Application	Typical Specimen Types
Oncology	Somatic mutation profiling	Tumor[b]
	Inherited cancer risk[a]	Blood[c]
	Minimal residual disease	Blood[d]
Constitutional	Inherited disease	Blood[e]
	Pharmacogenomics	Blood
	Prenatal testing	Blood, amniotic fluid[f], CVS[f]
	Newborn screening	Blood
	Identity testing	Bodily fluid, cells or tissue
Infectious disease	Diagnostic identification, prognostication	Varies depending on infectious agent and location of infection
	Quantitative, monitoring disease	

[a]*Can also be considered a constitutional subapplication.*
[b]*Tumor can be obtained from FFPE tissue or cells (cytology cell blocks), and cytology samples including cells from fine needle aspirate biopsies. Pathologic confirmation of tumor and adequate tumor quantity per requirements of the downstream genomic assay are critical to achieving success with this type of testing.*
[c]*Inherited cancer risk requires germline samples. This is classically obtained via peripheral blood samples but if blood is involved with the disease process, other samples, including but not limited to buccal swabs, may be used.*
[d]*Minimal residual disease testing is classically used in hematologic malignancy monitoring. This testing interrogates malignant cells in the peripheral blood.*
[e]*Inherited disease testing requires germline samples. Other sample types, including but not limited to buccal swabs, may be used instead of peripheral blood. Interpreting results from broad scale inherited disease testing such as exome or genome testing may be facilitated by testing a trio; Proband and both parents.*
[f]*The focus of prenatal testing is currently shifting from samples acquired by invasive means such as amniocentesis or chorionic villus sampling (CVS) to samples of cell-free fetal DNA in the maternal peripheral blood, so called noninvasive prenatal testing. The goal of obtaining amniotic fluid or samples of the placenta via CVS is to obtain DNA reflective of the fetus.*

failure for DNA extracted from FFPE tissue because it results in an insufficient quantity of fragments meeting the minimal fragment length required to proceed with downstream library preparation. DNA base damage is a more serious problem that can occur in tissue fixed in unbuffered formalin, where extended time at low pH can cause the formation of apurinic sites [27]. The latter is fortunately much less common in the clinical laboratory because most anatomic pathology laboratories use neutral buffered formalin.

Extraction of DNA from various sample types is routine in the clinical setting with many available commercialized solutions for this purpose.

Logistically, a laboratory will need to determine if manual or automated extraction better suits the testing workflow and volumes.

Three major forms of quality control may be employed to preclear samples for library preparation and downstream sequencing. First, gel electrophoresis methods are used to estimate the average length of DNA fragments. Second, spectroscopic absorbance measurements provide DNA yield and reveal the presence of contaminants that may adversely affect enzyme activity [28]. Third, the ability of DNA to be successfully amplified, a fundamental element in library preparation, can be assessed directly by real-time quantitative PCR (qPCR) amplification. By selecting a human reference gene and then designing nested amplicons of different length the range of sizes that can be amplified from the DNA extracted from FFPE tissue samples can be determined. Commercial kits are available to standardize this qPCR quality control step.

Library preparation can be a major source of sequencing errors in NGS data. While the library preparation method is unique to the sequencing instrument and the application, all library preparation methods have quantity input limits [24,29]. These limits are particularly relevant when analyzing samples that are composed of limited DNA quantity or compromised quality. Low-quantity samples run the greatest risk of yielding sequencing errors in large part due to the fact that they require greater PCR amplification. Base misincorporation by PCR enzymes makes a significant contribution to sequencing errors. Poor quality samples also run the risk of generating poor quality sequencing data by reducing the quantity of DNA that can be prepared into libraries for downstream NGS.

Different bioinformatics analysis methods are needed for identification of four classes of genomic variants: (1) single-nucleotide variants, (2) insertions and deletions, (3) structural alterations (e.g., translocations and gene fusions), and (4) copy number variations [30]. Traditionally, analyses of these variant types in a single clinical case would require use of different testing methods. The first two categories would typically be evaluated by sequencing. The latter two would be evaluated by cytogenetic methods such as fluorescence in situ hybridization or chromogenic in situ hybridization, or by molecular methods such as microarray technology or reverse transcription PCR (RT-PCR) assays in the case of transcript analysis of structural alterations. However, all these variants can be identified by NGS with different analytical methods for the sequencing data to identify these different classes of variants.

Other Applications of NGS

In addition to technologies directed at understanding changes at the DNA level, methods interrogating RNA expression for clinical use have emerged. RNA-based clinical assays are typically represented by expression profiles of genes associated with a particular disease state. For example, the OncoType DX test analyzes the expression of a panel of 21 genes by RT-qPCR with the readout predictive of the recurrence risk of breast cancer [31,32]. In addition, noncoding RNAs such as microRNA and long noncoding RNA expression are being linked to disease states and likely will be used in routine clinical testing in the future. The future of RNA-based testing is undoubtedly linked to NGS and recently prominent reviews of RNA sequencing technologies have been published [29,33].

Ensuring high-quality RNA is obtained from clinical samples for NGS testing can be difficult. As with DNA isolation from FFPE tissue, RNA isolated from these samples tends to be highly degraded. Fresh-frozen clinical samples are much preferred for RNA-based assays however current clinical models of tissue handling do not lend themselves well to this downstream sample type. It is also challenging to avoid RNA degradation during the isolation procedure since the enzymes that degrade RNA, ribonucleases, are ubiquitous.

New methods are emerging which seek to apply molecular diagnostic methods to analyze the very small amounts of cell-free DNA circulating in the bloodstream [34,35]. Clinical application of sequence analysis of cell-free circulating DNA has emerged for prenatal testing and evidence is mounting for clinical utility in tumor diagnostics. The degree to which circulating tumor DNA can replace the use of DNA obtained by tumor biopsy is still an open question, but the potential benefits are significant.

CLINICAL APPLICATIONS OF GENOMIC TESTING

The scope of genomic applications in the clinical setting is broad, but can be generalized into: (1) infectious disease, (2) constitutional or inherited conditions, and (3) oncology. Each of these categories contains additional subspecialized applications. The concept of differential applications is important because the application often dictates the optimal sample type and specifications (Table 17.2), as well as the approach to NGS testing as described above. If a sample is not composed of an adequate quantity of relevant genomic material, the test will not identify clinically relevant alterations that otherwise may exist in a qualified sample, thereby producing a false-negative result with potential for patient harm.

Downstream applications also have implications for test development and infrastructure design. To adequately implement genomic testing for multiple clinical applications, some level of customization based on the application should be expected. Some general elements for consideration in this process include assay design, local clinical practices, sample type and the related sample flow, patient consent and counseling, and reimbursement.

Development of a genomic medicine service for oncology includes implementation challenges common to all applications, but has the added complexity of the anatomic pathology specimen. Anatomic pathology cancer samples are precious and finite, and tissue processing can damage genomic material. Working with samples of suboptimal quantity and quality is a challenge for the genomic laboratory and coordinating the use of these samples with anatomic pathology adds complexity to a genomic testing workflow.

The effort to institute genomic testing to inform clinical care is ongoing. Retrospective review of individualized early adoption efforts revealed common implementation challenges [7]. This section will outline considerations from the viewpoint of the laboratory, the clinician, and the patient, to provide global perspective of what is required and what is to be considered when establishing a successful genomic medicine service.

Considerations for the Clinical Laboratory

The utility of a laboratory test is directly related to clinical use of the data to inform care. Developing and offering clinical genomic testing is resource intensive. Therefore, to avoid unnecessary time and effort, genomic testing should focus on applications that will inform clinical care. For the institution, launching a clinical genomic program is as much a cultural and political exercise as it is a scientific one [7]. A full-scale implementation plan requires effort beyond the laboratory validation for the assay. A complete business plan is usually required, as well as engagement of many stakeholders in the development of the testing applications to be implemented and the clinical pathways to ensure the appropriate use of the genomic information. Engagement of the institutional leadership and key political stakeholders early in this planning process is essential to gauge support and potential barriers to implementation such that they can be addressed in the planning stages.

As discussed earlier in the chapter, genomic medicine has several clinical applications. The implementation plan may focus solely on one application or it may outline how the laboratory intends to develop genomic testing over time for multiple applications. Each institution has its own strengths and areas of interest. It is wise to select a "pilot project" that aligns with the institutional vision and goals, and to engage with clinical champions of a new genomic approach. One of the greatest challenges identified in canvasing early adopters is the lack of appreciation by stakeholders outside of the laboratory (i.e., clinicians, institutional leadership, and payers) of the potential for genomics to improve patient care [7]. An implementation plan should include a firm scientific background to overcome cost concerns and a lack of institutional inertia that would prevent implementation of a genomic medicine program. At present, the compelling evidence supporting the "promise of genomic medicine" to improve patient care and the cost-effectiveness of clinical care is limited which remains a challenge to implementation [36].

Drafting a genomic testing plan requires many considerations. Smaller institutions with fewer resources may consider *partnering* and *outsourcing*. *Partnering* refers to establishing a collegial relationship with a group more experienced in clinical genomics. Early adopters are aware of the global benefit of increasing the number of groups performing high-quality clinical genomic testing. New adopters can model their genomic test development on the early adopter experience to expedite local implementation. *Outsourcing* the NGS sequence data generation and/or the bioinformatics needed to support clinical use of genomics is also a fundamental consideration because development and maintenance of clinically validated NGS testing and bioinformatics solutions require a significant investment of time and resources. Fortunately, many high-quality solutions are available for consideration making outsourcing a viable solution to this obstacle.

Early consideration of local clinical need is a fundamental step toward generating a validation plan. The scope of data made available by genomic testing is broad. Tailoring assays to meet expectations of clinicians will facilitate implementation of genomic testing and a genomic medicine program. For oncology, usually a targeted panel assay design is optimal to achieve a higher DOC. Considerations for the design of the cancer gene panel assay include the following issues. Should gene panels be developed according to tumor type (i.e., solid tumor, hematopoietic, sarcoma)? Should inherited cancer risk gene testing be performed by a separate inherited cancer risk panel that is developed with a companion informed consent process and ancillary supportive clinical services? Should the genes and coverage of a panel be customized for local practices or is a commercially designed assay acceptable? For a pharmacogenomics panel, should the assay be comprehensive or focus on genomic alterations that are specific to clinical scenarios such as pain management, cancer therapeutics, coagulation, or psychotherapy? For inherited disease, should the laboratory develop disease-specific gene panels, exome testing, or genome testing? If the latter two are preferred, what will be the process for addressing incidental findings and what is the approach for testing of minors? This list of considerations is not exhaustive but includes fundamental issues warranting discussion between laboratorians and clinicians.

Genomic technologies are still relatively new to most molecular pathologists and gathering information necessary to generate a validation plan may seem overwhelming. A validation plan will yield cost information central to the business plan. In an effort to develop clinical NGS testing, deciding which sequencer to acquire and the details of the assay to be developed can assist in determining additional equipment and reagent needs for assay development and performance and can facilitate outline of a draft workflow.

Determination of the sample types to be tested and the expected volume of testing will aid in determining the reagents required for validation and testing. Test validation and performance are often heavily dependent on regulatory requirements so it is essential to be informed in this regard. Excluding assays for inherited cancer risk, validation of oncology assays is complicated by the nature of anatomic pathology samples. Both FFPE tissue and various preparations of cytology samples should be considered for validation. As described above, both of these sample types come with technical challenges.

The laboratory should consider the sample types it may encounter clinically so that the same types of specimens can be analyzed during validation.

Drafting a detailed sample and testing workflow can be simplified into three main parts: test ordering, tracking, and resulting. "Ordering" warrants investigation of the current model of patient care and consideration to standardize a clinical genomic care pathway. The end goal is to ensure acquisition of high-quality samples in a fashion that does not delay or complicate clinical care and respects informed consent and patient autonomy. "Tracking" is commonly complicated because most laboratory information systems serving anatomic and clinical pathology are not designed to specifically track molecular/genomic workflows. When designing a genomic testing program, use of a laboratory informatics system specifically tailored to this type of testing is ideal, though may be costly.

"Reporting" as defined herein entails data interpretation and generation of final clinical reports. Data interpretation for most genomic-based assays generally requires what may be commonly referred to as a "dry bench" process. The "dry bench" is generally composed of sequence alignment, variant calling, variant annotation, and report generation for medical review. A laboratory must consider its resources and determine if solutions to these elements are best developed internally or if outsourcing is more economical. Increasingly, end-to-end solutions are becoming available on the market. Important considerations are the interface during assay validation between the wet bench and dry bench efforts and careful exploration of the variant annotation process and knowledge base that is relied upon for annotation and medical sign-out of results. In addition, seamless integration of information systems to avoid reconfiguration of complex genomic data is essential to generate comprehensive yet concise and accurate genomic reports for ease of use at the point of care.

An important consideration while ensuring fidelity of report format from the laboratory to the electronic health record (EHR) is searchability of the genomic data as it is critical for future reference in reporting and for research. A genomic medicine workflow will require many electronic interfaces of multiple systems supporting diverse users. The effort to create a seamless electronic workflow that supports efficiency in practice and optimizes patient safety is often underestimated, but should be investigated early in the implementation process. The National Human Genome Research Institute initiated the Clinical Sequencing Exploratory Research program (http://www.genome.gov/27546194) in 2010, which was extended in 2013, to support both the methods development needed for integration of clinical sequencing into patient care, as well as to study ethical, legal, and psychosocial issues of clinical genome sequencing.

Reimbursement is an additional consideration during efforts to implement clinical genomic testing. An institution must decide who will order testing and what group will take responsibility for pre/prior-authorization. There are a variety of reimbursement models and it is worthwhile to investigate different solutions to determine which will work locally. Successful reimbursement is a key element in creating a business plan that will garner support from stakeholders. As such, reimbursement is important to consider early during implementation planning.

Firm medical scientific evidence for the clinical usefulness of genomic test results will go a long way to convince stakeholders to support development and implementation of genomic medicine, yet supportive evidence currently is limited. To correct this limitation, evidence generation to understand the clinical usefulness of genomics in clinical care must be planned by those implementing genomic medicine programs in parallel with clinical practice. *Comparative effectiveness research* to assess the added value of genomics to clinical care and *implementation research* quantifying the impact of implementing genomic medicine programs on patient outcomes, healthcare costs, and clinician and patient satisfaction are needed [37]. Clinical evidence is also needed to determine which variants are actionable, in whom, and in what clinical situation. Information from programs practicing genomic medicine can generate an evidence base to support more widespread implementation and successful reimbursement [7].

Considerations for Patient Care Providers

The main challenge facing clinicians in integrating clinical genomics into practice is obtaining, interpreting, and managing results. Ideal reports are concise and include pertinent, current, supportive data for the clinical implications of a specific genomic test result. Development of a genomic medicine program is an opportunity to create a dedicated genomic medicine team of practitioners that can assist in clinical management that precedes and follows receipt of results. One important goal of implementation is to avoid impeding clinical workflow by having clinicians sift through and synthesize genomic information while discerning the clinical relevance of the results. An infrastructure item that can be leveraged to improve clinical care logistics is the electronic integration of the EHR with the laboratory information systems. Integration can facilitate efficient patient scheduling, and EHR status updates on the progress of the testing can decrease lost productivity to both the clinical and the laboratory staff.

Considerations for Patients

The patient perspective regarding genomics varies widely. Test menu development and informed consent can be tailored to safeguard patient autonomy. Tests and reporting can be designed so that results regarding germline information can be separated from non-germline information. A modular test menu can be achieved either with distinct panels or via bioinformatics solutions. Protection of genomic test results, consistent with protection of all healthcare information, is essential for patients to reduce patient fears regarding the potential misuse of their genomic information and to agree to genomic testing.

Genomic medicine programs must address care provider and patient education regarding testing and manage expectations accordingly. Including opportunities for counseling and informed consent in the clinical genomic care pathway workflow also solidifies a commitment to protect patient autonomy. A worthwhile consideration is to broaden standard informed consent procedures to address deposition of data into the EHR and/or research databases and collection of samples for biobanking and research.

Traditionally, the psychological impact of and expectations surrounding genomic testing has been addressed via genetic counseling and informed consent. It is predicted however that the wide use of genomic medicine services over the next decade may outstrip the supply of genetic counselors [38]. Therefore, investigation of innovative models of support and education are worthwhile [7]. Consideration should also be given to expansion of ancillary patient counseling services to include financial counseling. In the current and foreseeable future, reimbursement for genomic testing is uncertain, yet this is relatively expensive testing and patients should be adequately informed so they can compare potential benefits of obtaining genomic information to the out-of-pocket costs they may have to pay.

CONCLUSIONS

Since the completion of the Human Genome Project in 2003, the opportunities and applications of genomics in clinical care continue to broaden. Major technological advancements in parallel with decreasing costs of sequencing have permitted unprecedented access to genomic information. This has resulted in a massive knowledge base that continues to grow; especially as the clinical significance of non-protein-coding regions of the genome is being discovered. Understanding the medical significance of the genome however remains a challenge. Although most genomics work is pursued as a part of research programs, access to genomics for clinical care via CLIA-certified laboratories is increasing [39]. This chapter has reviewed the significant technological considerations underlying the genomic transformation of medicine. Common challenges to incorporating genomics into clinical care exist for the laboratory, the clinicians, and patients and this chapter has addressed major considerations for these various stakeholders.

REFERENCES

[1] Dulbecco R. A turning point in cancer research: sequencing the human genome. Science 1986;231:1055–6.

[2]   Collins FS. Shattuck lecture-medical and societal consequences of the Human Genome Project. N Engl J Med 1999;341:28–37.

[3]   Guttmacher AE, Collins FS. Genomic medicine—a primer. N Engl J Med 2002;347:1512–20.

[4]   Committee on Quality of Health Care in America, Institute of Medicine. Crossing the quality chasm: a new health system for the 21st century. Washington, DC: National Academy Press; 2001.

[5]   Salzberg SL, Pertea M. Do-it-yourself genetic testing. Genome Biol 2010;11:404.

[6]   Lupski JR, Reid JG, Gonzaga-Jauregui C, et al. Whole-genome sequencing in a patient with Charcot-Marie-Tooth neuropathy. N Engl J Med 2010;362:1181–91.

[7]   Manolio TA, Chisholm RL, Ozenberger B, et al. Implementing genomic medicine in the clinic: the future is here. Genet Med 2013;15:25–267.

[8]   Sanger F, Nicklen S, Coulson AR. DNA sequencing with chain-terminating inhibitors. Proc Natl Acad Sci USA 1977;74:5463–7.

[9]   International Human Genome Consortium. Initial sequencing and analysis of the human genome. Nature 2001;409:860–921.

[10]   International Human Genome Consortium. Finishing the euchromatic sequence of the human genome. Nature 2004;431:931–45.

[11]   Venter JC, Adams MD, Myers EW, et al. The sequence of the human genome. Science 2001;291:304–51.

[12]   Metzker ML. Sequencing technologies—the next generation. Nat Rev Genet 2010;11:31–46.

[13]   Hagemann IS, Cottrell CE, Lockwood CM. Design of targeted, capture-based, next generation sequencing tests for precision cancer therapy. Cancer Genet 2013;206:420–31.

[14]   Cottrell CE, Hussam AK, Bredemeyer AJ, et al. Validation of a next-generation sequencing assay for clinical molecular oncology. J Mol Diagn 2014;16:89–105.

[15]   Loman NJ, Misra RV, Dallman TJ, et al. Performance comparison of benchtop high-throughput sequencing platforms. Nat Biotech 2012;30:434–9.

[16]   Harris TD, Buzby PR, Babcock H, et al. Single-molecule DNA sequencing of a viral genome. Science 2008;320:106–9.

[17]   Eid J, Fehr A, Gray J, et al. Real-time DNA sequencing from single polymerase molecules. Science 2009;323:133–8.

[18]   Pfeifer JD. Clinical next generation sequencing in cancer. Cancer Genet 2014;206:409–12.

[19]   Yang Y, Muzny DM, Reid JG, et al. Clinical whole-exome sequencing for the diagnosis of Mendelian disorders. N Engl J Med 2013;369:1502–11.

[20]   Lee H, Deignan JL, Dorrani N, et al. Clinical exome sequencing for genetic identification of rare Mendelian disorders. JAMA 2014;312:1880–7.

[21]   Yang Y, Muzny DM, Xia F, et al. Molecular findings among patients referred for clinical whole-exome sequencing. JAMA 2014;312–1870.

[22]   Landrum MJ, Lee JM, Benson M, et al. ClinVar: public archive of interpretations of clinically relevant variants. Nucleic Acids Res 2016;44:D862–8.

[23]   Rehm HL, Berg JS, Brooks LD, et al. ClinGen—the clinical genome resource. N Engl J Med 2015;372(23):2235–42.

[24]   Chang F, Li MM. Clinical application of amplicon-based next-generation sequencing in cancer. Cancer Genet 2014;206:413–19.

[25]   Samorodnitsky E, Datta J, Jewell BM, et al. Comparison of custom capture for targeted next-generation DNA sequencing. J Mol Diagn 2015;17:64–75.

[26]   Paireder S, Werner B, Bailer J, et al. Comparison of protocols for DNA extraction from long-term preserved formalin fixed tissues. Anal Biochem 2013;439:152–60.

[27] Srinivasan M, Sedmak D, Jewell S. Effect of fixatives and tissue processing on the content and integrity of nucleic acids. Am J Pathol 2002;161:1961−71.

[28] Simbolo M, Gottardi M, Corbo V, et al. DNA qualification workflow for next generation sequencing of histopathological samples. PLoS One 2013;8:e62692.

[29] Li S, Tighe SW, Nicolet CM, et al. Multi-platform assessment of transcriptome profiling using RNA-seq in the ABRF next-generation sequencing study. Nat Biotechnol 2014;32:915−25.

[30] Abel HJ, Duncavage EJ. Detection of structural DNA variation from next generation sequencing data: a review of informatic approaches. Cancer Genet 2014;20:432−40.

[31] Cronin M, Sangli C, Liu ML, et al. Analytical validation of the Oncotype DX genomic diagnostic test for recurrence prognosis and therapeutic response prediction in node-negative, estrogen receptor-positive breast cancer. Clin Chem 2007;53:1084−91.

[32] Clark-Langone KM, Sangli C, Krishnakumar J, Watson D. Translating tumor biology into personalized treatment planning: analytical performance characteristics of the Oncotype DX Colon Cancer Assay. BMC Cancer 2010;10:691−702.

[33] Van Keuren-Jensen K, Keats JJ, Craig DW. Bringing RNA-seq closer to the clinic. Nat Biotechnol 2014;32:884−5.

[34] Fan HC, Gu W, Wang J, et al. Non-invasive prenatal measurement of the fetal genome. Nature 2012;487:320−4.

[35] Bettegowda C, Sausen M, Leary RJ, et al. Detection of circulating tumor DNA in early- and late-stage human malignancies. Sci Transl Med 2014;6:224.

[36] Scheuner MT, Sieverding P, Shekelle PG. Delivery of genomic medicine for common chronic adult diseases: a systematic review. JAMA 2008;299:1320−34.

[37] Eccles MP, Foy R, Sales A, et al. Implementation Science six years on—our evolving scope and common reasons for rejection without review. Implement Sci 2012;7:71.

[38] National Cancer Institute. Research-tested intervention programs. Available from: http://rtips.cancer.gov/rtips/programDetails.do?programId=277363.

[39] Gudgeon JM, Williams JL, Burt RW, et al. Lynch syndrome screening implementation: business analysis by a healthcare system. Am J Manag Care 2011;17:e288−300.

Chapter 18

Molecular Genetic Testing and the Future of Clinical Genomics[1]

S.H. Katsanis[1] and N. Katsanis[2]

[1]*Duke University, Durham, NC, United States,* [2]*Duke University Medical Center, Durham, NC, United States*

Chapter Outline

INTRODUCTION

The growing need for translation of genetic and genomic discoveries to health care has fueled a massive expansion of clinical genetic testing. New technologies are evolving to detect genetic variation in a patient, ranging from a single point mutation to a comprehensive atlas of rare and common variants. As providers, scientists, and policy advisors grapple with how to interpret and handle the onslaught of data, established and well-validated molecular technologies continue to play an important role in the quest for single disease-causing mutations. At the same time, the pace of technological development in genomics is rapid, leaving in its wake a series of policy and implementation challenges whose solutions will undoubtedly shape the field of genomic medicine.

---

1. A version of this chapter was first published in Nature Reviews Genetics in 2013, http://dx. doi.org/10.1038/nrg3493 by Nature Publishing Group [8].

Genomic and Precision Medicine. DOI: http://dx.doi.org/10.1016/B978-0-12-800681-8.00018-9

Genomics is transforming diagnostics. Genetic and genomic tests based on DNA are ordered across specialties from pediatrics through oncology. Regardless of the indication, the tools used for a clinical molecular genetic test to determine human variation are common—whether assessing a rare disease cause or assaying common genomic variants associated with a complex disorder or drug response.

Awareness and interest, coupled with the overwhelming amount of scientific discovery and frequent announcements of genomics and precision medicine discoveries have created a climate in which more testing is available; more practitioners are calling on molecular testing to aid them in routine practice; and more patients are requesting testing for family planning or risk assessment in their relatives. Fueling this growth is an increase in both the volume and detail of knowledge surrounding the genome and the expansion from diagnostic testing for single-gene disorders to predictive testing for complex, though more common, disorders.

This chapter describes the tools and applications of molecular genetics for detection of germline genome variation (i.e., excluding somatic variation). Distinct from research-based assays for variation, when ordering a clinical test, standards must be met to ensure accuracy of diagnostics. Laboratories testing for diagnosis, treatment, or management of a disorder must undergo clinical certification to ensure analytic validity, and some regulatory requirements evaluate clinical validity.

BASIC MOLECULAR TOOLS, TECHNOLOGIES, AND APPLICATIONS

A genetic test may take on many forms: a test may be genomic, interrogating dozens, hundreds, or thousands of genetic loci at once; or a genetic test may be specific to a particular variant. Chapter 16, Pharmacogenetics and Pharmacogenomics (Chung) outlines some of the ways that genomic information may be translatable to clinical care. In many scenarios, that translation represents the development of a genetic test that is applicable to clinical diagnosis, treatment, or management of a patient. Genetic testing applications span medical disciplines (e.g., pediatrics, cancer, pharmacology) and are applicable at all ages from embryo to adult (Table 18.1). Prenatal testing for inherited conditions is one of the more common genetic testing applications today, including testing for common trisomies (e.g., Down syndrome), which is now available through noninvasive methods [1]. Certain factors are considered before testing, including considerations for children and prenatal or embryonic testing [2].

In genomic medicine, the diagnosis of disease typically relies on a combination of clinical features and specific changes in biochemical and/or cellular markers. Due to the ease in which DNA can be assayed, it can serve as a robust disease marker. Indeed, genetic testing complements, and in some cases replaces, other methods of diagnosing single-gene disorders. The molecular genetic testing market, once viewed as a low value, small market

TABLE 18.1 Factors Considered in Selecting a Genetic Test

Description		Example	Preimplantation Genetic Diagnosis — Embryo/Blastocyst	Prenatal Testing — Fetus	Child	Adult
Newborn screening	Targeted tests for recessive genetic disorders	Phenylketonuria, cystic fibrosis, sickle cell anemia	NA	NA	Tests provided at birth vary by country and state/region	NA
Diagnostic testing	Confirmatory test or differential diagnosis testing for a symptomatic individual	Skeletal dysplasias, thalassemias, craniosynostoses	Specimen type and limited available amount for sampling may restrict platform selection (e.g., genome sequencing vs SNP/STR typing)	Turnaround time necessary may restrict platform selection	Where treatment is desired, turnaround time may restrict platform selection	
Carrier testing	Targeted testing for asymptomatic individuals potentially carrying one or more recessive mutation	Cystic fibrosis, thalassemias, Tay Sachs	Applied typically for rare disease, but applicable for other familial mutations		Carrier testing of minors is considered in context of individual pediatric cases	As indicated by clinical guidelines
Predictive testing	Tests for variants causing or associated with diseases/disorders with a hereditary component, usually with adult-onset symptoms	Most cancers, cardiovascular disease, diabetes	Some have discouraged genetic testing of asymptomatic minors for adult-onset conditions			As indicated by clinical guidelines

(Continued)

TABLE 18.1 (Continued)

| Description | | Example | Preimplantation Genetic Diagnosis | Prenatal Testing | Child | Adult |
			Embryo/ Blastocyst	Fetus		
Presymptomatic testing	Tests for variants causing or associated with diseases/disorders known to be inherited in the family, often with adult-onset symptoms	Huntington disease, hemochromatosis, Alzheimer disease	Some have discouraged genetic testing of asymptomatic minors for adult-onset conditions			As indicated by clinical guidelines
			Interpretation of VUS will depend upon presenting phenotypes in the family			
Pharmacogenetics	Targeted tests for variants associated with pharmaceutical dosage choice or adverse events	DNA tests for abacavir, warfarin, carbamazepine	Application not currently conducted, but theoretically feasible	Application not currently conducted, but conceivably applicable for screening treatment approaches in utero	Pharmacogenetic testing is considered in context of individual pediatric cases	As indicated by clinical guidelines

Adapted from Katsanis SH, Katsanis N. Molecular genetic testing and the future of clinical genomics. Nat Rev Genet 2013;14(6):415—426.

industry, is now seen as an industry with significant market and growth potential. Molecular genetic testing laboratories for diagnostics usually are CLIA-certified, but may be direct-to-consumer or research-based, and can be roughly divided into the following categories:

- *Commercial laboratories*: These laboratories tend to offer testing for a broad range of disorders including rare and common diseases. The tests offered may be laboratory-developed tests or kit-based tests.
- *Academic laboratories*: These laboratories remain affiliated with an academic institution or hospital. Many offer a narrow focus and specialize in molecular testing or testing (via a number of means) for a subset of disorders; a few offer a broad panel of molecular diagnostic services.
- *Hospital laboratories*: These laboratories operate within the health system as point-of-care services or within the hospital system. These laboratories tend to offer a few standard tests such as Factor V Leiden or kit-based tests that are broadly applicable and widely used.

In the United States, the Genetic Testing Registry (www.ncbi.nlm.nih.gov/gtr) provides guidance on available clinical genetic or genomic tests with limited information on the validity of the tests [3,4]. In Europe, EuroGentest (www.eurogentest.org) serves a similar function, also providing clinical utility "gene cards" that describe general best practice for particular disorders [5,6]. GeneReviews (www.ncbi.nlm.nih.gov/books/NBK1116) hosts peer-reviewed synopses on genetic conditions and includes test validity [7]. These resources and others enable evaluation of genetic tests in terms of the analytic validity, clinical validity, and clinical utility of the assays. Analytic validity is a measure of the ability of the test to detect a genetic/genomic variant and the clinical validity refers to the test's ability to predict the presence or absence of a clinical condition (Table 18.2) [8].

Diagnostic testing to confirm or rule out a genetic disorder in a symptomatic individual is one area that has quickly evolved in the last few decades from Southern blots in the 1980s, to polymerase chain reaction (PCR) screening in the 1990s, to gene sequencing in the current century. Genome and exome sequencing has emerged as the platform to detect a minute molecular change as a first-pass tool for diagnosing a genetic disorder [9]. This transition to sequencing genomes at ever-decreasing costs undoubtedly will change the future of diagnostics across all of medicine. But the basic molecular methodologies—hybridization, amplification, and Sanger sequencing—have been altered only in scale over the decades (Table 18.3). Most clinical genetic tests are focused on identifying patients' underlying pathogenic mechanisms for a disorder or trait of interest.

Sanger Sequencing

To evaluate the genome at the nucleotide level, the best approach is to sequence directly the region in question. This is typically done first by

TABLE 18.2 Evaluating Validity of Genetic Tests

Term		Definition	Complications in Molecular Tests	Calculation
Analytic validity Refers the accuracy and precision of the assay used for the test	Analytic sensitivity	Refers to the proportion of assays with the genotype who have a positive test result (false-negative rate of the assay)	• Allele drop out • Preferential amplification • Mosaicism	$\dfrac{\text{True positives}}{\text{True positives + False negatives}}$
	Analytic specificity	Refers to the proportion of assays without the genotype who have a negative test result (false-positive rate of the assay)		$\dfrac{\text{True negatives}}{\text{True negatives + False positives}}$
Clinical validity Refers to the accuracy with which a test predicts the presence or absence of a clinical condition or predisposition	Clinical sensitivity	Refers to the proportion of people with disease who have a positive test result (false negative rate of diagnosis)	• Variable penetrance • Variable expressivity	$\dfrac{\text{True positives}}{\text{True positives + False negatives}}$
	Clinical specificity	Refers to the proportion of people without disease who have a negative test result (false positive rate of diagnosis)		$\dfrac{\text{True negatives}}{\text{True negatives + False positives}}$
	Positive predictive value	Refers to the likelihood that a patient has the disease given that the test result is positive		$\dfrac{\text{True positives}}{\text{True positives + False positives}}$
	Negative predictive value	Refers to the likelihood that a patient does not have the disease given that the test result is negative		$\dfrac{\text{True negatives}}{\text{True negatives + False negatives}}$
Clinical utility		Refers to the value of the test for determining treatment, patient management, and family planning	• Depends upon health care system and environment	Subjectively determined based on reports supporting use and economic benefits
Personal utility		Refers to the value of the test for personal and family choices	• Depends upon personal vantage	Subjectively determined from an individual's perspective

Adapted from Katsanis SH, Katsanis N. Molecular genetic testing and the future of clinical genomics. Nat Rev Genet 2013;14(6):415–426.

TABLE 18.3 Clinical Genetic Testing Methodologies

Method	Common Point Mutations	Rare Point Mutations	Copy Number Variations	Uniparental Disomy[a]	Balanced Inversions/Translocations	Repeat Expansions	Primary Applications	Analytic Sensitivity/Specificity	Turnaround Time	Cost[b]
Sanger gene sequencing	X	X					Confirmation of genetic diagnosis or variant found through genome sequencing/screening (e.g., Treacher Collins syndrome diagnosis)	High	Average–high	Average
Target mutation detection (PCR)	X	X[c]				X	Screening population/batches for a particular genotype (e.g., cystic fibrosis carrier testing)	High	Low	Low
FISH			X		X		Sub-cytogenetic variation (e.g., Angelman syndrome)	Low	Low	Low
Array CGH			X	X			Macro/micro-cytogenetic variation (e.g., a new referral or challenging diagnostic case)	Average	Average	Average
Southern blot/MLPA			X			X	Sub-chromosomal rearrangements (e.g., Fragile X syndrome)	High	High	Low
Genome-wide SNP microarrays	X		X				Screening population/batches for thousands of genotypes (e.g., cardiovascular disease risk assessment)	Low	Low	Low

(Continued)

TABLE 18.3 (Continued)

Method	Common Point Mutations	Rare Point Mutations	Copy Number Variations	Uniparental Disomy[a]	Balanced Inversions/Translocations	Repeat Expansions	Primary Applications	Analytic Sensitivity/Specificity	Turnaround Time	Cost[b]
Panel/pathway sequencing	X	X					Assessing known variants in candidate genes for a particular condition (e.g., long QT syndrome)	Average/low	Average	Average
Linkage analysis (STRs/SNPs)	X		X[c]				Efficient indirect testing of a variant in a family	Low	Low	Low
Genome/exome sequencing	X	X	X[d]				A new referral or challenging diagnostic case	Low	High	High
Key: Categorical assignments in this row are subjective and vary according to context of the tests being ordered and the laboratory conducting the tests. The "low," "average," and "high" are presented to simplify and compare platforms generally								*low: <80%, average: 80–98%, high: >98%*	*low: within a week, average: 1 week–1 month, high: >1 month*	*low: <$400, average: $400–$2000, high: >$2000*

[a] UPD can be detected by any method if both parents are genotyped. However, only the indicated approaches will detect UPD in absence of the parental genetic samples.
[b] Costs of the testing will vary widely from one laboratory to the next; however, these estimates are based on the charge of the test from a sampling of laboratories, not on the costs of consumables or the reimbursed amount.
[c] Familial mutations or genomic rearrangements can be assayed.
[d] CNV detections are improving in NGS, but are more efficient in genome than exome sequencing, and are of limited reliability for clinical diagnostics.
Adapted from Katsanis SH, Katsanis N. Molecular genetic testing and the future of clinical genomics. Nat Rev Genet 2013;14(6):415–426.

amplification (via PCR) of a fragment of DNA, usually in the range of 500 bp. Subsequently, the amplified DNA fragment is sequenced in both directions (5' to 3') using fluorescently labeled nucleotides. The resulting sequence can be read using a fluorescent detection machine through capillary electrophoresis. Bidirectional sequencing and variations on this approach, validated for a number of years, have rendered Sanger sequencing the "gold standard" for clinical diagnostics. This direct approach has high analytic validity, though long reads can deteriorate quality for base-calling and minute specimen may produce PCR artifacts [10].

The value of direct Sanger sequencing of a particular exon of a gene or small region of a genome is the ability to combine a clinical indication for a candidate gene with a high detection rate for point mutations and small variants [8]. It is applied ideally to clinical cases with a strong candidate and in genetic disorders with few causative genes. For example, direct sequencing of the fibroblast growth receptor genes for a suspected skeletal dysplasia is a swift, accurate, and inexpensive way to confirm highly penetrant causative variants for achondroplasia [11].

Given that the Sanger sequencing process is labor intensive and low throughput, and that the cost of interrogating each nucleotide has remained high (and stable) in recent years, primarily because of rising personnel and administrative overhead costs, Sanger sequencing is impractical and inefficient for testing many genes, or when a strong candidate is not clear. In these cases, massively parallel sequencing may be a better approach to narrow the clinical testing to specific candidate regions of interest. When broader genome-wide testing with lower specificity is used as a first-pass screening tool (e.g., exome sequencing, microarrays), targeted Sanger sequence analysis is used to confirm a mutation site. Despite the high sensitivity, however, Sanger sequencing cannot detect structural changes including chromosome rearrangements and copy number variations (CNVs), so it is insufficient for diagnosis for many genetic disorders [8].

Targeted Mutations

Many of the commonly used molecular tests (e.g., Factor V Leiden test for thrombosis, many pharmacogenetic variants) are simple mutation detection (or genotyping) assays, designed to target a particular variant for a particular trait. This approach is most valuable for interrogating one particular nucleotide of interest for screening a large number of samples. For example, a particular pharmacogenetic variant may be tested on multiple individuals at once efficiently via a genotyping assay. Throughput of multiple samples via PCR and detection of a variation (e.g., fluorescent detection, gel separation) offers high confidence to detect variants in a laboratory conducting the same test in high volume. In addition, a direct PCR assay may be used to test common disease-causing repeat expansions, such as those in myotonic dystrophy or Fragile

X syndrome, which may be tested by direct amplification of the repeated fragment [12].

Structural and Chromosomal Variation

While detecting nucleotide-level changes are important, many genomic changes on the macro-scale require cytogenetic approaches to detect CNV and genomic rearrangements. The resolution of cytogenetic variation has improved significantly with innovation in chemistry and microscopy, most notably through the development of multi-probe fluorescent *in situ* hybridization (FISH) and comparative genomic hybridization (CGH). Gradually, traditional cytogenetic methods of banding and FISH are being phased out in the clinic in favor of array CGH, which utilizes probes to detect chromosomal and genomic rearrangements and deletions with greater precision than FISH [8]. Depending on design and probe density, array CGH can offer resolution from whole chromosomes to deletions and duplications a few kilobases in size [8,13]. Array CGH can detect rearrangements (except balanced translocations) and uniparental disomy (UPD) with high sensitivity. On the other hand, the specificity is limited: A number of submicroscopic genomic rearrangements have been found that hold unclear significance to the clinical phenotype of a patient. Databases cataloguing such CNV are being developed, such as DatabasE of Chromosomal Imbalance and Phenotype in Humans using Ensembl Resources (DECIPHER, decipher.sanger.ac.uk) and International Standards for Cytogenomic Arrays Consortium (ISCA Consortium, www.iscaconsortium.org) [14,15].

The forefather in detecting submicroscopic genomic rearrangements, the Southern blot, also is undergoing a renaissance. Southern blots involve the restriction fragmentation of a genome, blotting of the fragments run on a gel, and hybridization of the regions of interest to the blot. Once widely used in combination with restriction fragment length polymorphisms (RFLPs) for molecular diagnoses, "Southerns" continue to be used to detect small genetic changes as well as large repeat variants not amenable to PCR amplification (e.g., trinucleotide repeat expansions) [12]. In recent years, however, multiplex ligation-dependent probe amplification (MLPA) assays have replaced Southerns for some applications. MLPA can detect CNVs, mosaic mutations, methylation status and can be used to confirm structural anomalies detected by FISH or CGH [16,17]. In most cases MLPA cannot detect balanced genomic rearrangements such as translocations or inversions, which is a significant limitation given how common these events are in human genetic disease [18].

Microarrays

Microarray-based genotyping can be divided into three main applications: array CGH to detect structural anomalies (see discussion above), phenotype-specific microarray panels, and genome-wide panels of single-nucleotide polymorphisms (SNPs) [8]. Phenotype-specific panels containing SNP alleles of interest for specific phenotypes, such as panels for retinal degeneration, may be used in clinical testing for a condition with multiple causative or associated genes [19,20]. This approach is more efficient than Sanger sequencing in interrogating multiple genes and inexpensive in comparison to genome sequencing. However, the continuous discovery of novel causal alleles and genes, as well as variable penetrance and expressivity of known mutations, limits the ability of this approach for finding causative alleles [8].

Large-scale genome-wide SNP genotyping offers a single cost-efficient platform to assess thousands of variants, often risk variants for common genetic disorders [21]. This platform allows for predictive and presymptomatic testing as a multiplex platform for a host of conditions, including certain cancers and pharmacogenetic tests, as well as for ophthalmologic, cardiac, renal, and neurological disorders (among others) [8]. Particular loci may be evaluated with high analytic validity, but the limited scope of variant detection in current SNP panels confines analysis to predefined points in the genome. Most SNP-based diagnostics are probabilistic, not deterministic, with variable degrees of clinical validity, since arrays identify a limited range of variants [8].

Indirect Testing

Indirect molecular tests are those that test for variation in the genome near or linked to a causal variant, which is applied most typically to test for familial variants that are difficult to directly test. Despite the emergence of new technologies to interrogate causal variants in a patient, indirect methodologies continue to play a role in diagnostics, primarily for the diminished costs over direct methods [8]. Indirect methods such as linkage analysis using SNPs and short tandem repeats (STRs) are appropriate for certain genetic anomalies or to work up pedigrees. Techniques including denaturing gradient gel electrophoresis and single-strand conformation polymorphism that have been mostly phased out in the United States and Europe are still utilized in developing regions with limited resources [22,23].

Genome Sequencing

Building upon the Sanger sequencing technology, next-generation sequencing (NGS) employs powerful massively parallel assays to efficiently sequence many genes of interest, the whole exome, or the whole genome. Targeted exon capture prior to genome sequencing enables analysis of the majority of the coding regions of the genome (the exome), comprising about 1% of the human genome. Whole-genome sequencing refers to the evaluation of almost all of the euchromatic human genome, excluding the heterochromatic regions that are too repeat-dense to reliably read using current technologies. Exome sequencing is a fast and accurate discovery approach for some mutations causing single-gene disorders, and the clinical cost is well below that of Sanger sequencing of some larger genes, making exome sequencing economically the best choice as a first run when a genetic cause is unknown [24–27].

Clinical genome and exome sequencing is available from multiple clinical laboratories and is expanding rapidly. However, interpretation of exome results is limited to the variants that are clearly causative or clinically actionable. Below we discuss the challenges and considerations for interpreting genomic variants identified through genome sequencing. Clinicians at major academic centers routinely counsel for and order exome sequencing for unexplained genetic disorders and to order genome sequencing for cancer diagnostics and other indications.

Cost considerations notwithstanding, the primary practical barrier to use of genome sequencing in clinical settings is the limited ability of the technology to reliably detect the absence or presence of mutations [8]. Different NGS platforms have been shown to deliver results of variable quality, with some instruments more accurate at individual base calls and others covering a broader range of the genome [28,29]. Targeted exome approaches, for example, restrict the clinical sensitivity in comparison to the greater depth of coverage of genome sequencing, which ultimately limits the interpretive scope. Currently, it is not possible to obtain high-quality sequence from the entire human genome, or even the euchromatic genome, sufficient for exhaustive clinical interpretation [8].

Further, to parse sequencing data efficiently, genome sequencing is best applied to not just the proband, but also both unaffected parents in a trio to efficiently ascertain *de novo* (new) and inherited mutations under limited information with regard to mode of inheritance. For efficiency, many clinical exome sequencing services sequence only the proband and confirm variants of interest in the parents. In either case, access to the biological relatives remains valuable in interpretation of genetic variation. To refine further the vast amounts of data obtained from a single sequencing run, confirming the integrity of a mutation detected by NGS by Sanger sequencing in probands and family members is typical. However, there remain acute interpretive problems that are dependent on the scope of the initial genome analysis.

VARIANT INTERPRETATION

In the past, genetic data largely served a confirmatory role in the clinic, rather than driving diagnosis. The knowledge of pathogenic lesions typically leads to population-based arguments about possible patient outcomes. A major challenge in personalizing medicine is to convert pathogenic genetic data into a primary diagnostic tool that, in combination with clinical observation and biometric data, can shape clinical decisions and long-term management in a proactive way. While most emerging clinical genome sequencing paradigms are focused on phenotypes in order to probe its genetic architecture in detail, a broader approach will contribute meaningfully to what is a core question to precision medicine: Should we sequence every patient admitted to a hospital and if so, how do we interpret those data to maximize clinical utility?

Causal Disease Variants

In a clinical diagnostic setting, detected variants are assigned to one of six categories of results:

1. *Pathogenic*: Sequence variation is previously reported and is a recognized cause of the disorder (e.g., *CFTR* delF508).
2. *Likely pathogenic*: Sequence variation is previously unreported and is of the type which is expected to cause the disorder (e.g., *de novo*, nonsense).
3. *Likely benign*: Sequence variation is previously unreported and is probably not causative of disease.
4. *Benign*: Sequence variation is previously reported and is a recognized neutral variant.
5. *Uncertain significance*: Sequence variation is not known or expected to be causative of disease, but is found to be associated with a clinical presentation [30].

Assigning one of these categories of variants is subject to interpretation based on prior cases of particular variants, population frequencies, clinical findings, and possibly case-specific research data.

While these categories are valuable for labeling causative variants, many genome variants are important to human function but are neither disease-causing nor benign. A variant may be considered protective from disease or related to drug response. Or a variant may alter protein function but not lead to disease.

The challenge of interpreting sparsely documented genetic mutations is not a new concept applicable only to genome sequencing; rather, interpreting variants of uncertain significance (VUS) in a clinical setting has been a challenge since the outset of molecular diagnostics. The nonuniformity of interpretive criteria is exemplified by the fact that, in some instances, individual laboratories responsible for assigning significance of a molecular finding are based largely on experience of that laboratory in the gene of interest. Especially in rare disease testing, a laboratory specializing in testing a particular gene or syndrome will hold institutional history on incidences of rare unpublished variants [8]. The sheer volume of data from genome sequencing requires a sophisticated and transparent exchange of variants associated with phenotypes or clinical indications. Disease-centric mutation databases have morphed into human disease variant databases valuable for documenting clinical variation, such as the Human Gene Mutation Database (HGMD, www.hgmd.cf.ac.uk), and the hand-curated database ClinVar (www.ncbi. nlm.nih.gov/clinvar) [31]. The fragmentation of historic mutation databases necessitates a concentrated effort to develop standards for interpreting variation and consistent and informative reporting. The variation database pilots likely will expand into the broader, focused exchange necessary to facilitate interpretation of both rare and common variation across varying platforms and laboratories around the world [8].

Variants of Uncertain Significance

A human genome harbors thousands of variants, many of them *de novo* and hundreds that likely alter gene function [32]. Research laboratories use a host of genetic filters to assess and prioritize variants for a particular condition. For alleles previously not associated with human pathology or for which there is limited biological insight, *in silico* prediction algorithms (such as Polyphen, VAAST, ESEfinder) are commonly applied and in some cases incorporated into clinical reports [33−36]. Many of the informatics-based filters are of modest predictive value, so research laboratories rely upon a combination of informatics, literature searches, and *in vivo* modeling to interpret each variant. Laboratories also may enrich for specific variants, analyze multiple family members, examine concordance in computational algorithms, and parse the morbid human and mouse genomes for variation [35,37]. Rare alleles may be compared to human disease gene and model organism databases [38−41]. A clinical laboratory, however, may not have access to the same tools or systematic processes to evaluate tens, hundreds, or thousands of variants in a proband.

Even in the context of a single-gene disorder, variant interpretation is problematic since a significant fraction of alleles have poor predictive value and modifier alleles can have profound phenotypic effects. A research laboratory may deploy physiologically relevant functional assays to assess the biological effect of a novel variant on the protein and its relationship to the phenotype [42−44]. Such tools exist for disorders of mitochondrial function and a handful of other conditions [45,46], but such approaches in the research realm are not easily translated into standard clinical laboratory protocols. Not only are these assays expensive, labor intensive, difficult to automate, and can be challenging to interpret in the context of human mutation, but also they are outside the scope of existing regulatory guidelines [8].

As genome sequencing enters the clinical realm, we must work out how to communicate relevant findings to best inform clinical practice. Genetic variants that appear to precipitate a phenotype may also depend on environmental factors, modifier genes, epigenetics, and the additive/synergistic effects from multiple variants [47]. Even simple genetic test results can be misunderstood in clinical translation. Thus, communicating complex genomic results with a range of interpretations is challenging. If large-scale genomics is to succeed at translating precision medicine into the clinic, it is incumbent upon all stakeholders to develop appropriate and flexible approaches for long-term interpretation, review, and communication of clinically relevant genetic information. Open and longitudinal data sharing is an essential component for ensuring maximum benefit from genome sequencing data for patient management.

PATIENT ACCESS TO GENETIC TESTS

Genome and exome sequencing have appropriately generated substantial debate in the medical community with regard to the delivery and impact of the information on clinicians, patients, and society in general. Decreasing costs of and increased accessibility to the tools of biomedical research and individual health information are challenging the traditional paradigms of patient care. Until recently, physician and patient information exchange has been asymmetrical, if not paternalistic: Patients are expected to adhere to regimens prescribed by a physician. However, it is clear that people with Internet access will seek medical information online, refuting the idea that patients want or need only a small amount of information or nothing more than a prescriptive regimen [48]. We also know that the rise of crowd-sourced patient Web sites (e.g., www.PatientsLikeMe.com) fulfills a need that is not otherwise being met by the traditional health care system [49,50].

As large-scale data are developed on individual patients, how to handle the data for long-term patient management is undetermined. Perhaps only patients should hold sequence variants or perhaps variants should go into electronic medical records. Clinical laboratories are obligated to maintain raw data from genetic tests in addition to the test results, but perhaps they also have an obligation to provide said data to patients requesting the data [51]. Sequencing results also may identify mutations relevant to medical care (e.g., breast cancer, Alzheimer's) that is not the focus of the original test requisition [52]. Clinical professional guidelines instruct clinical laboratories to return any secondary findings found in particular genes [53]. Studies measuring distress in individuals learning their genetic risk of Alzheimer's disease demonstrated minimal harm in disclosing unexpected results [54]. Ultimately, the duty to inform patients of predictable risks could be influenced by the legal pressure and threat of malpractice [55,56].

Whether to test for or disclose genotypes for adult-onset disorders to a minor child and/or her parents requires a delicate balance, weighing the benefit of protecting information to foster autonomous decision making against the value in allowing parents to assess risks and benefits specific to their child [57]. In any case, failure to disclose genetic information that may be actionable could have dire consequences both to the child and the parties withholding the information.

CONCLUSIONS

Genetic and genomic tests for micro- or macro-molecular variation is integral to bringing genomic variation research discoveries to precision diagnostics and treatment. The molecular tools are evolving rapidly, particularly with the decreasing costs of genome analysis, but retain some core tenants for clinical application. Regardless of the assay used for detecting a given variant, interpreting variants in terms of the presenting condition of the patient is not only the bottleneck but is also key to devising the genetic architecture comprising a person's condition. Each human carries thousands of variants that define their uniqueness—the challenge in precision medicine is determining which are causative of a condition, which are contributing to a condition, which are secondary to the condition, and which are benign.

GLOSSARY TERMS, ACRONYMS, AND ABBREVIATIONS

Array comparative genomic hybridization (array CGH) A microarray-based method of identifying differences in DNA copy number by comparing a sampled genome to a reference genome.

Copy number variation (CNV) A structural genomic variation that results in copy number changes in a specific chromosomal region. Usually, there are two copies of each locus, but if, for example, duplications or triplications occur the number of copies will increase.

Denaturing gradient gel electrophoresis (DGGE) An electrophoresis assay that uses a chemical gradient to denature the sample as it moves across an acrylamide gel to separate molecular fragments.

Epigenomics Describes a heritable effect on chromosome or gene function that is not accompanied by a change in DNA sequence, but rather by modifications of chromatin or DNA.

Exome The collection of protein-coding regions (exons) in the genome. This is the portion of the genome that is translated into proteins. As exons comprise only 1% of the genome and contain the most easily understood, functionally relevant information, sequencing of only the exome is an efficient method of identifying many variants that are likely to affect a trait.

Fluorescent *in situ* hybridization (FISH) A molecular and cytogenetic method using a fluorescently labeled DNA probe to detect a particular chromosome or gene using fluorescence microscopy.

Laboratory-developed test (LDT) A type of *in vitro* diagnostic test that is designed, developed, and used within a single laboratory.

Large-insert clones Large haplotype fragments that are inserted into, for example, bacterial artificial chromosomes.

Linkage analysis A statistical method for identifying a region of the genome that is implicated in a trait by observing which region is inherited from the parental strain carrying the trait in offspring that carry the trait.

Multiplex ligation-dependent probe amplification (MLPA) A molecular technique involving the ligation of two adjacent annealing oligonucleotides followed by quantitative PCR amplification of the ligated products, allowing the detection of deletions, duplications, and trisomies and characterization of chromosomal aberrations in copy number or sequence and SNP or mutation detection.

Next-generation DNA sequencing (NGS) Non-Sanger-based high-throughput DNA sequencing technologies. Compared to Sanger sequencing, NGS platforms sequence as many as billions of DNA strands in parallel, yielding substantially more throughput and minimizing the need for the fragment-cloning methods that are often used in Sanger sequencing of genomes.

Noninvasive prenatal diagnosis Method of obtaining a prenatal diagnosis by detecting fetal cells circulating in maternal blood.

Oligonucleotide arrays Hybridization of a nucleic acid sample to a very large set of oligonucleotide probes, which are attached to a solid support, to determine sequence, to detect variations, or for gene expression or mapping.

Penetrance The proportion of individuals with a given genotype who display a particular phenotype.

Preimplantation genetic diagnosis (PGD) *In vitro* method of identifying genetic defects in *in vitro* fertilization embryos prior to maternal transfer and implantation.

Restriction fragment length polymorphisms (RFLPs) Variations between individuals in the lengths of DNA regions that are cut by a particular endonuclease.

Sanger DNA sequencing A method used to determine the nucleotides present in a fragment of DNA. It is based on the chain-terminator method developed by Frederick Sanger, but currently uses labeling of the chain-terminator dideoxynucleotides, which allows sequencing in a single reaction.

Short tandem repeat (STR) A DNA sequence containing a variable number of highly polymorphic, tandemly repeated short (2−6 bp) sequences.

Single-strand conformation polymorphism (SSCP) An electrophoresis method to detect variants by separating amplified single-stranded molecules in a non-denaturing polyacrylamide gel.

Single-nucleotide polymorphism (SNP) A difference in the nucleotide composition at a single position in the DNA sequence.

Uniparental disomy (UPD) Inheritance of both homologues of a pair of chromosomes from only one parent.

Variants of uncertain significance (VUS) An alteration in the sequence of a gene, the significance of which is unclear.

REFERENCES

[1] Allison M. Genomic testing reaches into the womb. Nat Biotechnol 2013;31(7):595−601.

[2] Duncan RE, Savulescu J, Gillam L, Williamson R, Delatycki MB. An international survey of predictive genetic testing in children for adult onset conditions. Genet Med 2005;7(6):390−6.

[3] Javitt G, Katsanis S, Scott J, Hudson K. Developing the blueprint for a genetic testing registry. Public Health Genomics 2010;13(2):95−105.

[4] Rubinstein WS, Maglott DR, Lee JM, Kattman BL, Malheiro AJ, Ovetsky M, et al. The NIH genetic testing registry: a new, centralized database of genetic tests to enable access to comprehensive information and improve transparency. Nucleic Acids Res 2013;41 (Database issue):D925−35.

[5] Dierking A, Schmidtke J. The future of Clinical Utility Gene Cards in the context of next-generation sequencing diagnostic panels. Eur J Hum Genet 2014;22(11):1247.

[6] Javaher P, Kaariainen H, Kristoffersson U, Nippert I, Sequeiros J, Zimmern R, et al. EuroGentest: DNA-based testing for heritable disorders in Europe. Community Genet 2008;11(2):75−120.

[7] Pagon RA, Adam MP, Ardinger HH, Bird TD, Dolan CR, Fong C-T, et al. GeneReviews®, University of Washington, Seattle; 1993−2016.

[8] Katsanis SH, Katsanis N. Molecular genetic testing and the future of clinical genomics. Nat Rev Genet 2013;14(6):415−26.

[9] Pasche B, Absher D. Whole-genome sequencing: a step closer to personalized medicine. JAMA 2011;305(15):1596−7.

[10] SenGupta DJ, Cookson BT. SeqSharp: a general approach for improving cycle-sequencing that facilitates a robust one-step combined amplification and sequencing method. J Mol Diagn 2010;12(3):272−7.

[11] Krakow D, Rimoin DL. The skeletal dysplasias. Genet Med 2010;12(6):327−41.

[12] Wallace AJ. Detection of unstable trinucleotide repeats. Methods Mol Med 1996;5:37−62.

[13] Fruhman G, Van den Veyver IB. Applications of array comparative genomic hybridization in obstetrics. Obstet Gynecol Clin North Am 2010;37(1):71−85.

[14] Swaminathan GJ, Bragin E, Chatzimichali EA, Corpas M, Bevan AP, Wright CF, et al. DECIPHER: web-based, community resource for clinical interpretation of rare variants in developmental disorders. Hum Mol Genet 2012;21(R1):R37−44.

[15] Wapner RJ, Martin CL, Levy B, Ballif BC, Eng CM, Zachary JM, et al. Chromosomal microarray versus karyotyping for prenatal diagnosis. N Engl J Med 2012;367 (23):2175−84.

[16] Hills A, Ahn JW, Donaghue C, Thomas H, Mann K, Ogilvie CM. MLPA for confirmation of array CGH results and determination of inheritance. Mol Cytogenet 2010;3:19.

[17] Kozlowski P, Jasinska AJ, Kwiatkowski DJ. New applications and developments in the use of multiplex ligation-dependent probe amplification. Electrophoresis 2008;29 (23):4627−36.

[18] Talkowski ME, Rosenfeld JA, Blumenthal I, Pillalamarri V, Chiang C, Heilbut A, et al. Sequencing chromosomal abnormalities reveals neurodevelopmental loci that confer risk across diagnostic boundaries. Cell 2012;149(3):525−37.

[19] Hageman GS, Gehrs K, Lejnine S, Bansal AT, Deangelis MM, Guymer RH, et al. Clinical validation of a genetic model to estimate the risk of developing choroidal neovascular age-related macular degeneration. Hum Genomics 2011;5(5):420−40.

[20] Zanke B, Hawken S, Carter R, Chow D. A genetic approach to stratification of risk for age-related macular degeneration. Can J Ophthalmol 2010;45(1):22−7.

[21] Schaaf CP, Wiszniewska J, Beaudet AL. Copy number and SNP arrays in clinical diagnostics. Annu Rev Genomics Hum Genet 2011;12:25−51.

[22] Massaro JD, Wiezel CE, Muniz YC, Rego EM, de Oliveira LC, Mendes-Junior CT, et al. Analysis of five polymorphic DNA markers for indirect genetic diagnosis of haemophilia A in the Brazilian population. Haemophilia 2011;17(5):e936−43.

[23] Pereira Fdos S, Matte U, Habekost CT, de Castilhos RM, El Husny AS, Lourenco CM, et al. Mutations, clinical findings and survival estimates in South American patients with X-linked adrenoleukodystrophy. PLoS One 2012;7(3):e34195.

[24] Need AC, Shashi V, Hitomi Y, Schoch K, Shianna KV, McDonald MT, et al. Clinical application of exome sequencing in undiagnosed genetic conditions. J Med Genet 2012;49 (6):353−61.

[25] Nguyen MT, Charlebois K. The clinical utility of whole-exome sequencing in the context of rare diseases—the changing tides of medical practice. Clin Genet 2014;88(4):313−19.

[26] Park JY, Kricka LJ, Fortina P. Next-generation sequencing in the clinic. Nat Biotechnol 2013;31(11):990−2.

[27] Yang Y, Muzny DM, Reid JG, Bainbridge MN, Willis A, Ward PA, et al. Clinical whole-exome sequencing for the diagnosis of Mendelian disorders. Nq Engl J Med 2013;369(16): 1502−11.

[28] Quail M, Smith ME, Coupland P, Otto TD, Harris SR, Connor TR, et al. A tale of three next generation sequencing platforms: comparison of Ion torrent, pacific biosciences and illumina MiSeq sequencers. BMC Genomics 2012;13(1):341.

[29] Sulonen AM, Ellonen P, Almusa H, Lepisto M, Eldfors S, Hannula S, et al. Comparison of solution-based exome capture methods for next generation sequencing. Genome Biol 2011;12(9):R94.

[30] Richards S, Aziz N, Bale S, Bick D, Das S, Gastier-Foster J, et al. Standards and guidelines for the interpretation of sequence variants: a joint consensus recommendation of the American College of Medical Genetics and Genomics and the Association for Molecular Pathology. Genet Med 2015;17(5):405−24.

[31] Stenson PD, Mort M, Ball EV, Howells K, Phillips AD, Thomas NS, et al. The Human Gene Mutation Database: 2008 update. Genome Med 2009;1(1):13.

[32] MacArthur DG, Balasubramanian S, Frankish A, Huang N, Morris J, Walter K, et al. A systematic survey of loss-of-function variants in human protein-coding genes. Science 2012;335(6070):823−8.

[33] Houdayer C, Dehainault C, Mattler C, Michaux D, Caux-Moncoutier V, Pages-Berhouet S, et al. Evaluation of in silico splice tools for decision-making in molecular diagnosis. Hum Mutat 2008;29(7):975−82.

[34] Cartegni L, Wang J, Zhu Z, Zhang MQ, Krainer AR. ESEfinder: a web resource to identify exonic splicing enhancers. Nucleic Acids Res 2003;31(13):3568−71.

[35] Adzhubei IA, Schmidt S, Peshkin L, Ramensky VE, Gerasimova A, Bork P, et al. A method and server for predicting damaging missense mutations. Nat Methods 2010;7(4):248−9.

[36] Yandell M, Huff C, Hu H, Singleton M, Moore B, Xing J, et al. A probabilistic disease-gene finder for personal genomes. Genome Res 2011;21(9):1529−42.

[37] Kumar P, Henikoff S, Ng PC. Predicting the effects of coding non-synonymous variants on protein function using the SIFT algorithm. Nat Protoc 2009;4(7):1073−81.

[38] Amberger J, Bocchini C, Hamosh A. A new face and new challenges for Online Mendelian Inheritance in Man (OMIM(R)). Hum Mutat 2011;32(5):564−7.

[39] Blake JA, Bult CJ, Kadin JA, Richardson JE, Eppig JT. The Mouse Genome Database (MGD): premier model organism resource for mammalian genomics and genetics. Nucleic Acids Res 2011;39(Database issue):D842−8.

[40] Robinson PN, Kohler S, Bauer S, Seelow D, Horn D, Mundlos S. The Human Phenotype Ontology: a tool for annotating and analyzing human hereditary disease. Am J Hum Genet 2008;83(5):610−15.

[41] Sprague J, Bayraktaroglu L, Clements D, Conlin T, Fashena D, Frazer K, et al. The Zebrafish Information Network: the zebrafish model organism database. Nucleic Acids Res 2006;34(Database issue):D581−5.

[42] Merveille AC, Davis EE, Becker-Heck A, Legendre M, Amirav I, Bataille G, et al. CCDC39 is required for assembly of inner dynein arms and the dynein regulatory complex and for normal ciliary motility in humans and dogs. Nat Genet 2011;43(1):72−8.

[43] Rosenthal N, Brown S. The mouse ascending: perspectives for human-disease models. Nat Cell Biol 2007;9(9):993−9.

[44] Siddiqui SS, Loganathan S, Krishnaswamy S, Faoro L, Jagadeeswaran R, Salgia R. *C. elegans* as a model organism for in vivo screening in cancer: effects of human c-Met in lung cancer affect *C. elegans* vulva phenotypes. Cancer Biol Ther 2008;7(6):856−63.

[45] Pelak K, Shianna KV, Ge D, Maia JM, Zhu M, Smith JP, et al. The characterization of twenty sequenced human genomes. PLoS Genet 2010;6:9.

[46] Zaghloul NA, Liu Y, Gerdes JM, Gascue C, Oh EC, Leitch CC, et al. Functional analyses of variants reveal a significant role for dominant negative and common alleles in oligo-genic Bardet-Biedl syndrome. Proc Natl Acad Sci USA 2010;107(23):10602−7.

[47] Majewski J, Schwartzentruber J, Lalonde E, Montpetit A, Jabado N. What can exome sequencing do for you? J Med Genet 2011;48(9):580−9.

[48] van Uden-Kraan CF, Drossaert CH, Taal E, Smit WM, Moens HJ, Siesling S, et al. Health-related Internet use by patients with somatic diseases: frequency of use and charac-teristics of users. Inform Health Soc Care 2009;34(1):18−29.

[49] Swan M. Crowdsourced health research studies: an important emerging complement to clinical trials in the public health research ecosystem. J Med Internet Res 2012;14(2):e46.

[50] Wicks P, Massagli M, Frost J, Brownstein C, Okun S, Vaughan T, et al. Sharing health data for better outcomes on PatientsLikeMe. J Med Internet Res 2010;12(2):e19.

[51] Evans BJ, Dorschner MO, Burke W, Jarvik GP. Regulatory changes raise troubling ques-tions for genomic testing. Genet Med 2014;16(11):799−803.

[52] Jarvik GP, Amendola LM, Berg JS, Brothers K, Clayton EW, Chung W, et al. Return of genomic results to research participants: the floor, the ceiling, and the choices in between. Am J Hum Genet 2014;94(6):818−26.

[53] Green RC, Berg JS, Grody WW, Kalia SS, Korf BR, Martin CL, et al. ACMG recommen-dations for reporting of incidental findings in clinical exome and genome sequencing. Genet Med 2013;15(7):565−74.

[54] Green RC, Roberts JS, Cupples LA, Relkin NR, Whitehouse PJ, Brown T, et al. Disclosure of APOE genotype for risk of Alzheimer's disease. N Engl J Med 2009;361(3):245−54.

[55] Marchant GE, RMilligan RJ, Wilhelmi B. Legal pressures and incentives for personalized medicine. Pers Med 2006;3(4):391−7.

[56] McGuire AL, Knoppers BM, Zawati MH, Clayton EW. Can I be sued for that? Liability risk and the disclosure of clinically significant genetic research findings. Genome Res 2014;24(5):719−23.

[57] Robertson S, Savulescu J. Is there a case in favour of predictive genetic testing in young children? Bioethics 2001;15(1):26−49.

Chapter 19

Bringing Genomics to Medicine: Ethical, Policy, and Social Considerations

Laura Lyman Rodriguez and Elyse Galloway
National Human Genome Research Institute, National Institutes of Health, Bethesda, MD, United States

Chapter Outline

INTRODUCTION

... [the] complex world of genetic and genomic research, where torrents of information have meaning that may or may not be established or widely accepted and that rotates on an axis of incomplete public policies and regulations, is not susceptible to easy, linear solutions... [1]

The above quote is focused on ethics and policy dilemmas faced within the genomics research context, but it is equally applicable to the realm of genomic medicine. The scientific progress and technological feats accomplished during these early days of the "genome era" have been extraordinary and empowering. Innovative research teams have generated a map of the structures within the genome that control gene expression, a rich catalog of common and rare human genome variants, and through their strong data sharing practices have created broadly accessible and growing data repositories that already contain genomic data for over one million individuals [2]. Today, multidisciplinary research teams are adding to these resources by building a living dictionary of the functional and clinical significance for individual genomic variants. Through these and other resources, the power

Genomic and Precision Medicine. DOI: http://dx.doi.org/10.1016/B978-0-12-800681-8.00019-0

of genomics to understand health and disease has become an engine for research and discovery.

This genomic potential is entering the clinical realm with increasing regularity as well [3]. In the United Kingdom, for example, the National Health Service England is exploring how to integrate genomic medicine into healthcare delivery and has proposed to reorganize the clinical infrastructure supporting genomic laboratory services to accommodate the growing clinical potential [4]. As the science advances and the clinical possibilities grow, we are at a challenging juncture in the evolution of genomic medicine. The potential to improve care is real, but the knowledge about how to do so is still emerging, and the systems and standards to support implementation are not yet developed, or in some cases even conceived.

As genomic medicine transitions from discovery to knowledge application, it is as important to pursue considerations about the ethics, policy, and social research questions posed by its realization as it is to advance the science. This chapter will explore some of the high-level issues influencing the progression from concept to practice, but other chapters in this volume also explore a few of the complex challenges in greater depth.

THE NATURE OF GENOMIC DATA AND GENOMIC EXCEPTIONALISM

Genomic technology is a powerful tool with great potential to inform clinical care, but the information generated pertains not only to individuals, but to families, groups, and ancestral populations. In light of this broad context, consideration of approaches to manage privacy and use of genomic data should include individual perspectives and preferences layered with any cultural significance and the potential for group risks and benefits [5]. Furthermore, genomic information, with important caveats, is stable and does not vary with time. However, the interpretation of genomic information is evolving rapidly as more is learned through research, and also while genetic information can be definitive for a diagnosis or provide a marker through which to track heritable disorders, in most cases (e.g., for common diseases such as cancer or asthma), genomic information will represent a predictive tool to inform practitioners about patient susceptibility to disease [6]. As we gain a deeper understanding of the genomic bases for disease, the predictive quality of genomic information might become as important to some as its diagnostic qualities are to others [7]. The differential meaning of genetic information, and differences in how individuals or cultures might interpret or feel about the information, contribute to the sensitive nature of the data and the array of opinions about how to respect, manage, and safeguard it.

Throughout the short history of the genomics field, there has been great debate as to whether or not genetic and genomic information are *exceptional* and require different treatment relative to other health or personal information. Proponents of this approach argue that genetic/genomic information by virtue of its uniqueness to the individual and its conveyance of future health risks is substantively different from other information about an individual and must be protected through laws and policies specific to these characteristics [8]. In contrast, opponents of this approach assert that genomic information is no different than other health information and should be managed and used by researchers and health care providers within the same frameworks established to oversee other health information [9]. The challenge is that both perspectives, and the range of perspectives in-between, have merit.

It is important to keep in mind that for nearly all the rationales for treating genetic and genomic information differently, there are examples of other medical data types that pose similar risks for patients [9]. Therefore, the potential utility of genomic data to inform medical decision making should not be unduly constrained by these concerns, but rather respectfully and appropriately managed. Recent discussions about genetic exceptionalism have offered an analogy to how physicians manage the necessary, but sensitive, need to physically examine their disrobed patients [10]. This analogy highlights the fact that the genome is not merely a quantitative measure, but also conveys information that is descriptive and revealing of vulnerabilities, including those that are shared among families. Similar to the special respect that is afforded to patients when physicians view their naked bodies, genomic data should be obtained, analyzed, and managed intentionally and with clear informed consent from patients.

Many patients are concerned about the confidentiality of their genetic and genomic data, and the potential for its misuse [11]. From a legal perspective, genetic and genomic data are often covered by state health privacy laws, resulting in a complicated patchwork of protections that can confuse patients more than assure them that the confidentiality of their data will be protected [12]. The Genetic Information Nondiscrimination Act (GINA) of 2008 is a federal law that carves out nuanced protections for the appropriate use of genetic information in health insurance and employment settings and provides a baseline of consistent privacy protections [13]. While some contend that this propagates exceptionalism, an alternative view is that GINA provides an acknowledgment within a broad area of law that genetic information is complex, representing both physical data and informational context, and should be managed appropriately across all of its dimensions. Indeed, GINA's provisions are implemented largely through amending other laws, such as the Health Insurance Portability and Accountability Act, which protects health privacy [14]. Genomic information is unique to an individual, it is sensitive due to its multidimensional context, but it is not exceptional.

CONSIDERING RACE AND ANCESTRY

The relationship between race and genomics research is, not surprisingly, fraught with sensitivities. While race may represent some biological information about ancestry, it is actually a complex product of social, biologic, and environmental factors [15]. Socioeconomic status, education, health behavior, clinical characteristics, and social disparities may be associated with race and the differences in the incidence of diseases and their outcomes. As such, the clinical application of genomic information across our ethnically diverse population raises important considerations and delicate nuances for practitioners to reflect upon.

In 2005, BiDil became the first FDA-approved therapeutic to include race within the drug label [16]. However, important questions remain about the generalizability of claims about race in drug safety or efficacy. The concept of race and ethnicity as a reliable biological surrogate has been well refuted [17]. There are known differences in frequencies of some genetic variants among ancestral populations, but there are also known to be substantial differences among variant frequencies within certain subpopulations, in some instances more than those observed between populations [18]. A major hurdle for genomic medicine going forward is the relative dearth of genomic studies that include non-European populations [19]. This gap in our knowledge base creates a critical weakness in our ability to assess the validity of genomic contributors to disease outside of European ancestry populations [20]. To address this, it is imperative that research studies identify methods to recruit more diverse study populations, both in targeted efforts to over-represent groups known to have differential health outcomes for certain disease (such as African-Americans with high blood pressure) [21], as well as broader inclusion of diversity within any genomic health study.

In clinical practice, there is contention among physicians about the use of race in healthcare decision making and the potential for assumptions about genomic differences among racial groups to contribute to existing health disparities [21]. Some assert that along with the context of pyscho-socioeconomic information, there is benefit to including race in the assessment of a patient's presentation. Likewise, the inclusion of genetic information in assessments of observed group differences is important in cases where health outcomes persist despite controlling for socioeconomic status, access to care, and other nongenetic factors [18]. Studies indicate that physicians, regardless of their own race, view indicators such as medical history and family history as the most important information in medical decision making, but that when physicians experience anxiety with regard to clinical uncertainty, they are more likely to rely on race as a factor to consider [22,23]. The effect of this on health outcomes for patients is not known. Future work to further characterize how physicians consider race and ethnicity in formulating clinical decisions, and how genetic information might inform any biases that may be introduced, will be important to pursue.

Given the potential for genomic medicine to fuel a transformation toward precision medicine across clinical domains, it is important that clinicians appreciate the context that race and ancestry can provide for a patient, but recognize the need to be discerning in implementing genomic findings within their patient populations. Misuse or overinterpretation of race and ancestry in medical applications of genomic information could diminish the value of and severely undermine the public's confidence in genomic medicine.

PATIENT EMPOWERMENT

Disruptive innovations in genomic technologies, in particular low-cost DNA sequencing, are thrusting genomic information into the clinical realm at breakneck speed [24]. The same genomic technologies that physicians and other health professionals are considering as cost-effective options to inform clinical decisions have similarly penetrated the consumer market. Direct-to-consumer (DTC) companies such as 23andMe and Ancestry.com have brought access to genomic sequence data to the individual patient or curious member of the public. While both companies have strong research arms, their consumer-based focus has changed the context in which interested individuals view their own genomic information and, importantly, their desire to understand and make decisions (health or otherwise) based upon the information they receive from these companies. The resulting democratization of access to genome sequence information empowers individuals to consider the potential implications of their specific variants for their health, even at a time when the available information is limited and often not yet clinically validated.

Concurrent with the rise of DTC genomic data providers, there is a paradigm shift happening in how patients (and research participants) wish to engage with information about themselves. Patient organizations and citizen science initiatives alike have enabled and recruited participants to "patient-powered" research and clinical trials. In many cases, patients receive information back from these studies on an individual and an aggregate level, empowering their ability to seek out clinical options through research and to identify with patient communities. Indeed, there are compelling examples of motivated patients catapulting forward the definition and characterization of their diseases, and the search for diagnostic or therapeutic options through social media and access to genomic technology [25,26]. As instances of these patient-driven success stories continue to emerge, the public will become more comfortable with and confident in genomic medicine, and expect their providers to be as well.

The rise in patient–participant driven access to genomic information offers exciting possibilities for research and clinical care [27]. However, there is concern about the readiness of health care providers and resources for clinical support systems to be able to integrate this type of information into care [28]. As discussed in subsequent chapters, the reality of these shifts in patient engagement will need to be addressed by clinical infrastructures and educational resources to support both patients and providers if genomic medicine is to become a trusted component within our clinical tool box.

EVOLVING OVERSIGHT STRUCTURES

Genomics research has been closely linked to robust technology development since the Human Genome Project. The so-called "next-generation sequencing" (NGS) technology resulted in a net decline of 400-fold in costs to produce an individual's full genome sequence between 2008 and 2014 [29]. As the potential for "affordable" clinical DNA sequencing platforms has become plausible, genomics has become one of the fastest growing sectors of the biotechnology marketplace [30]. During this burst of development many companies have come and gone, innovating across the genome sequencing pipeline to provide "services" from testing supplies and sequencing instruments, to clinical testing and data generation, to data analysis and interpretation services, and even solutions to manage the massive data storage challenge (Fig. 19.1). These developments inform, but are distinct and developing somewhat independently from the actual clinical encounter between patient and provider. This active "push" on the technology front generates a tension between the desire to wait for the broad uncertainty in the knowledge base, the relevant regulatory structures, and the available clinical support mechanisms to settle out and the recognition that the market is already moving genomics into the clinic [32]. Specific issues related to the regulation of genomic technologies are discussed in another chapter, but because this regulatory uncertainty is substantively affecting the way that this area of technology, and to a related degree medicine, develops [33] it is important to note here.

As genomic technologies move from research tool to clinical care, expectations change with regard to the regulatory controls overseeing their development and ensuring patient safety. In addition, the rapid evolution in genomic testing modalities is challenging the development of definitive (and reasonably stable) testing or device parameters needed to satisfy regulatory expectations within a single laboratory, let alone across the clinical laboratory community [34]. Add to that that each "test" produces up to three billion bits of data that convey potential health information, most of which are not relevant to the clinical question(s) at hand, and in most cases the evidence to support their clinical validity is still emerging. This reality of evolving technology and emerging clinical evidence renders it difficult, at best, to define standards for analytic quality, clinical validity, and in turn any consensus on appropriate standards through which to establish professional practice guidelines.

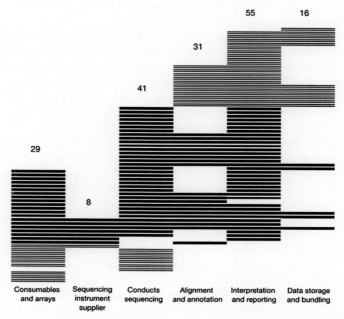

FIGURE 19.1   As NGS has come to the scientific forefront, an array of companies have contributed to the genome sequencing pipeline, providing "services," such as sequencing instrument supplies, data interpretation, and data management and storage. While the development of such services is extremely beneficial to the growth, implementation, and usage of NGS technologies, the fragmented state of the current system poses challenges to healthcare's application of genomic medicine [31].

Much of the current debate centers on the pathways through which to provide appropriate oversight without stifling the rapid innovation that is at the heart of genomic medicine. A primary concern is that FDA enforcement, by virtue of its need to strictly assess analytic and clinical validity, will "lock-down" genomic testing technologies, and that advances will stagnate [30]. It has also been questioned whether the FDA has jurisdiction in this area, or if the provision and interpretation of specific genomic analyses is part of the practice of medicine [35]. The reality is that at this juncture in our technology development and our understanding of genomic contributions to health and disease both might be true.

In the research realm, there are substantial efforts to bring the genomics community together around standard principles for genomics research that will advance clinical applications, as well as technical standards to promote "interoperability" across studies and facilitate data aggregation and sharing. Launched in 2013, the Global Alliance for Genomic Health, which included 374 organizational members from 36 countries representing academic, biotechnology, pharmaceutical, government, and patient perspectives as of 2014, has undertaken many of these efforts [36].

Community-driven approaches to create a framework by which translational and clinical genomics research can effectively proceed toward the realization of genomic medicine applications will serve as an important complement to government-based oversight that is focused on patient safety and efficacy of new treatments or devices brought to market. In the United States, FDA is actively seeking public input on how to accomplish this latter role in a manner that does not hamper on-going innovation or disrupt the pace of the fast-moving genomics arena [37]. The dialogs within the community and among the community and the responsible government entities will be important to monitor in the coming years, and will surely have ramifications for how genomic medicine technologies are developed in the future.

THE NEED FOR TRANSPARENCY

Despite the ambiguity in the oversight pathway for genomic testing technologies, the advent of high-throughput platforms for testing, the decline in NGS costs, and the proliferation of research data about thousands of individual genomic variants has resulted in a dramatic rise in the number of genetic and genomic tests on the market and an increase in the number of laboratories performing these tests. The allure of cutting-edge technology has also contributed to an emphasis on genomic services as a strategic business model, heightening the competition within the market. An outcome of these dynamics was a quagmire of information about available genetic tests with little organization, consistency, or depth to support clinicians and patients seeking information. The National Institutes of Health (NIH) Genetic Testing Registry (GTR) was established in 2011 to begin to meet this need [38]. As of 2015, GTR provided the public with information on over 25,000 individual tests. The information is voluntarily provided by the testing laboratories offering the services, and GTR further provides links to research information or other reports within the scientific literature about the registered tests [39]. The entry of GTR into the testing landscape provided an important measure of transparency to genetic testing services for the public.

Since 2013, NIH has launched two intitiatives to provide additional and requisite support for physicians and patients (and clinical laboratories) seeking to interpret genomic information. The first, ClinVar, provides a free, public archive of reported human variants along with any reported relationship with a phenotype and the supporting evidence available [40]. ClinVar does not provide any annotation of the assertions submitted, but rather makes transparent the information available about the sources of those submissions and of the claims made. A later complementary initiative, the Clinical Genomics Resource (or ClinGen), is in the early stages of building a knowledge base of variants, phenotypes and functional effects, and other clinical information to develop a consensus approach to determining the clinical relevance of genetic variants and disseminating that information to the public [41]. When operational, ClinGen will deposit all variant information it receives within ClinVar. The two programs have worked together to develop a rating system to indicate different quality levels for the validation information available about a given variant, as well as any relationship to observed phenotypes [42] (Fig. 19.2). ClinGen investigators are partnering with ClinVar in other ways as well [42], and will work with professional organizations to assist with any clinical practice guideline development, as requested. Such resources provide critical information sharing tools needed to accumulate the knowledge essential to inform clinical interpretations about a patient's given genetic variants and broader genomic background.

Despite these initiatives, market forces that focus on information control threaten the ultimate success of genomic medicine. Currently, some clinical laboratories use patients' test results to construct proprietary databases of human genomic variants and observed relationships to disease. An example where this is profoundly affecting the ability of the clinical laboratory community to provide equitable access to service for patients centers on variant data for the *BRCA1* and *BRCA2* genes [43]. Resource efforts such as

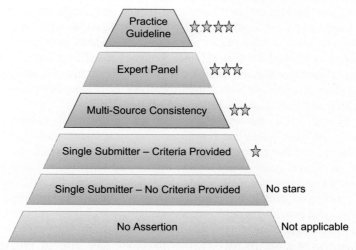

FIGURE 19.2   Variants submitted to ClinVar receive a star rating based on the evidence available to support its clinical validity. ClinGen and ClinVar investigators collaborated to develop the criteria for this system. This ranking system ranges from no stars (no criteria for published assertion provided) to four stars (professional practice guidelines issues pertaining to use of the variant in care). The ClinGen network, which includes more than 300 supporters, is also organizing expert panels in different "domains" (e.g., cardiovascular, metabolic) to assess the evidence available for variants with reported functional significance. Those variants listed in ClinVar that have received expert panel review (by ClinGen panels or other expert panels meeting established parameters) are eligible for a three-star rating in the resource.

ClinGen and ClinVar will stimulate and support transparency and promote broad access to critical information about particular variants observed in patients, and thereby diminish the number of variants of unknown significance (VUS). However, they cannot require data sharing from genetic testing laboratories, so the data within it will be incomplete. Grass-roots organizations, such as the *Free the Data* movement, are appealing directly to patients to share their testing information to improve the capacity for quality care for all through the creation of an openly accessible reference database [38]. Unfortunately, without a legal mandate for clinical laboratories to share this type of information, the time lost for patients whose physicians do not have access to critical information about the genomic variants they possess may bring a high price.

RETURN OF RESULTS

As the number of genes and gene variants with reported connections to disease has skyrocketed, the debate about the return of genomic information to individuals has intensified. This subject is the subject of other chapters in this volume, but there are substantive policy issues to be considered as well as the technical and clinical challenges. Much of the recent policy debate has focused on what an appropriate framework to return results should be, how to define an "incidental" or "secondary" result that might have potential health implications for a patient, and what the obligations (and liability) of physicians or clinical laboratories are to provide updates over time as new information is learned. Of course, since genomes are shared within families there are also important and complex questions about any potential obligations to share information with family members, or responsibilities to return information to families after a patient's death [44].

In March 2013, the American College of Medical Genetics and Genomics (ACMG) issued the first practice recommendations to tackle this issue, calling for the return of 57 (later adjusted to 56) specific incidental findings in clinical genome and exome sequencing [45]. Later that year, the President's Commission for the Study of Bioethical Issues (PCSBI) issued a report defining an ethical framework through which to consider these types of findings and provided specific recommendations for how to proceed in the context of research, clinical care, and DTC scenarios [46] (Table 19.1).

While the original ACMG recommendations took several provocative positions in their effort to advance the dialog, they have since modified their original proposal and are substantially aligned with the ethical construction for managing secondary results proposed by the PCSBI. Ongoing modification of the ACMG position is to be expected and is consistent with the rapid and dynamic changes in the available understanding of genomic information's clinical relevance. One of the more controversial aspects of the

TABLE 19.1 Recommendations for the Ethical Management of Incidental and Secondary Findings in the Clinical Context—Presidential Commission for the Study of Bioethical Issues

Informed Consent	Empirical Data	Clinical Judgment in Managing Incidental Findings
Clinicians should inform patients that incidental and secondary findings are potential results of the tests or procedures being conducted, engaging patients in shared decision making about the scope of findings that will be communicated and ultimately respecting patient preference.	Reliable comparative and cost-effectiveness analyses should be conducted by federal agencies and other interested parties to evaluate bundled tests versus sequential, discrete diagnostic tests, aiding in creating strong clinical practice guidelines, informing laboratory and payer practices and policies, and enhancing patient outcomes.	Medical educators must emphasize and cultivate a culture of thorough communication with patients, as well as a culture of professional judgment in ordering and conducting tests and interventions necessary for addressing health concerns related to their patients.
Decision aids and graphical representations should be used by clinicians to communicate difficult information about incidental and secondary findings and describe a patient's absolute risk.		Professional and public health organizations should provide guidance to clinicians based on standards for proposed screening programs to assist the management of incidental findings.

In the Presidential Commission for the Study of Bioethical Issue's 2013 report, *Anticipate and communicate: ethical management of incidental and secondary findings in the clinical, research, and direct-to-consumer contexts*, recommendations 6–10 addressed context-specific issues for the clinical arena [47].

ACMG recommendations, which has been retained through the updates issued to date, is the call to manage pediatric and adult patient secondary findings equivalently [48]. ACMG asserts that genome sequencing presents a distinct context from that of the traditional candidate gene predictive test models considered previously and argues that ignoring risk for potentially preventable harm to the child as an adult or to family members would not be in the best interest of the child [49]. Whether in the pediatric or adult context, the return of genomic results and secondary findings will be an important discussion to monitor over the next few years.

A new theme within the current debate is the change in patient/research participant/consumer expectations to learn their genome sequence data if it is generated (and the capacity to order it directly), and the growing appreciation for individual preferences and capacities to "handle" the uncertainty intrinsic to genomic information. Among the reasons that patients have provided to explain their interests in receiving genetic and genomic results is the notion of "personal utility" [50]. Personal utility can stem from just knowing the information, or it could pertain to the use of the information for other life decisions such as seeking long-term care insurance or modifying retirement plans, etc. There may be no action to be taken by a patient in response to genomic results, but for many being offered the choice to receive it holds substantial value and is an important demonstration of respect [31].

Given the complexities of interpreting genetic test information, the pace of genomic discoveries, and the number of VUS identified another policy challenge is the lack of clarity about any responsibility for tracking updates to a patient's genomic profile. Patients and their providers have a role, as do consulting medical geneticists and clinical laboratories [51]. Professional practice guidelines will be important to provide recommended standards for care and to define the respective roles and obligations. Going forward, it will also be critical to prioritize infrastructure needs such as electronic health records systems, clinical decision support, and provider education resources to enable quality patient care over time by all care givers.

CONCLUSIONS

As we explore how to apply genomic information and the suite of technologies that generate it for clinical purposes, "research" and "care" are amalgamating. This blending of purpose is being shaped not only by the science, but also by shifts in "consumer" demand to have access to the benefits of research technology in the clinic. It is especially important at this time for clinicians to engage in the full consideration of how genomic science is employed to improve patient care. Issues to anticipate range from the technical, e.g., what specific genome variants might or might not indicate about a given patient's risk for disease or response to treatment, to reflecting on the appropriate ethical framework for the management of genomic information relative to other health information, e.g., how and when to share genomic information with patients or family members. Looking ahead, new research initiatives will extend our knowledge base and our information resources to guide patient care. In fact, exciting new proposals, such as the Precision Medicine Initiative, will not only advance genomic medicine, but will integrate genomic medicine strategies with other developments in health care, including mobile health and nascent learning healthcare infrastructures [52]. With time, consensus will develop to define the evidence appropriate for using genomic information in clinical practice, standards for demonstrating safety and efficacy of genomic technologies will stabilize, and best practices for how to communicate with patients about the associated complexities will emerge. For now, the field must harness the promise of genomic technology to improve health, but it must do so while seeking a deliberative progression between research knowledge and clinical implementation that is always mindful of the individuals for whom clinical care is provided.

REFERENCES

[1]   Fisher R. GIM. 2012;14(4).

[2]   DbGaP-NCBI. National Center for Biotechnology Information. U.S. National Library of Medicine. Available from: http://www.ncbi.nlm.nih.gov/gap; 2015.

[3]   Manolio TA, Chisholm RL, Ozenberger B, Roden DM, Williams MS, Wilson R, et al. Implementing genomic medicine in the clinic: the future is here. Genet Med 2013;15: 258−67.

[4]   Rabesandratana T. U.K.'s 100,000 genomes project gets £300 million to finish the job by 2017. American Association for the Advancement of Science; 2014. Available from: http://news.sciencemag.org/biology/2014/08/u-k-s-100000-genomes-project-gets-300-million-finish-job-2017.

[5]   McEwen JE, Boyer JT, Sun KY. Evolving approaches to the ethical management of genomic data. Trends Genet 2013;29(6):375−82.

[6]   Biesecker L. Opportunities and challenges for the integration of massively parallel genomic sequencing into clinical practice: lessons from the ClinSeq project. Genet Med 2012;14:393−8.

[7]   The integration of genetic technologies into health care and public health: a progress report and future directions of the Secretary's Advisory Committee on Genetics, Health, and Society. Department of Health and Human Services; 2009.

[8]   Evans JP, Burke W, Khoury M. The rules remain the same for genomic medicine: the case against "reverse genetic exceptionalism". Genet Med 2010;12:342−3.

[9]   Evans JP, Burke W. Genetic exceptionalism. Too much of a good thing? Genet Med 2008;10:500−1.

[10] Sulmasy DP. Naked bodies, naked genomes: the special (but not exceptional) nature of genomic information. Genet Med 2015;17:331−6.

[11] Martin Jean. Privacy and confidentiality. Handbook of global bioethics. Netherlands: Springer; 2014. p. 119−37.

[12] Presidential Commission for the Study for Bioethical Issues. Privacy and progress in whole genome sequencing. 2012.

[13] Hudson KL, Holohan MK, Collins FS. Keeping pace with the times—the Genetic Information Nondiscrimination Act of 2008. N Engl J Med 2008;358:2661−3.

[14] New rule protects patient privacy, secures health information: enhanced standards improve privacy protections and security safeguards for consumer health data (2013, January 17). Retrieved February 18, 2015, from http://www.hhs.gov/news/press/2013pres/01/20130117b.html.

[15] Bamshad M. Genetic influences on health: does race matter? JAMA 2005;294(8):937−46.

[16] Brody H, Hunt LM. BiDil: assessing a race-based pharmaceutical. Ann Fam Med 2006;4 (6):556−60.

[17] Cooper RS, Kaufman JS, Ward R. Race and genomics. N Engl J Med 2003;348:1166−70.

[18] Burchard EG, Ziv E, Coyle N, Gomez SL, Tang H, Karter AJ, et al. The importance of race and ethnic background in biomedical research and clinical practice. N Engl J Med 2003;348:1170−5.

[19] Bustamante CD, Burchard EG, De La Vega FM. Genomics for the world: medical genomics has focused almost entirely on those of European descent. Other ethnic groups must be studied to ensure that more people benefit, say. Nature 2011;475(7355):163−5.

[20] Haga SB. Impact of limited population diversity of genome-wide association studies. Genet Med 2010;12:81−4.

[21] Barr DA. The practitioner's dilemma: can we use a patient's race to predict genetics, ancestry, and the expected outcomes of treatment? Ann Intern Med 2005;143(11):809−15.

[22] Cunningham BA, Bonham VL, Sellers SL, Yeh HC, Cooper LA. Physicians' anxiety due to uncertainty and use of race in medical decision making. Med Care 2014;52(8):728−33.

[23] Snipes SA, Sellers SL, Tafawa AO, Cooper LA, Fields JC, Bonham VL. Is race medically relevant? A qualitative study of physicians' attitudes about the role of race in treatment decision-making. BMC Health Serv Res 2011;11:183.

[24] McCarthy JJ, McLeod HL, Ginsburg GS. Genomic medicine: a decade of successes, challenges, and opportunities. Sci Transl Med 2013;5:189.

[25] Mnookin B. One of a kind. The New Yorker 2014.

[26] Winslow, R. Teen helped research her own disease: rare cancer may be linked to gene mutation. The Wall Street Journal 2014.

[27] Kaye j, Currne L, Anderson N, Edwards K, Fullterton S, Kanellopoulou N, et al. From patients to partners: participant-centric initiatives in biomedical research. Nat Rev Genet 2012;13:371−8.

[28] McCarthy JJ. Driving personalized medicine forward: the who, what, when, and how of educating the health-care workforce. Mol Genet Genomic Med 2014;2:455−7.

[29] Wetterstrand K.A. DNA sequencing costs: data from the NHGRI Genome Sequencing Program (GSP), www.genome.gov/sequencingcosts [accessed 15.02.15].

[30] Rawlins M. Biotechnology: genomics and us. Nature 2013;498:34.

[31] Curnuttte MA, Frumovitz KL, Bollingrt JM, McGuire AL, Kaufman D. Development of the clinical next-generation sequencing industry in a shifting policy climate. Nat Biotechnol 2014;32:980−2.

[32] Biesecker LG, Burke W, Kohane I, Plon SE, Zimmern R. Next generation sequencing in the clinic: are we ready? Nat Rev Genet 2012;13(11):818−24.

[33] Kaufman D, Curnutte M, McGuire AL. Clinical integration of next generation sequencing: a policy analysis. J Law Med Ethics 2014;42:5−8.

[34] Javitt GH, Carner KS. Regulation of next generation sequencing. J Law Med Ethics 2014;42:9−21.

[35] Ray T. Legislators question FDA, stakeholders on how LDT oversight impacts industry, patients, innovation. 2014. Retrieved February 18, 2015, from: https://www.genomeweb.com/clinical-genomics/legislators-question-fda-stakeholders-how-ldt-oversight-impacts-industry-patient.

[36] Members. Global Alliance for Genomics and Health. Available from: http://genomicsand health.org/members#members-top; 2015.

[37] Optimizing FDA's regulatory oversight of next generation sequencing diagnostic tests— preliminary discussion paper. U.S. Food and Drug Administration; 2015.

[38]   Rubinstein WS, Maglott DR, Lee JM, Kattman BL, Malheiro AJ, Ovetsky M, et al. The NIH genetic testing registry: a new, centralized database of genetic tests to enable access to comprehensive information and improve transparency. Nucleic Acids Res 2013;41 (Database issue):D925–35.

[39]   National Library of Medicine. 2015. About GTR®. Available from: http://www.ncbi.nlm. nih.gov/gtr/docs/about/ [accessed 15.02.15].

[40]   Introduction-ClinVar-NCBI. National Center for Biotechnology Information. U.S. National Library of Medicine. Available from: http://www.ncbi.nlm.nih.gov/clinvar/intro/; 2015.

[41]   Rehm HL, Berg JS, Brooks LD, Bustamante CD, Evans JP, landrum MJ, et al. ClinGen— the Clinical Genome Resource. N Engl J Med 2015;372:2235–42.

[42]   About—ClinGen | Clinical Genome Resource. Available from: http://clinicalgenome.org/ about/; 2015.

[43]   Nguyen S, Terry SF. Free the data: the end of genetic data as trade secrets. Genet Test Mol Biomarkers 2013;17(8):579–80.

[44]   Appelbaum PS, Parens E, Waldman CR, Klitzman R, Fyer A, Martinez J, et al. Models of consent to return of incidental findings in genomic research. Hastings Cent Rep 2014;44:22–32.

[45]   Green RC, Berg JS, Grody WW, Kalia SS, Korf BR, Martin CL, et al. ACMG recommendations for reporting of incidental findings in clinical exome and genome sequencing. Genet Med 2013;15(7):565–74.

[46]   President's Commission for the Study of Bioethical Issues. Anticipate and communicate: ethical management of incidental and secondary findings in the clinical, research, and direct-to-consumer contexts. U.S. Department of Health & Human Services; 2013.

[47]   Presidential Commission for the Study of Bioethical Issues. Context specific recommendations—clinical recommendations. In: Anticipate and communicate: ethical management of incidental and secondary findings in the clinical, research, and direct-to-consumer contexts. 2013. p. 9–13.

[48]   ACMG Board of Directors. ACMG policy statement: updated recommendations regarding analysis and reporting of secondary findings in clinical genome-scale sequencing. Genet Med 2014;17:68–9.

[49]   McGuire AL, Joffe S, Koenig B, Biesecker BB, McCullough LB, Blumenthal-Barby JS, et al. Ethics and genomic incidental findings. Science 2013;340:1047–8.

[50]   Foster MW, Mulvihill JJ, Sharp RR. Evaluating the utility of personal genomic information. Genet Med 2009;11:570–4.

[51]   Aronson SJ, Eugene H, Clark BM, Varugheese M, Baxter S, Babb JL, et al. Communicating new knowledge on previously reported genetic variants. Genet Med 2012;14:713–19.

[52]   FACT SHEET: President Obama's Precision Medicine Initiative. The White House. Office of the Press Secretary. Available from: http://www.whitehouse.gov/the-press-office/2015/01/30/fact-sheet-president-obama-s-precision-medicine-initiative; 2015.

Chapter 20

Educational Issues and Strategies for Genomic Medicine: For the Public and for Providers

Alanna Kulchak Rahm and Michael F. Murray
Genomic Medicine Institute, Geisinger Health System, Forty Fort, PA, United States

Chapter Outline

INTRODUCTION

Since the human genome project was completed in 2003, genomic medicine has held the promise to advance our understanding of human health and disease and is the cornerstone of precision medicine. The new Precision Medicine Initiative (https://www.whitehouse.gov/precision-medicine) launched by President Obama in 2015 builds on advances in genomics, bioinformatics, social media, mobile technology, and patients who participate as active partners in research to "give everyone the best chance at good health" [1]. In order to realize the promise of genomic medicine, healthcare professionals, particularly primary care providers, need training in human genetics that facilitates the

Genomic and Precision Medicine. DOI: http://dx.doi.org/10.1016/B978-0-12-800681-8.00020-7

incorporation of genomics into practice. In addition, substantive public engagement will also be required in the new era of precision medicine, requiring informational and educational strategies targeting the public as well [2].

Therefore, in this chapter we will discuss different methods for engaging the public and providers in education about genomic medicine. We will discuss current public and provider understanding of genomics, methods for engaging the public and providers in education, and new and different learning opportunities to help the public and providers become more engaged, active, and confident in understanding and utilizing genomic medicine to improve health.

EDUCATION FOR THE PUBLIC

The effectiveness of genomic medicine rests on the assumption that providing personalized genetic risk information will motivate individuals to use this information to improve their health. While some individuals may have first-hand experience with genetics through a visit to a genetic counselor, a genetic test, or a genetic disease in the family; most individual consumers, physicians, and policymakers have little experiential knowledge of genetics. Therefore, much of the public's exposure to genomic medicine is through second-hand personal stories or stories disseminated through media/Internet outlets [3]. Improving the public's knowledge of genetics is an important first step toward the effective use of genomic medicine as it becomes more integrated into everyday prevention and medical care [2].

Smerecnick and colleagues [4] applied Roger's knowledge framework to understanding the public's knowledge and needs related to genetics. In his *Diffusion of Innovations* [5], Everett Rogers distinguished between three types of knowledge necessary to adopt innovations: (1) awareness knowledge, (2) how-to knowledge, and (3) principles knowledge. Table 20.1 shows how these different types of knowledge relate to public understanding of genetics and education needs around genomic medicine.

While individuals can make decisions about health without detailed principles knowledge of genetics, awareness knowledge without how-to knowledge is less likely to lead to initiation or sustained use of genomic information in health or health behaviors. However, principles knowledge may help improve the sustained use of the how-to knowledge due to the deeper understanding, which may then lead to the ability to influence others [5]. Participants in a National Human Genome Research Institute (NHGRI) meeting used the concepts of *genetic health literacy* and *genetic science literacy* to describe the different knowledge needs of the public. *Genomic science literacy* relates to principles knowledge in that it encompasses knowledge of basic genetics and genomics, as well as the mathematical knowledge to support risk comprehension. *Genetic health literacy*

TABLE 20.1 Genomic Definition of Different Knowledge Types in Roger's Knowledge Framework

Knowledge Type	Genomic Definition
Awareness knowledge	Awareness that a genetic risk factor is associated with a given disease or condition
How-to knowledge	Practical knowledge of how the genetic factor influences disease risk, and how that influences health, and what can be done to modify risk. This can also be represented by the term "genetic *health* literacy"
Principles knowledge	Theoretical knowledge of genetics and the technical mechanisms through which a certain gene interacts with other factors to develop disease, and how alteration in that gene impacts disease development. This can also be represented by the term "genomic *science* literacy"

relates to how-to knowledge, in that it encompasses an individual's capacity to gather, understand, and use genomic information to make decisions about health [6].

In general, the public's principle's knowledge of genetics is limited and may include many misperceptions about impact of genetics on health or conditions in which genetics may play a role in risk [4,6,7]. Individuals are also less likely to attend to or retain such knowledge, except during times when that information is most salient or relevant to that individual (such as when a sibling is diagnosed with cancer) [4]. Studies consistently find that while members of the public may have familiarity with the word "gene" or other terms related to genetics, this does not always translate to understanding of the underlying concepts. The public may also be aware of genetic risk factors associated with disease and that genetic predisposition does not necessarily mean disease is a certainty, however, this awareness often varies by disease. Finally, awareness and how-to knowledge also appear to vary greatly with sociodemographics as individuals with higher education levels, women, and whites all seem to have greater awareness and how-to knowledge [4,6,7].

Health Education Interventions for the Individual

One-on-one education in the healthcare environment can be an effective way of improving individual knowledge. Genomic medicine interventions delivered by trained clinicians can create awareness, provide how-to knowledge, and are more likely to assist with principles knowledge, as these encounters often happen during those "teachable moments" when the issue is more likely to be salient to the individual. Genetic counselors and other trained genomics clinicians provide individualized education during a clinical encounter that is described as "the process of helping people understand and adapt to the medical, psychological, and familial implications of genetic contributions to disease" [8]. Genetic counseling has been shown to improve use of preventive screening behavior for women with family history of breast cancer [9], and in cases of Hereditary Breast and Ovarian Cancer or Lynch Syndrome, can facilitate surgical decision making [10] and family communication about risk [11]. Such education in the clinical environment can also affect psychosocial adjustment to genomic information by decreasing cancer worry and anxiety [12].

However, while education delivered by healthcare providers during the healthcare encounter can improve knowledge, educational strategies that do not include ascertainment of preconceived notions and misperceptions are likely to be unsuccessful [4]. As individuals come to healthcare encounters with improved awareness of and familiarity with genetics from outside sources, misconceptions may hinder the ability for healthcare providers to improve how-to knowledge or principles knowledge. This has already been seen in the study of personalized medicine and direct-to-consumer genetic testing cohorts, where the utilization of genetic counseling services to help interpret complex information is low, with most participants claiming perceived understanding of the results [13]. Use of clinical decision aids and other decision support interventions are frequently used as methods of eliciting prior perceived risk, perceived understanding, as well as values and preferences. Use of decision aids can also improve subjective knowledge of the condition (principles knowledge) and/or how-to knowledge about risk and risk perception [14].

Collection of family health history and utilization of such information within the healthcare environment is another intervention that has been shown to improve how-to knowledge and genomic health literacy by helping individuals make decisions about their health in the context of familial risk, even when theoretical and scientific genetic knowledge (principles knowledge) is low [15]. Currently, family history is underutilized in healthcare due to many reasons, such as lack of appropriate documentation tools, clinical decision support for providers, or insufficient interpretation [15,16]. However, family health history remains one of the least expensive genomic medicine tools and, if utilized more effectively, holds promise for improving individual understanding and utilization of genomics within the healthcare setting. Recently, patient education regarding family history collection was shown not only to improve the quality of family health history collected, but also improved risk stratification and care recommendations [17].

Educating the Public Through Mass Media

As stated above, the general public has limited personal experience with genetics. Mass media (news stories, magazine articles, television shows, Internet) can be an effective way to improve public knowledge about medical subjects by exposing the public to information and education when there is no other way for the public at large to have direct experience with the subject [18]. Mass media can bring to the forefront previously unknown issues for the public to deliberate and debate by alerting the public to "what to think about" [19]. Keeping a subject in the forefront of the public's mind through repeated and varied representation in the media can improve the different types of knowledge, including principles knowledge by increasing the saliency (or personal relevance) to individuals viewing the media message. Celebrity narratives especially, have not only increased awareness of specific illnesses or conditions, but have also promoted public health policy and inspired individuals to take appropriate preventive actions [20].

One caveat to increasing media messages about genomic medicine is that individuals may be primarily exhibiting awareness knowledge when indicating they have knowledge about a subject. However this "knowledge" an individual may be professing is often an expression of familiarity through exposure rather than an indication of any specific theoretical or technical knowledge of the subject [21]. The public does recall exposure to media messages about genetics, however, such information is usually limited to high-exposure subjects such as breast cancer, prenatal testing, or cloning [22]. Recent study of the "Angelina effect"—the news coverage of actress Angelina Jolie Pitt revealing prophylactic surgery due to carrying a mutation in the *BRCA1* gene—also indicated that while most individuals surveyed were aware of her double mastectomy, very few (<10%) of those surveyed accurately understood the difference between Angelina Jolie Pitt's risk of breast cancer relative to a woman's risk without a *BRCA* mutation [20].

Mass media facilitate public understanding by utilizing symbolic or metaphorical language to shape context and make points quickly. Once the media has put forth the issue into the public eye, the use of metaphors and symbolic language help the public understand what the issue is, why it is important, and why they should be concerned (or not) about it by likening the more abstract and unknown subject (like genomic medicine) to everyday concerns and issues [18,23]. However, the information presented in the media may not always be correct or framed in such a way to promote the appropriate understanding of how to use the information for personal health. Mass media is also more likely to present genetics in an overly optimistic light [3,23], however, alternate frames from the medical perspective that promote caution and how to utilize genomic information carefully and appropriately to improve personal health are also presented in the media. When presented with direct-to-consumer genetic testing, members of the general public discussed a desire to seek more assistance or information in order to utilize the technology appropriately. They also discussed examining their own personal values and expectations prior to any decision to pursue such genetic testing—actions expected of an "informed consumer" [21].

Other new media tools for education to promote public knowledge of genomic medicine include podcasts, Facebook, Twitter, blogs, YouTube, TED talks, among others [6]. However, when used by scientists, researchers, and health professionals, care must be taken to ensure language that is accessible and understandable to the general public is utilized [23]. This means not only using lower age or grade-level appropriate language, but also working with target audiences to assess how different representations of information are perceived and understood through their cultural and linguistic lens. As the landscape of digital media evolves, members of the public also influence the visibility of media stories through blogging, reposting, and retweeting through personal channels as well. The more media messages and stories can be tailored culturally and linguistically to the target population, the more effective they will be at raising awareness and improving how-to and principles knowledge.

Community and Deliberative Engagement Forums

Another way to help the public become aware of and knowledgeable about an issue with little personal salience, like genomic medicine, is through community or deliberative engagement forums. These forums can last hours to days and involve purposefully engaging members of the public to learn about and discuss a complicated, values-heavy issue to gain informed perspective and meaningful insights into the way people think about the complex topic [22]. In the realm of genetics, these forums have been used to develop public governance structures for biobank programs [24]. Such forums are also useful in understanding how different populations already understand genomics, existing general misconceptions, and value of genomic medicine to the population. Such forums also highlight the necessity of linguistically and culturally congruent education. For example, through these forums researchers may learn that certain populations may only connect "environment" with the physical world and not changeable health behaviors such as smoking or diet [6]. Other populations may value the specific education on "their" diseases and how their population is different [21]. Linguistic differences, such as the simple use of a term like "overweight" may be so incongruent to a culture, that dissemination of any knowledge about the issue would be impossible until meaningful messages that are important to the target population are determined [5,25].

Educating Through Involvement: Citizen Scientist/Crowdsourcing

When individuals in the public are motivated to participate in genomic science, the promise of genomic medicine is more likely to be realized. Individual stories and research studies repeatedly show individuals using genomic medicine to improve their health. Francis Collins, current director of the National institutes of Health, changed his lifestyle to lose weight and exercise after genetic testing revealed an increased risk for diabetes [26]. Internet technology is increasingly utilized by many scientific disciplines as a way to engage the public in science [27]. Citizen scientist programs, crowdsourcing, or other methods of involving members of the public as partners in scientific investigations are becoming more common and easier to facilitate as technology becomes more widely available. As the culture changes to embrace the idea that greater engagement between the public and research will result in greater engagement of the public in health, these types of programs may become more prolific. Online communities interested in genomics and health research already exist including, Patients Like Me, 23andMe, the Genetic Alliance, GenomeConnect, and others. Participants in these communities can connect to other like individuals, share experiences, connect with researchers, and enter into data-sharing partnerships with the intent of facilitating and accelerating research in areas that are important to the public [28]. Another involvement of the public in the science of genetics has been through unique museum-based programs, including the traveling exhibit Genome: Unlocking Life's Code (https://www.genome.gov/27554054) and the Genetics of Taste Lab at the Denver Museum of Nature and Science (http://www.dmns.org/science/museum-scientists/nicole-garneau/the-genetics-lab). In the Genetics of Taste Lab, research questions were crowdsourced and citizen scientists are trained to consent, collect, and process buccal samples from museum patrons [29]. While sampling exhibit patrons provides the broad population necessary to answer the research question, the citizen scientists involved in the research are acquiring how-to and principles knowledge that is improving their overall genetic literacy through answering patron questions, consenting, and processing genetic samples.

Citizen scientist programs also help science focus on what the public is most concerned about, such as health and wellness. Tapping into the excitement and interest of individuals motivated to contribute to science has been called an "unfulfilled potential in research" [30]. As more individuals are using smart devices to collect information about themselves, their health, and their environment, this data can be a treasure trove of data for scientists, as well as a golden opportunity to engage individuals in learning in a way that has the potential to improve both how-to and principles knowledge of genetics, as well as to improve the overall health of the public at large.

EDUCATION FOR HEALTHCARE PROVIDERS

Approaches to Knowledge Acquisition and Education for Providers

The old aphorism in medical education regarding a standard approach to learning complex information, expressed as "see one, do one, teach one," can be applied (with caveats) to genomics and personalized medicine. The idea of learning through observation first, then supervised action with its inherent trial and error, and finally teaching is a powerful model. In the modern era with powerful electronic resources available, and with many settings where there are no local experts on new knowledge, the "see one" aspect can be done through printed and audiovisual resources, and the "do one" can be accomplished through peer-review practice-improvement opportunities. The "teach one" aspect is often up to the learner, and contingent upon their own confidence with the topic, but there is certainly a lot of need to play it forward in this content arena.

Healthcare professional learners are often motivated by questions from patients about issues that apply to that individual's health and wellness. An interesting spin on that motivation has occurred in a number of settings where providers themselves have undergone genomic testing and then examined their own results as a prompt for understanding the opportunities and the limits of our understanding of genomes [31−33]. This learning exercise, which makes the provider simultaneously experience both the patient and provider perspectives, simultaneously achieves many complex learning goals. The exercise of starting with your own genome as the learning tool has the potential to become a standard approach to learning to see one, do one, teach one in the years ahead, especially for primary care and other non-genetics specialist providers. A team at Temple University in Philadelphia recently added a new group of subjects to the list of potential candidates for genomic testing as a teaching exercise; they carried out whole-exome sequencing on the cadavers used in medical student anatomy and this provided a longitudinal opportunity for students to explore the value and implications of genomic testing in medicine [34].

In addition to self-examination of sequence, many providers learn best from peers who can help them define the learning agenda and then fulfill the learning goals. In this they often look to the professional organizations that they know and trust. With this practice in mind, the National Human Genome Research Institute sponsored the launch of the ISCC in 2013. This group acts as a gathering site for different professional organizations in healthcare who seek to create and share learning resources with their membership [35]. In addition to gathering resources they have published a framework for professional competencies in the area of genomic medicine to allow organizations to chart their own course within genomic medicine [36]. The competency framework allows considerable flexibility for provider organizations to decide what the priorities are for their group, but at the same time identifies family history, genomics testing, genome-specific therapies, somatic genomics, and microbial genomics as core areas to address.

The ISCC's mission continues to broaden in innovative ways such as including linguistics through a working group on "speaking genetics," which is exploring language in communication and education around genomics in all spheres. In addition, there is continued expansion in more traditional arenas such as the important work going forward to facilitate interactions between healthcare providers and related organizations such as insurers.

Sources of Provider Education

There is a growing of resources available to any provider who seeks education in the rapidly changing area of Genomic Medicine. Attempts to list all of the relevant or important educational sources is bound to be incomplete, however Table 20.2 lists readily available and free online sources that are created by experts in the field and are reliable.

TABLE 20.2 Some of the Publically Available Reference Sources for Providers

Resource	Description	Web site
Clinical Pharmacogenetics Implementation Consortium (CPIC) guidelines	CPIC guidelines on gene–drug pairs are peer-reviewed, published, and simultaneous posted with supplemental information/data and updates	https://www.pharmgkb.org/page/cpic
GeneReviews	A library of over 600 expert-authored, peer-reviewed chapters presented in a standardized format and focused on clinically relevant and medically actionable information on the diagnosis, management, and genetic counseling for specific inherited conditions	http://www.ncbi.nlm.nih.gov/books/NBK1116/
Genetics and Genomics Competency Center (G2C2)	A clearinghouse for Genetics and Genomics Resources for use in the classroom or in a practice. It includes information on peer-reviewed resources including Web sites, books, articles, and other materials	http://g-2-c-2.org//
Genetics Home Reference	A resource from the national library of medicine, that is, provides information about the effects of genetic variations on human health	http://ghr.nlm.nih.gov
Genetic Testing Registry (GTR)	The GTR is a clearing house for information regarding	http://www.ncbi.nlm.nih.gov/gtr/

(Continued)

TABLE 20.2 (Continued)

Resource	Description	Web site
	available genetic tests, conditions/phenotypes, genes, and laboratories	
MedGen	MedGen is the National Center for Biotechnology Information's portal to information related to Medical Genetics. Terms from the multiple sources are aggregated into concepts, each of which is assigned a unique identifier and a preferred name and symbol	http://www.ncbi.nlm.nih.gov/medgen
OMIM (Online Mendelian Inheritance in Man)	An Online Catalog of Human Genes and Genetic Disorders	http://www.omim.org
Talking Glossary of Genetic Terms	A resource maintained by the National Institutes of Health's NHGRI	http://www.genome.gov/Glossary/

Detailed descriptions of everything from definitions of terms, to pharmacogenomics, to description of uncommon genetic syndromes can be found among these references with links to the primary literature.

The Provider as a Guide for Interpreting Available Information

The direct-to-consumer genetics companies as well as online software programs for self-interpretation of genomic findings point to a model for genomic sequencing and interpretation that starts outside of the healthcare system and then engages with providers at a point after data has been generated and some conclusions have been drawn. This evolving model will likely take years before we see any preferred path for large segments of healthcare consumers. Regardless of the details it seems certain that providers will often plan the role of guide in this setting and will need to understand the details of patient information outside of the healthcare system [37]. The patients who come to them with clear notions about what their genomic data calls for will be best served by providers who can help them put their observations and conclusions in the greater context of healthcare delivery so that best choices and outcomes can be obtained.

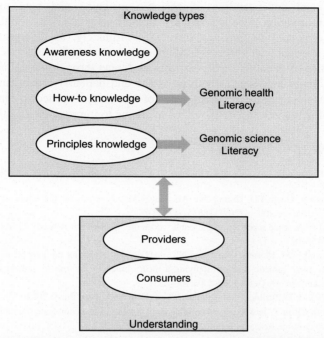

FIGURE 20.1   Knowledge types and relation to improving understanding of genomic medicine.

CONCLUSIONS

Clearly, a multi-faceted approach to educating both the public and providers about genomic medicine is required (Fig. 20.1). It is not enough to simply raise awareness. How-to knowledge and principles knowledge must be improved through genomic health and genomic science literacy improvement of the general public and non-genetics healthcare providers. It is also imperative to understand what the individual and collective communities currently understand about genomics and how to utilize genomic information to manage their health. In a workshop held in 2011 by the NHGRI, multiple stakeholders agreed that efforts must be made to understand what the public wants to know, what they feel would be useful, and the context in which they would like to accept this education [6].

Individual health education through interventions provided by healthcare providers, healthcare systems, and public health programs can help increase individual exposure to and understanding of genomic medicine. Other approaches, such as mass media messages, deliberative engagement forums, and citizen scientist programs or crowdsourcing are additional approaches that can be useful in improving public and provider how-to and principles knowledge of genomic medicine.

A variety of methods of engaging and educating the public and providers in genomic medicine are vital. Focused efforts aimed at increasing provider education and understanding of genomic medicine and how to utilize genomic information in collaboration with their patients are essential to realizing the potential of genomic medicine. Ultimately, as demonstrated in Fig. 20.1, a realistic and broad-based appreciation of the limits as well as the opportunities of genomic medicine will only come through the collaborative efforts of many individuals (both providers and patients) and organizations working to improve the different types of knowledge needed to realize the promise of genomics and personalized medicine to improve the health of all.

REFERENCES

[1] Collins FS, Varmus H. A new initiative on precision medicine. N Engl J Med 2015;372 (9):793–5.
[2] Collins FS, Green ED, Guttmacher AE, Guyer MS. A vision for the future of genomics research. Nature 2003;422(6934):835–47.
[3] Condit C. Science reporting to the public: does the message get twisted? CMAJ 2004;170 (9):1415–16.
[4] Smerecnik CM, Mesters I, de Vries NK, de Vries H. Educating the general public about multifactorial genetic disease: applying a theory-based framework to understand current public knowledge. Genet Med 2008;10(4):251–8.
[5] Rogers EM. Diffusion of innovations. 5th ed. New York: Free Press; 2003.
[6] Hurle B, Citrin T, Jenkins JF, et al. What does it mean to be genomically literate? National Human Genome Research Institute Meeting Report. Genet Med 2013;15(8):658–63.
[7] Condit CM. Public understandings of genetics and health. Clin Genet 2010;77(1):1–9.
[8] Resta R, Biesecker BB, Bennett RL, et al. A new definition of Genetic Counseling: National Society of Genetic Counselors' Task Force report. J Genet Couns 2006;15(2):77–83.
[9] Watson M, Kash KM, Homewood J, Ebbs S, Murday V, Eeles R. Does genetic counseling have any impact on management of breast cancer risk? Genet Test 2005;9(2):167–74.
[10] Schwartz MD, Lerman C, Brogan B, et al. Impact of BRCA1/BRCA2 counseling and testing on newly diagnosed breast cancer patients. J Clin Oncol 2004;22(10):1823–9.
[11] Hughes C, Lerman C, Schwartz M, et al. All in the family: evaluation of the process and content of sisters' communication about BRCA1 and BRCA2 genetic test results. Am J Med Genet 2002;107(2):143–50.
[12] Bish A, Sutton S, Jacobs C, Levene S, Ramirez A, Hodgson S. Changes in psychological distress after cancer genetic counselling: a comparison of affected and unaffected women. Br J Cancer 2002;86(1):43–50.
[13] Darst BF, Madlensky L, Schork NJ, Topol EJ, Bloss CS. Perceptions of genetic counseling services in direct-to-consumer personal genomic testing. Clin Genet 2013;84(4):335–9.
[14] Stacey D, Legare F, Col NF, et al. Decision aids for people facing health treatment or screening decisions. Cochrane Database Syst Rev 2014;1 CD001431.
[15] Guttmacher AE, Collins FS, Carmona RH. The family history—more important than ever. N Engl J Med 2004;351(22):2333–6.
[16] Cross DS, Rahm AK, Kauffman TL, et al. Underutilization of Lynch syndrome screening in a multisite study of patients with colorectal cancer. Genet Med 2013;15(12):933–40.
[17] Beadles CA, Ryanne WR, Himmel T, Buchanan AH, Powell KP, Hauser E, et al. Providing patient education: impact on quantity and quality of family health history collection. Fam Cancer 2014;13(2):325–32.

[18] Mccombs ME, Shaw DL. The agenda-setting function of mass media. Public Opin Q 1972;36(2):176−87.

[19] Borzekowski DL, Guan Y, Smith KC, Erby LH, Roter DL. The Angelina effect: immediate reach, grasp, and impact of going public. Genet Med 2014;16(7):516−21.

[20] Lanie AD, Jayaratne TE, Sheldon JP, et al. Exploring the public understanding of basic genetic concepts. J Genet Counsel 2004;13(4):305−20.

[21] Rahm AK, Feigelson HS, Wagner N, et al. Perception of direct-to-consumer genetic testing and direct-to-consumer advertising of genetic tests among members of a large managed care organization. J Genet Counsel 2012;21(3):448−61.

[22] Siegel JE, Heeringa JW, Carman KL. Public deliberation in decisions about health research. Virtual Mentor 2013;15(1):56−64.

[23] Condit CM. How geneticists can help reporters to get their story right. Nat Rev Genet 2007;8(10):81520.

[24] Olson JE, Ryu E, Johnson KJ, et al. The Mayo Clinic Biobank: a building block for individualized medicine. Mayo Clin Proc 2013;88(9):952−62.

[25] Knierim SD, Rahm AK, Haemer M, et al. Latino parents' perceptions of weight terminology used in pediatric weight counseling. Acad Pediatr 2014.

[26] Wadman M. Francis Collins: one year at the helm. Nature 2010;466(7308):808−10.

[27] Shuttleworth S, Frampton S. Constructing scientific communities: citizen science. Lancet 2015;385(9987):2568.

[28] Lee SS, Crawley L. Research 2.0: social networking and direct-to-consumer (DTC) genomics. Am J Bioethics 2009;9(6−7):35−44.

[29] Garneau NL, Nuessle TM, Sloan MM, Santorico SA, Coughlin BC, Hayes JE. Crowdsourcing taste research: genetic and phenotypic predictors of bitter taste perception as a model. Front Integr Neurosci 2014;8:33.

[30] Naci H, Ioannidis JP. Evaluation of wellness determinants and interventions by citizen scientists. JAMA 2015;314(2):121−2.

[31] Zierhut H, McCarthy Veach P, LeRoy B. Canaries in the coal mine: personal and professional impact of undergoing whole genome sequencing on medical professionals. Am J Med Genet A 2015;167A(11):2647−56 PMID: 26219924.

[32] Salari K, Karczewski KJ, Hudgins L, Ormond KE. Evidence that personal genome testing enhances student learning in a course on genomics and personalized medicine. PLoS One 2013;8(7):e68853 PMID: 23935898.

[33] Sanderson SC, Linderman MD, Zinberg R, et al. How do students react to analyzing their own genomes in a whole-genome sequencing course? Outcomes of a longitudinal cohort study. Genet Med 2015;17(11):866−74 PMID: 25634025.

[34] Gerhard GS, Paynton B, Popoff SN. Integrating Cadaver Exome Sequencing Into a First-Year Medical Student Curriculum. JAMA 2016;315(6):555−6.

[35] Manolio TA, Murray MF, Inter-Society Coordinating Committee for Practitioner Education in Genomics. The growing role of professional societies in educating clinicians in genomics. Genet Med 2014;16(8):571−2 PMID: 24503779.

[36] Korf BR, Berry AB, Limson M, et al. Framework for development of physician competencies in genomic medicine: report of the Competencies Working Group of the Inter-Society Coordinating Committee for Physician Education in Genomics. Genet Med 2014;16(11):804−9 PMID: 24763287.

[37] Murray MF. Why we should care about what you get for "only $99" from a personal genomic service. Ann Intern Med 2014;160(7):507−8 PMID: 24514942.

Chapter 21

Regulation of Genomic Technologies

Elizabeth Mansfield, Katherine Donigan, Steven Tjoe and Adam C. Berger
US Food and Drug Administration, Silver Spring, MD, United States

Chapter Outline

The US Food and Drug Administration (FDA) regulates medical devices under the authority of the Federal Food, Drug, and Cosmetic Act (FD&C Act) [1]. Medical devices include instruments, in vitro diagnostic (IVD) reagents, and components, parts, and accessories that meet the statutory definition of a device. The Code of Federal Regulations (CFR) defines IVD products as "those reagents, instruments, and systems intended for use in the diagnosis of disease or other conditions, including a determination of the state of health, in order to cure, mitigate, treat, or prevent disease or its sequelae."(See 21 CFR 809.3) The term "in vitro," in this context, refers to devices used in the collection, preparation, and examination of specimens taken from the human body. Although these devices are termed "diagnostics," the regulatory definition encompasses the much wider spectrum of information that IVDs can deliver, such as risk, prognosis, or prediction. Tests intended for clinical laboratory use created using genomic technologies are medical devices under FDA's regulatory authority and encompassed by the definition of IVD products. Under the FD&C Act and its implementing regulations, including 21 CFR 809.3, IVD products are devices, although the clinical laboratory community generally refers to them as tests. The words "test," "IVD," and "device" may be used interchangeably throughout the remainder of this document.

Genomic and Precision Medicine. DOI: http://dx.doi.org/10.1016/B978-0-12-800681-8.00021-9

Diagnostic technologies have increased in sophistication and accuracy over the centuries, moving from, for example, van Leeuwenhoek's use of the microscope to identify single-celled microorganisms in the 17th century to today's molecular analyses that can detect medically relevant changes at the level of the genome. The levels of sophistication, granularity, and precision that genome-based IVDs can deliver have occurred primarily as a result of technologies developed since the 1970s. The ability to detect or sequence specific DNA regions led to the development of tests that could identify pathogenic changes in DNA structure, evolving from technologies such as Southern blots and Maxam—Gilbert DNA sequencing to polymerase chain reaction (PCR) methodologies that could amplify specific DNA segments for direct analysis of genetic changes or further processing.

In the 1980s, FDA authorized the first DNA-based tests which were those that identified microorganisms in patient samples using probes that could specifically recognize the organism's DNA. Examples include DNA hybridization tests for *Mycobacterium* spp., *Chlamydia*, *Neisseria*, and other microorganisms. Significant developments in human DNA-based IVDs came soon after with the ability to amplify short segments of known sequences of DNA using PCR, although fluorescence in situ hybridization probe kits were the first FDA-cleared DNA tests for use in detecting human genomic alterations. More complex tests that detect genomic sequence and structural changes, using real-time (quantitative) PCR, array-based DNA and RNA tests, and other types of qualitative and quantitative technologies, expanded the ability to examine DNA changes by, in some cases, enabling the query of thousands to millions of points in the genome simultaneously. However, these technologies still relied on known sequence information. It was not until the development of rapid, automated shot-gun chain-termination sequencing, and the accompanying technology to assemble sequence fragments into contiguous DNA sequences, that DNA sequences could be generated in large scale without requiring a priori knowledge of sequence information. This large-scale sequencing technology, which enabled the Human Genome Project, has further evolved into the technology we know today as "next-generation" sequencing (NGS). NGS enables rapid, broad, and deep sequencing of nearly any DNA from a portion of a gene, entire exomes, or (nearly) a whole genome and is now being used clinically for a number of purposes. NGS as a technology is not limited to qualitative DNA testing; it also can be used to sequence RNAs, to identify transcript isoforms and to assess relative RNA expression levels [2]. Here, the discussion focuses on qualitative DNA sequencing, as this use is currently most clinically available for diagnostic use.

The technologies being used in today's genomic IVD tests have several technical advantages, including parsimonious use of sample to yield information about many variants at once, a relatively low cost per variant detected, and, in some cases, the developer does not need to predefine the variant(s) to be detected. These multiple measurement tests are currently being used in several types of clinical queries, particularly in the "personalized" or "precision" medicine domain, for example, to determine if a tumor has one or more of several genetic mutations that render it susceptible to certain therapies or to ascertain a patient's HLA type for tissue transplant. Human genomic nucleic acid sequencing is inherently multiplex and has been used either to detect one, a few, or many genetic differences from a "reference" sequence [3], usually as an indicator of disease, risk for disease, or likelihood of response to therapy. Genomic copy number arrays have been used to measure the number of copies of genomic sequences within a patient's genome, in cases in which autosomal copy numbers less than or greater than two (the "normal" diploid state) may be pathogenic. Copy number determination primarily uses arrays of mapped single-nucleotide polymorphisms (SNPs), although increasingly, NGS is being used for this application. Other nucleic acid tests assess the clinical state of the patient by examining gene "signatures" and developing a "score" using quantitative RNA expression patterns (or sometimes other measurements) and algorithms that may, for example, indicate a prognosis or aid in clinical decision making.

INTENDED USE AND CLASSIFICATION

For genomic tests, as for traditional diagnostic tests, in determining regulatory classification of such tests, FDA assesses the intended use of the test as well as the technological characteristics of the test. Under its medical device framework, IVDs, including genomic tests, are classified into one of three classes—Class I, Class II, and Class III—that correspond to the level of risk that the IVD presents to patients and the public. In general, Class I devices are subject to the least regulation, and Class III devices are subject to the most regulation. Class I IVDs present the lowest level of risk and generally do not require premarket review by FDA. Class II IVDs present a moderate level of risk and are generally subject to premarket review. Class III IVDs present the highest level of risk and are subject to premarket approval (PMA) and other regulatory controls to ensure these tests can be used safely and effectively. In determining classification, FDA evaluates the amount of regulation that provides a reasonable assurance of the device's safety and effectiveness (i.e., the amount of regulation needed so that the device's probable benefits outweigh any probable risks of injury or illness). For IVDs, a reasonable assurance of safety and effectiveness is generally demonstrated through a reasonable assurance of analytical validity, clinical validity, and safety under the device's conditions of use. In general, for genomic tests, the type of technology used has not been a determining factor in classification; rather the intended use and controls required to mitigate the occurrence or the consequences of an incorrect result for a patient when the test is used as intended have been the primary determinants. Thus, technological factors such as multiplexing or sequencing chemistry may have little bearing on a genomic test's classification. Note that a device that consists of multiple devices would be subject to CDRH's policy that, when a device combines devices from different classes, it is *classified in the highest of the predicate device classifications* unless the combined devices are regulatable as separate articles, e.g., they are detachable (FDA. Replacement reagent and instrument family policy, http://www.fda.gov/RegulatoryInformation/Guidances/ucm079185.htm).

The term "intended use" means the general purpose of the device or its function and encompasses the indications for use. The term indications for use, as defined in 21 CFR 814.20(b)(3)(i), describes the disease or condition the device will diagnose, treat, prevent, cure, or mitigate, including a description of the patient population for which the device is intended. The indications for use statement for genomic tests have usually identified what is being measured (e.g., mutations in *Kras* codons 12 and 13), specified the disease or other condition of interest, or included the clinical reason for measurement (e.g., to determine if a cancer patient will benefit from a particular therapy). Additionally, indications for use statements have often included many other specifications (e.g., the type of sample to be tested, the population for which the test is designed (the target population), and the clinical setting in which the test is performed, if other than a general setting).

Prescription use IVDs provide medically relevant information to healthcare providers for decision making about a patient's clinical management, so their impact on the patient in terms of risk is not usually direct, that is, the act of generating test results is not expected to cause patient harm (although direct health hazards to the patient and test operator due to possibility of disease transmission from needle sticks, for example, are recognized). However, incorrect or misinterpreted results can lead directly to patient harm, including inappropriate or unnecessary follow-up procedures, such as additional testing and/or treatment. If, for example, a test fails to detect a true *Kras* mutation, the healthcare provider might treat the patient with a drug that he or she will not respond to. Not only would the patient suffer any toxicities of the drug, but he/she could miss out on a treatment to which he or she might respond. When the patient stands to suffer serious or irreversible harm (or even death) from application of an inaccurate result to clinical management, or when the inaccurate result could critically delay necessary treatment, the test is usually considered to be high risk. When the potential harm from an incorrect test result is less severe due to its adjunctive nature, the limited usefulness of the information, the ease of detection of an incorrect result, or other mitigations, the test may be considered to be moderate risk or even low risk. It is important to understand that the actual performance (e.g., accuracy, precision) of an IVD is not a factor in determining risk.

Classification, once established, generally applies to all IVDs within the device type. However, infrequently, new information becomes available and has caused FDA to propose reclassification. This has typically resulted in down-classification (e.g., reclassification from Class II to Class I) and considerations have included new information pertaining to risk mitigation, extensive regulatory and clinical experience with a particular IVD type that acts as a mitigating factor, or changes in the way an IVD result is used in clinical practice.

Classification, among other considerations, determines the applicable path for regulatory submission. High-risk genomic tests have been generally considered to be as Class III devices. Class III devices generally will require PMA of an application. A PMA application requires information such as the indications for use of the device, description of the device and how it operates, description of and data from analytical evaluations, and description of and data from any clinical evaluations that were required to demonstrate that the test can achieve an acceptable diagnostic performance for its intended use (See 21 CFR 814.20). FDA's evidentiary standard for premarket review of IVDs, as for all devices, is valid scientific evidence (See 21 CFR 860.7(c) (2). Valid scientific evidence includes "evidence from well-controlled investigations, partially controlled studies, studies and objective trials without matched controls, well-documented case histories conducted by qualified experts, and reports of significant human experience with a marketed device, from which it can fairly and responsibly be concluded by qualified experts that there is a reasonable assurance of the safety and effectiveness of a device under its conditions of use. [Remainder omitted for brevity, but relates to variable requirements for evidence and what is not regarded as valid scientific evidence.]), and when a PMA is required, an applicant must provide in the PMA valid scientific evidence to demonstrate a reasonable assurance that the device is safe and effective for its intended uses (See 21 CFR 860.7(d)(1) and (e)(1)). Under FDA's medical device regulations, there is reasonable assurance that a device is safe when it can be determined, based on valid scientific evidence, that the probable benefits to health from use of the device for its intended uses and conditions of use outweigh any probable risks and where the valid scientific evidence adequately demonstrates the absence of unreasonable risk of illness or injury associated with the use of the device for its intended uses and conditions of uses (See 21 CFR 860.7(d)(1) for complete information on safety determination). A reasonable assurance of safety and effectiveness is a standard that is tailored to the specific intended uses and conditions of use in question, including the indicated target population. For IVDs, a reasonable assurance of safety and effectiveness is generally demonstrated through a reasonable assurance of analytical validity, clinical validity, and safety under the device's conditions of use. See section "Performance Characteristics" for more information on these characteristics.

When a genomic test is considered to be moderate risk, it is usually classified as Class II. If FDA has previously granted marketing authorization to another genomic test with the same intended use and technological characteristics as a new genomic test (i.e., a predicate), a 510(k) premarket notification is typically the appropriate pathway for the new genomic test. The term 510(k) derives from the section (510) and subsection (k) of the FD&C Act that requires premarket notification prior to introduction of a device into interstate commerce or commercial distribution for human use. Devices which are granted marketing authorization through the 510(k) program are "cleared" by FDA. The current standard for 510(k) clearance is a finding that the new device is "substantially equivalent" to a predicate device. "Substantially equivalent" or "substantial equivalence" means that the new device has the same intended use as the predicate device and that the applicant can demonstrate that the device has either the same technological characteristics as the predicate device or has different technological characteristics and the information submitted to FDA does not raise different questions of safety and effectiveness than the predicate device and demonstrates that the device is at least as safe and effective as a legally marketed device [4].

In the case of a new device for which there is no predicate, the new device will automatically be placed into Class III [5]. If FDA believes that the risks associated with a new device is low-moderate risk and can be sufficiently mitigated by regulatory controls applicable to Class I or Class II devices (i.e., general controls or a combination of general and special controls), FDA may identify the device as a likely candidate for the FDA's de novo classification pathway, instead of requiring the submission of a PMA for the new device [6]. Alternatively, a submitter may submit a de novo request for classification when the submitter believes their device to be eligible. If the submitter of a de novo request for classification demonstrates that general controls, or a combination of general controls and special controls, are sufficient to provide a reasonable assurance of safety and effectiveness, FDA will grant the de novo request for classification and issue a written order classifying the specific IVD and IVD type in Class I or Class II. The IVD that is granted marketing authorization may serve as a predicate for future 510(k) submissions. As above, for IVD tests, a reasonable assurance of safety and effectiveness is generally demonstrated through a reasonable assurance of analytical validity, clinical validity, and safety under its conditions of use.

Genomic IVDs could thus be classified as Class I, Class II, or Class III. Most genomics IVD regulatory applications received by FDA to date have either been Class III or Class II, accounting for the specific risks and mitigations applicable to the genomic test.

PERFORMANCE CHARACTERISTICS

FDA's premarket review of IVDs through the 510(k), de novo, or PMA submission pathways includes an evaluation of the performance characteristics of the IVD. Performance characteristics include analytical performance—various measures of how well a test measures or detects a particular analyte or analytes—and clinical performance—a measure or measures of how well the device result correlates with the disease or condition it is intended to detect, as well as safety under the conditions of use described in the labeling of the IVD. To demonstrate analytical performance, metrics such as analytical accuracy, precision, sensitivity, specificity, and other measures that are important for the particular test in question are provided in a premarket submission.

The analytical characteristics of importance may vary based on whether the test is quantitative or qualitative, the technology used in measurement or detection, and other factors. IVD clinical performance is generally described in terms of diagnostic sensitivity and specificity, positive and negative predictive values, and other metrics that are relevant to the particular clinical use of the test. In many cases, there is no predetermined analytical or clinical performance "bar" that must be met; rather the requirements for the actual performance of the test are set by what the performance needs are for the test in order to deliver medically useful results and whether there are existing tests (predicates) with defined performance to which it can be compared.

The intended use of an IVD generally informs the types of data and information FDA will request to assess the performance of a particular device. For example, for an IVD intended to detect specific mutations in codons 12 and 13 of the *Kras* gene, FDA expected that the sponsor would demonstrate that the test could actually detect each of the claimed mutations and would not detect mutations that do not exist (Summary of Safety and Effective Data at http://www.accessdata.fda.gov/cdrh_docs/pdf11/P110030B.pdf). This is the accuracy metric, usually demonstrated by comparison to a standard, a reference method, or best available method, using at least some proportion of clinical samples, at a demonstrated mutant allele frequency (or sensitivity) that is reasonable for its intended use. The extent of use of clinical samples and the lowest consistently detectable mutant allele frequency have historically been the subject of discussions in meetings between FDA and the sponsor under FDA's presubmission program (Additional information on FDA's pre-submission program is available at http://www.fda.gov/downloads/medicaldevices/deviceregulationandguidance/guidancedocuments/ucm311176.pdf). Other examples of potentially applicable metrics from past FDA reviews of genomic tests are listed in Table 21.1, but this list is

TABLE 21.1 Additional Performance Metrics of Interest

Performance Metric	Characteristic of Interest	Example of Study
Repeatability	Ability to obtain the same result upon repeated measurements under the same conditions	Run same sample multiple times on the same instrument, with the same reagent lot, by the same operator
Reproducibility	Ability to obtain the same result under varied conditions	Run same sample multiple times under different conditions, e.g., different instruments, chips/arrays/flow cells different reagent lots, different operators, different sites

(Continued)

TABLE 21.1 (Continued)

Performance Metric	Characteristic of Interest	Example of Study
Specificity: Interference	Change in result in the presence of an endogenous or exogenous potential interfering substance	Run samples with endogenous/exogenous interfering substances present and absent
Specificity/selectivity: Cross-reactivity	Change in result in the presence of potentially cross-reactive substance	Search of relevant genome sequence for cross-reacting sequences. Run samples in the presence of potentially cross-reacting substance
Stability of sample and reagents	Change in result as a consequence of sample storage or reagent age	Run samples with different duration or condition of storage. Run samples with reagent lots of different ages
Demultiplexing	Ability to accurately demultiplex mixed samples	Analyze system output accuracy from pooled samples with known sequence/variants indexed with all applicable barcode sequences
Limit of detection/limit of quantitation	Establish the lower limit at which the test gives accurate results	Run natural or diluted samples at progressively lower concentration (allele fraction) until change in result/failure to detect signal occurs. Measure repeatedly in this range to establish the lowest concentration (limit of detection) where correct result/true signal occurs more than 95% of the time
Precision (quantitative measurements)	Establish the quantitative performance of the test at different concentrations of analyte	Run samples covering range of measurement repeatedly to obtain output values
Linearity (quantitative measurements)	Establish the linearity and range of linearity of the test	Measure a sample above the upper limit of linearity range and dilutions of this sample repeatedly; calculate deviations of the measured concentrations from the best fitted straight line

provided as an example and is not exhaustive. Each technology and test will be assessed as to the critical performance factors needed to demonstrate "a reasonable assurance of safety and effectiveness" or "substantial equivalence."

For tests utilizing technologies that by their design measure numerous analytes from a single sample (e.g., arrays, NGS), some traditional analytical performance metrics are impractical or impossible to demonstrate at the per analyte level. For example, in sequencing a gene, each base is considered an analyte for which the system could generate a true or a false result, and given the a priori possibility that any of the four bases could be present at any position, it would be impossible in practice to demonstrate the test performance on a per analyte basis as each gene could have anywhere from hundreds to millions of bases. This poses challenges for both the development and the regulatory review of tests that measure large numbers of analytes. Therefore, FDA has taken what it believes to be a practical approach of requesting that performance be demonstrated in different ways, depending on the complexity of the test. Genomic tests that report on limited numbers of prespecified changes, e.g., a test for *Kras* mutations at codons 12 and 13, have been required to demonstrate analytical validation on a per analyte basis (Summary of Safety and Effective Data at http://www.accessdata. fda.gov/cdrh_docs/pdf11/P110030B.pdf). Genomic devices that could be used to measure many analytes whose state are unknown a priori, for example, whole-exome sequencing, have been required to demonstrate performance accuracy by detecting *types* of mutations it is claimed the test can detect, such as single nucleotide variants, insertions and deletions of various sizes, and copy number variants (http://www.accessdata.fda.gov/cdrh_docs/ reviews/K123989.pdf). Performance across a number of genomic contexts, and in samples where genome sequence information is well characterized, has been needed. For tests that query an intermediate number of analytes, validations that combine the "per analyte" and "per analyte type" approaches have been needed. The term "intermediate" is contextual in nature as the number of analytes that would be required to be validated would vary based on the intended use and performance characteristics of each individual test. Validation approaches depend on the specific claims of, and technology behind a test; accordingly, different approaches may be appropriate for other types of genomic tests, such as arrays to detect SNPs or to measure relative or (nearly) absolute levels of RNA or DNA.

As technology evolves, FDA is challenged to determine which validation requirements are important to demonstrate that a genomic device is "reasonably safe and effective," and how those requirements can be implemented in a practical manner. FDA has taken the least burdensome approach to the demonstration of analytical performance by the developer, taking into account the critical characteristics of a given test. For example, to provide regulatory authorization for an NGS test that claimed to detect essentially any change in the cystic fibrosis (*CF*) gene, analytical validation included demonstration of performance characteristics for known, common *CF* mutations, as well as demonstrating that repeatedly sequencing a "normal" *CF* gene would yield the correct sequence consistently (http://www.accessdata.fda.gov/cdrh_docs/reviews/K132750.pdf). To authorize marketing of an NGS instrument and its associated reagents and software, FDA requested that the submitter demonstrate performance for different types of changes across the genome, in a number of genomes, where the changes were in various genomic contexts (Decision Summaries for these products at http://www.accessdata.fda.gov/cdrh_docs/reviews/K123989. pdf; http://www.accessdata.fda.gov/cdrh_docs/reviews/K133136.pdf). This approach queried the ability of the device to perform across various genomic changes at demonstrated levels of accuracy.

Additional information about the specific issues addressed for regulatory clearance or approval for specific devices can be found on FDA's Web site (A periodically updated list of nucleic acid-based tests cleared or approved by FDA at http://www.fda.gov/MedicalDevices/ProductsandMedicalProcedures/ InVitroDiagnostics/ucm330711.htm).

COMPLEX TECHNOLOGY CONSIDERATIONS

As genomic technologies increase in complexity and are able to measure more and more analytes from one sample, demonstrating analytical performance for each possible analyte becomes infeasible. For example, as discussed above, even in the relatively simple case of sequencing of a single gene to determine if there are deleterious mutations, there are hundreds to millions of individual DNA bases, each of which is an analyte on its own. It would be a nearly impossible task to demonstrate that each and every possible change in DNA sequence could be detected, as there are no clinical samples or artificial constructs to cover the entire spectrum of each possible base at each position in the gene, not to mention the substantial amount of validation testing that would require. FDA seeks to apply the level of regulation necessary to provide reasonable assurance that an IVD, including a genomic IVD, is safe and effective for its intended use. Accordingly, in conducting premarket review of an IVD, FDA requests information that is necessary for FDA to make a determination of whether the standards for marketing authorization are met.

Multiplex genomic testing, or simultaneous measurement of multiple analytes generated by a common process from the same sample in a single run, is often used to distinguish the existence of one analyte among a large possible set. For example, multiplex tests for infectious agents in respiratory viral panels are intended to detect which one or more infectious agents are present in a patient sample, rather than running individual tests for each possible infectious agent. Other types of multiplex testing can include those that are designed to query a sample for multiple possible SNPs that may inform health status. These multiplex tests have required that each possible analyte be validated; may only have required that a subset of analytes that are reasonably expected to be representative of all the other analytes in terms of performance be validated (with a justification as to why this is sufficient); or have required other validation metrics such as the complex agreement strategy used for validation of chromosomal copy number detection tests (http:// www.accessdata.fda.gov/cdrh_docs/reviews/k130313.pdf). Which of these possibilities was appropriate was determined based on scientific and clinical questions, such as how critical it was to distinguish one virus from another, whether validating by type of mutation/variant was convincingly justified as representative, or whether the relevant "analyte" was capable of being and was specified beforehand. For DNA sequencing, even when the mutation of interest in a specific gene will not necessarily be known prior to testing, as can occur with heritable genetic diseases and in other conditions, one may know what the most common mutations are for that gene, and can usually validate the sequencing test around those mutations. However, there may be many rarer mutations, or even private (confined to a single family) mutations that are deleterious. In the case of a single-gene sequencing test for *CF*, where many mutations are known, and it is expected that additional rare mutations are yet to be discovered, FDA requested analytical validation using common mutations and accepted the ability of the test to consistently and correctly call wild-type sequence at all other positions in the gene as sufficient to establish test accuracy (http://www.accessdata.fda.gov/cdrh_docs/ reviews/K132750.pdf).

NGS, which can be used to sequence nearly the entire genome, poses similar challenges for demonstrating analytical performance since validating each analyte becomes impractical as the number of bases queried increases. Indeed, in FDA's decision to authorize marketing of an NGS sequencing device, the Agency accepted a demonstration that the instrument, as configured, could detect various types of mutations, including single nucleotide variants, insertions/deletions, polynucleotide repeats, etc., contained within a wide range of genomic contexts (e.g., GC-richness, different chromosomes), as compared to highly characterized genomic reference sequences (http:// www.accessdata.fda.gov/cdrh_docs/reviews/K123989.pdf). FDA used a flexible regulatory approach in authorizing the marketing of this device, examining the overall abilities of the system, rather than the explicit ability to detect particular genomic changes. It is important to note that the validation testing under this classification regulation (21 CFR 862.2265) only establishes the instrument's general capabilities and does not establish the instrument's capabilities or suitability with respect to any specific claims.

The analytical and clinical performance of an NGS system can also be highly dependent on the informatics tools applied to the reads generated from the sample/reagent/instrument configuration. These informatics tools have multiple functions, and generally include so-called variant callers, which can simultaneously identify different types of variants at a high accuracy. To date, there is no single "ideal" caller for all types of variants; rather different callers have been optimized for the types of variants of greatest interest to the user and have differing false-positive/negative rates that can depend on the number of times a particular segment of DNA has been sequenced and other factors. FDA is currently working internally and partnering with other governmental agencies to develop and make available high-quality reference materials, analytical tools, and methods that will enable more accurate identification of genomic alterations.

The claims for a genomic test have been required to be supported by a demonstration that the entire system, including the informatics, can produce the desired result. If the claims for an NGS test have not included certain variant types, then FDA has not requested performance information for those excluded variant types to be provided. In addition, FDA recognizes that materials, tools, and methods developed by third parties may constitute part of a test system and has recognized that their validation may be needed as part of the test system.

LOOKING TO THE FUTURE

IVD genomic tests run the gamut of complexity from simple PCR detection of single base pair changes to the highly complex, computationally driven tests such as NGS for (nearly) whole-genome sequencing. FDA's regulation of IVD genomic tests ensures that such tests produce meaningful results for patients, while enabling developers to innovate at the speed of technological and scientific advancement. Accordingly, with FDA regulation of IVD genomic tests, patients have access to the most technologically advanced high-quality testing.

As NGS in particular becomes increasingly used for all types of genomic testing, FDA has begun to craft a streamlined regulatory approach that takes into account the variety of system configurations and types of tests that could be created and used. This approach is part of the "Precision Medicine Initiative" (PMI), which is a federal effort designed to generate new medical knowledge and uses through research, technology, and policies directed specifically at individualizing clinical care and therapy. The PMI, announced on January 30, 2015, includes a complex set of activities and goals, some of which direct FDA to develop flexible approaches to evaluate analytical performance, assess informatics performance, and evaluate clinical validity in part through reliance on community-developed consensus standards and high-quality, well-curated databases of clinically valid genotype–phenotype associations. This approach is intended to allow iterative technological change and promote rapid incorporation of valid scientific evidence into testing, while providing accurate and meaningful test results for patients based on current knowledge. FDA issued a preliminary discussion paper (http://www.fda.gov/downloads/MedicalDevices/NewsEvents/WorkshopsConferences/UCM427869.pdf) and held a public meeting (http://www.fda.gov/MedicalDevices/NewsEvents/WorkshopsConferences/ucm427296.htm) to discuss initial concepts for how the Agency might go about developing this flexible and adaptive approach to genetic test regulation.

Recently FDA published two draft guidance documents proposing new approaches to test design, development, and validation (http://www.fda.gov/ScienceResearch/SpecialTopics/PrecisionMedicine/default.htm). FDA has committed to continuing engagement with the public on regulation of new technologies through future workshops and opportunities for input.

For its PMI responsibilities, FDA has focused its regulatory efforts on three areas—analytical standards, informatics, and the use of genomic databases. The first prong under FDA's consideration is "standards-based" analytical validation activities, whereby a laboratory may be considered as in compliance with certain process standards, using standard reference materials. Such a laboratory would thus be expected to have the knowledge and materials to develop specific types of NGS tests, with the eventual regulatory goal of using alternative approaches to demonstrate analytical validity, such as conformity with FDA-recognized standards for supporting or assuring analytical validity. The second prong is a cloud-based, open-source informatics platform and toolkit called precision FDA that would allow developers to benchmark their selected informatics pipelines against various well-characterized genomic sequences, and to demonstrate performance over various types of genomic contexts such as insertions/deletions of various sizes and in various genomic backgrounds such as differing GC content of sequence or repeat structures. This platform would allow test developers to maintain their confidential data or share it with others in a secure environment; participants will be able to "play in a community sandbox" to test ideas, compare tools, modify publicly accessible tools to meet their own needs, and work collaboratively to develop better bioinformatics tools for the community. The third prong is use of appropriate, high-quality, well-curated clinical databases as sources of valid scientific evidence. The aggregation of data from many sources in public databases where phenotype and genotype data are appropriately adjudicated and rated as to quality of evidence could have the potential to enable more rapid development of the necessary evidence base, could provide publicly available evidence assessments, and could alleviate the need for each sponsor to perform individual studies to demonstrate the clinical validity of tests that examine the same analyte for the same intended use.

The development of FDA's portion of the PMI is an ongoing effort and represents the Agency's commitment to creating a flexible and practical regulatory environment that will ensure that patients have access to cutting edge technologies that produce meaningful results.

CONCLUSIONS

FDA has over 30 years of experience regulating genomic technologies, encompassing many different types of technologies and tests. Underlying this regulation are three essential principles: risk classification based on intended use, assurance that the device accurately and reliably measures the analytes it is intended to measure, and assurance that the test result has an appropriately strong correlation to the disease or condition in question. New technologies and rapid generation of extremely complex datasets call for flexible approaches to oversight; however, the three essential elements listed above are basic tenets that will likely continue to form the basis of these adaptive approaches since they are vital to understanding the value and quality of tests. As genomic technologies have evolved, the regulatory requirements for demonstrating a reasonable assurance of safety and effectiveness have likewise evolved to ensure that the right questions are asked, the right data are provided, and access to high-quality innovative tests is available as rapidly as possible.

DISCLOSURE

The information and questions contained in this document are not binding and do not create new requirements or expectations for affected parties, nor is this document meant to convey FDA's recommended approaches or guidance.

REFERENCES

[1] Federal Food, Drug and Cosmetic Act, Section 201(h); United States Code, Section 321(h).
[2] SEQC/MAQC-III Consortium. A comprehensive assessment of RNA-seq accuracy, reproducibility and information content by the Sequencing Quality Control Consortium. Nat Biotechnol 2014;32(9):903−14.
[3] Zook JM, Chapman B, Wang J, Mittelman D, Hofmann O, Hide W, and Salit M. Integrating human sequence data sets provides a resource of benchmark SNP and indel genotype calls. Nat Biotechnol 2014;32(3):246−51.
[4] See Section 513(i)(1)(A) of the FD&C Act (21 USC 360c(i)(1)(A) for the statutory basis for substantial equivalence.
[5] Section 513(f)(1) of the FD&C Act (21 USC 360c(f)(1)).
[6] Section 513(f)(2) of the FD&C Act (21 USC 360c(f)(2)). See also the information provided by the FDA on the de novo process at http://www.fda.gov/AboutFDA/CentersOffices/OfficeofMedicalProductsandTobacco/CDRH/CDRHTransparency/ucm232269.htm.

Chapter 22

Developing the Value Proposition for Personalized Medicine

Shelby D. Reed[1], Eric Faulkner[2,3] and David L. Veenstra[4]

[1]*Duke University, Durham, NC, United States, [2]Evidera, Bethesda, MD, United States,
[3]University of North Carolina at Chapel Hill, Chapel Hill, NC, United States, [4]University
of Washington, Seattle, WA, United States*

Chapter Outline

A diverse set of costs and benefits—across all stakeholders—are relevant to quantifying the value of personalized medicine. Patients value an accurate diagnosis, reaching treatment goals more quickly, avoidance of toxicities, and improved health outcomes. Payers value avoidance of unnecessary treatments and reduced medical expenditures. Physicians value greater confidence that they are ordering the treatment strategy with the best benefit-risk profile for a given patient, and health systems value approaches that optimize clinical pathways and budget requirements for care delivery. Biopharmaceutical companies value improved treatment outcomes, higher success rates, and faster drug development. Scientists value greater understanding of human disease and variability in treatment responses between individuals (Table 22.1).

Questions about the value of personalized medicine typically examine whether the expected benefits from altered medical management due to test results are worth the cost of testing. In a health care system with limited resources, opportunity costs representing alternative ways this money could be spent to improve health outcomes are relevant. Cost-effectiveness analysis

TABLE 22.1 Value Drivers for Personalized Medicine—Stakeholder Perspectives

Stakeholders	Value Drivers
Patients	Diagnosis, appropriate treatments, avoidance of toxicities, improved health, and "value of knowing"
Payers	Avoidance of unnecessary treatments and reduced medical expenditures; defined patient populations
Physicians	Greater confidence in treatment strategy, "value of knowing"
Health systems	Optimizing clinical pathways and budget requirements for care delivery
Biopharmaceutical companies	Improved treatment outcomes, higher success rates, and faster drug development
Scientists	Greater understanding of human disease and variability in treatment responses between individuals

is the most widely used framework to assess comparative costs and effects of health care interventions. From the perspective of the health care system, the overall value of personalized medicine comprises many focused cost-effectiveness assessments of individual tests used in specific clinical scenarios, of which some demonstrate good value for money and some do not. Distinguishing between good and poor value in leveraging this test information is of increasing interest to governments, private payers, and providers. As a result, costs associated with personalized medicine testing are increasingly scrutinized, particularly as a future state emerges that involves testing hundreds to thousands of biomarkers versus a single marker or limited panel testing today. Broad reimbursement and market access for a new technology is often considered the "fourth hurdle" (after the first three hurdles of safety, efficacy, and quality) in the full execution of translational medicine, and the development process for a new personalized medicine is often not considered complete without a formal evaluation of its cost-effectiveness [1].

Standard methodological approaches have successfully been applied to evaluate the cost-effectiveness of numerous personalized medicine tests, particularly for companion diagnostics developed in the context of a clinical trial. Recently, however, questions have emerged about the extent to which existing methods can characterize the value of personalized medicine—specifically, how to account for the scope of potential benefits and risks from broader testing, including all downstream consequences [2,3]. The challenges parallel scientific advances from the identification of one or more biomarkers to stratify patients for prognosis or treatment-related benefits and risks to strategies that use individual patients' genomic, transcriptomic, proteomic, metabolomic, and other "omics" profiles to tailor treatment regimens. The advent of whole-genome sequencing presents a host of methodological issues, such as how to handle incidental findings and associated costs, how to value knowledge about specific genetic findings that have no clinical implications, or whether to account for potential value to family members.

To explicate the issues around the value of increasingly individualized care, we begin with an overview of cost-effectiveness analysis and the factors determining the value of personalized medicine, using examples in cancer care. We then identify limitations of the traditional cost-effectiveness analysis framework and methodological issues that will arise with increasing use of multiple "omics" profiles to guide individualized selection of treatments.

COST-EFFECTIVENESS ANALYSIS

An important feature of cost-effectiveness analysis is that it is comparative [4]. When evaluating a personalized medicine test, direct comparisons of both costs and health outcomes are necessary. To estimate the incremental cost-effectiveness ratio (ICER) requires one to model its impact on treatment decisions and to estimate the difference in mean costs, $\hat{C}$, and the difference in mean effectiveness, $\hat{E}$, incurred by patients managed with the test versus patients managed without the test:

$$ICER = \frac{\hat{C}_{\text{test}} - \hat{C}_{\text{no test}}}{\hat{E}_{\text{test}} - \hat{E}_{\text{no test}}}.$$

When health benefits subsequent to testing are greater than with standard care without testing, the denominator of the ICER (i.e., ΔE) increases and the economic value of the testing strategy improves. Preferred measures of effectiveness are life-years or quality-adjusted life-years (QALYs) to facilitate comparisons across other types of tests and medical interventions. When total costs are lower with testing relative to standard care and clinical outcomes are improved, integrating the test into practice is considered to be "economically dominant." Although claims that personalized medicine saves money are plentiful, two recent reviews of 59 and 84 cost–utility analyses (i.e., cost-effectiveness analyses that use QALYs as the measure of effectiveness) by Phillips et al. and Hatz et al. [5,6] both found that only about 20% of tests demonstrated cost savings. More frequently, costs are higher with testing along with concomitant improvements in health outcomes, and decision makers are charged with judging whether associated health gains are worth the additional cost. The review by Hatz et al. [6] revealed that the median ICERs in base-case analyses were lower for screening tests ($8497/QALY (2008 $US)) and tests for disease prognosis ($10,150) compared to tests to stratify patients for treatment response/nonresponse ($37,308) or to identify patients more likely to experience adverse events with specific treatments ($39,196). In other words, personalized medicine is often not cost saving because testing leads to downstream health care interventions that are cost increasing, despite providing improved health outcomes.

Although ICERs are computed using just four parameters, the derivation of each requires synthesis of information on a test's analytic and clinical validity; impact on clinical decision making and health outcomes (i.e., clinical utility); and cost of testing, clinical management, and medical events. Each must be estimated over a time horizon that is sufficient to capture all downstream cost and health consequences. In the United States, where patients are often not enrolled in commercial health insurance plans over long periods of time, they may look to shorter time horizons than payer counterparts in other markets that take a long-term societal view (e.g., Australia, Canada, the United Kingdom). However, it is standard practice in the competitive marketplace for health insurance to provide coverage for interventions that have long-term benefits, such as statin therapy.

FACTORS DETERMINING COST-EFFECTIVENESS

Numerous factors must be jointly considered when assessing the value of a personalized medicine strategy, including the characteristics of the patients identified for testing, costs directly associated with testing, as well as the test's impact on clinical decision making, patient behavior, and subsequent health outcomes (Table 22.2) [7]. Even with a low-cost test, if only a small proportion of patients tested warrant a change in standard therapy, the clinical benefits of the treatment change for this group must be sufficient to offset the total cost of testing the broader patient population. On the other hand, if standard care comprises a costly drug regimen (e.g., targeted oncology treatments or treatment combinations), a test that could identify just a small fraction of patients more likely to benefit from a less expensive treatment could be highly cost-effective or cost-saving.

Because personalized medicine tests have no direct impact on health outcomes, their cost-effectiveness relates to their clinical utility—the extent to which test results modify treatment decisions and the impact those treatments have on QALY gains. Personalized medicine tests likely to yield greater clinical utility include those where short-term surrogate markers (e.g., blood pressure) are not available to clinicians to monitor response to therapy; when an adverse drug-related event predicted by a test is severe, costly, or prolonged (e.g., abacavir hypersensitivity); when the lack of drug response leads to major clinical events (e.g., antiplatelet therapy to reduce risk of vascular events); when the baseline risk of a clinical event is high (e.g., risk of myocardial infarction after coronary revascularization); and when there is a

TABLE 22.2 Factors Influencing the Value of a Personalized Medicine Testing Strategy

Factors Associated With Testing

- Cost required to identify candidate patients
- Test performance/positive and negative predictive value, as this influences treatment decisions
- The extent to which testing is ordered in appropriately selected patients
- Costs directly associated with testing, including:
  - Time for counseling and obtaining consent
  - Acquiring the sample(s) for testing
  - Laboratory costs to run the test, including assay costs
  - Cost of reporting test results
  - Interpreting the test result and documenting in medical records
  - Potential follow-up tests
  - Additional office visits to discuss results and implications with the patient and caregiver
- The relative sizes of the groups identified by the test (e.g., prevalence of a genetic or somatic biomarker) that would be considered treatment responders or benefit from a specific care approach
- Prognostic and/or predictive power associated with test results

Factors Associated With Treatment Patterns With and Without Testing

- Extent to which providers change treatment decisions as a result of the test/follow test results
- Cost of the treatment and downstream costs avoided (e.g., improved health outcomes or patient-centric outcomes) or incurred (e.g., costs associated with adverse drug effects)
- Identification of the appropriate standard of care (may vary by market)
- The fidelity with which the treatment is delivered (e.g., appropriate dosing, adherence to therapy), the extent to which treatment strategies change based on test results
- The relative cost of alternative treatment strategies including monitoring, management of side effects
- The relative magnitude of treatment effects associated with alternative treatments

Factors Associated With Subsequent Outcomes

- Costs associated with toxicities and clinical events
- Impact of toxicities and clinical events on quality of life and survival (i.e., QALYs)
- Timing of toxicities and clinical events

significant proportion of patients in whom testing leads to a departure from standard treatment.

For many existing tests, however, few studies report on how frequently treatment decisions are changed based on test results and even fewer report on differences in health outcomes between test versus no test strategies [8]. The lack of data on clinical utility is often cited as a barrier to wider implementation and reimbursement for diagnostic tests [9,10]. In the afore-mentioned review of cost–utility analyses, the authors found that all six tests that met Centers for Disease Control and Prevention criteria for demonstrating clinical utility also had cost–utility analyses performed. Among tests with "likely" clinical utility, only one-fifth had associated cost–utility data.

DECISION ANALYTIC MODELS

Decision analytic models are often developed by health technology product manufacturers to investigate the potential cost-effectiveness of a personalized medicine strategy before its impact on clinical decision making or clinical utility is known. These models can incorporate data from a variety of sources, encompass the decision problem including patient benefits and health care costs, and quantify uncertainty [11]. They are the most common approach to conducting cost-effectiveness studies. Decision models enable stakeholders to assess value drivers and the maximum willingness to pay per QALY threshold at which the strategy is cost-effective. In early cost-effectiveness analyses of genetic profiles in the setting of estrogen receptor-positive, node-negative breast cancer, analysts assumed that all patients would forgo chemotherapy when assay results suggested low risk of recurrence and that all other patients would receive chemotherapy [12–14]. These studies predicted that genetic profiling would reduce overall medical costs. However, as genetic profiling was adopted in practice, retrospective and prospective cohort studies revealed that receipt of chemotherapy was not perfectly in accord with patients' risk levels as determined by genetic profiling [15,16]. When these "real-world" treatment patterns were integrated into decision models, the cost-effectiveness of genetic profiling for early-stage breast cancer was less attractive than the early models predicted based on ideal treatment patterns although, by most standards, these tests were still considered as good value [17,18].

Detractors may use these results to criticize "early-stage" decision models as hypothetical and overly optimistic. Such examples reflect that we are still in the early days of evidence generation for a variety of test applications (e.g., risk assessment, prediction, diagnosis, treatment selection, and monitoring), each of which has different associated value drivers that are not yet consistently considered in practice. However, the development of transparent decision models requires that analysts make explicit all of a model's parameter values and assumptions. Early modeling efforts provide an analytic framework to evaluate whether a test set at specific prices could be cost-effective across various scenarios [19,20]. Threshold analyses can determine the minimum extent to which clinical decisions would have to change to obtain good value for money with a test. Decision models can also examine the cost-effectiveness of a testing strategy in specific subgroups defined by age, gender, or comorbid conditions or the impact of specific model parameters that may vary across individuals like the impact of anxiety or disease symptoms on quality of life [21].

VALUE OF INFORMATION ANALYSIS

With more advanced modeling efforts, decision models can estimate the value of conducting further research. Modelers can assign distributions in place of point estimates for model parameters and apply Monte Carlo simulation in a probabilistic sensitivity analysis to examine how uncertainty across a model's parameters jointly impacts the results. With a few additional steps, analysts can generate formal value of information (VOI) estimates representing the expected value of ascertaining additional information to improve the precision of a model's parameters and reduce the likelihood of incorrect conclusions about a test's cost-effectiveness [22,23]. VOI analyses often reveal that a model's findings are largely dependent on the values of one or a few parameters. The net value of decreasing uncertainty with additional data can then be compared with the cost of research to acquire it. This information can help prioritize data collection efforts by public research funding agencies, clinical trial networks, or product manufacturers as recently demonstrated for funding future studies in cancer genomics [24,25]. Although VOI analysis is just beginning to be used in the health care sector, it offers significant promise in personalized medicine in particular because of the paucity of direct evidence of clinical utility (e.g., from randomized clinical trials).

PERSONALIZED MEDICINE IN CANCER

There is little question that personalized medicine is transforming cancer care. Use of genomic information for acquired somatic mutations in tumors that can guide treatment selection is already standard of care for several cancers. In addition to the scientific rationale for developing targeted drug therapies, personalized care has high value potential owing to the high cost of many new cancer therapies, significant drug-associated risks and toxicities, and potential for effective treatments to delay disease progression and death.

Economic evaluations of personalized medicine strategies in cancer run the gamut from those focused solely on cost offsets associated with testing, to those examining the cost-effectiveness of a drug versus standard care within the targeted subgroup (without consideration of patients in the nontargeted group), to highly complex models examining multiple sequential testing and treatment strategies [26].

The introduction of monoclonal antibodies, including the anti-vascular endothelial growth factor (VEGF) antibody bevacizumab and anti-epidermal growth factor receptor (EGFR) antibodies cetuximab and panitumumab (and more recently, regorafenib and aflibercept), represented a significant advance in treating metastatic colorectal cancer. Clinical studies had shown higher response rates among individuals with KRAS wild-type tumors, but the cost of treatment was high. Several economic evaluations have confirmed that KRAS mutation testing prior to treatment with monoclonal antibodies saves costs [27−29]. As expected, this relatively low-cost test for a reasonably prevalent biomarker (∼60% of the population could be identified as responders) that could limit the number of patients who received a costly drug was rapidly adopted and reimbursed by most payers [9]. Acceptance of this personalized medicine scenario was influenced by several payer-relevant factors, including the ability to avoid providing the drug to a substantial portion of the tested population that will not benefit (i.e., 40%), the number needed to test was deemed reasonable, and the unit cost of available testing platforms for KRAS was small in comparison to the cost of the innovator oncology therapy. In such cases, a complex decision model may be viewed as unnecessary, but there are additional issues to consider such as the cost-effectiveness of the targeted treatment, costs of subsequent lines of treatment, and potential clinical benefits forgone in the nontargeted subgroup who did not receive the treatment.

When the monoclonal antibodies were compared to lower-cost standard chemotherapy regimens (primarily combinations of 5-fluorouracil, leucovorin, oxaliplatin, irinotecan) for first-line therapy in metastatic colorectal cancer, the ICERs generally surpassed the relatively well-accepted threshold in the United States of $100,000 per QALY even when their use was limited to individuals without KRAS and BRAF mutations [27,30,31]. Thus, although KRAS testing itself was shown to be a cost-saving strategy from a payer perspective (compared to use of monoclonal antibodies in all patients), the economic value of the test combined with targeted therapy was unfavorable because the targeted therapy itself was not cost-effective. This is not uncommon for new cancer drugs due to drug pricing for many branded products, with one recent example being an evaluation of combination therapy including pertuzumab in HER2-positive breast cancer finding that the addition of the drug would not be cost-effective even under the most favorable conditions [32]. Another analysis of monoclonal antibodies in metastatic colorectal cancer reveals the importance of the comparator. When costs and outcomes of monoclonal antibodies were compared, bevacizumab was found to represent good value. In this study, bevacizumab was less costly than the other monoclonal antibodies but differences in outcomes were uncertain due to the lack of head-to-head studies [30].

Some economic evaluations reveal that it may be cost-effective to forgo personalized medicine testing and treat all patients with the treatment of interest, whereas others suggest testing followed by targeted therapy is not cost-effective and standard care is preferred. In one early modeling study of EGFR testing in non-small-cell lung cancer (NSCLC), uniform use of erlotinib without testing was shown to be preferable to targeted use of erlotinib as long as the willingness to pay per QALY threshold was less than $150,000/

QALY [33]. These findings were consistent with a trial-based economic evaluation in which the ICER was $94,600 (2007 Canadian dollars) per life-year gained across all patients with advanced NSCLC, although cost-effectiveness was improved in some subgroups [34]. An example where neither targeted treatment nor uniform treatment with a new drug was considered a good value was an early model-based study of olaparib, a PARP inhibitor used as maintenance therapy in recurrent ovarian cancer. Despite greater effectiveness of olaparib among patients with BRCA 1/2 mutations [35], a cost-effectiveness analysis using progression-free survival (PF-LYS) as an outcome found that neither global use of olaparib ($234,000/ PF-YLS) nor the test-olaparib strategy ($193,000/PF-YLS) represented good value relative to no maintenance therapy [36].

EVOLVING ISSUES IN ECONOMIC EVALUATIONS IN CANCER

The inconsistent findings on cost-effectiveness of targeted cancer agents underscore the need to examine value on a case-by-case basis and the complexity of modeling considerations [37]. However, the rate at which new personalized medicine tests and targeted therapies are entering the market is faster than publication of high-quality economic evaluations. These studies are demanding in terms of data requirements, time, and technical expertise required. The daunting task of synthesizing this information is evidenced by private payers' reliance on subscriptions to health technology assessments (HTAs) to review new personalized medicine strategies although an acknowledged limitation is their lack of cost-effectiveness analyses [38]. Payers are addressing these challenges using specific approaches for evaluating companion diagnostics in oncology [39].

Understanding of genomic alterations associated with tumor development and metastasis is increasing. Cancers historically classified by organ involvement are now classified by molecular characteristics, an approach that may: (1) involve dozens to hundreds or thousands of markers or (2) have application across multiple cancer types potentially altering clinical trial and evidence development approaches [40–42]. The next era in genomic profiling for cancer therapy will present new challenges for analysts conducting cost-effectiveness analyses and payers who want to ensure that medical expenditures are used wisely. Analysts may confront sparse data from randomized trials on comparative safety and efficacy of drug combinations in highly specific groups. Even after cancer patients have experienced disease progression following several courses of chemotherapy, genetic profiling can identify additional drugs that may offer therapeutic benefit [43]. Evidence of a potential biologic mechanism may place pressure on payers for coverage of additional indications, even if evidence to support its use in a specific cancer is lacking, requiring development of adaptive modeling approaches and/or alteration of evidence requirements, including those involving postmarket real-world evidence, to address core evidence gaps.

MARKET ACCESS AND ADOPTION

Personalized medicine also has special considerations for market access, which not only includes evaluation and reimbursement by HTA and payer bodies, but also includes other access gatekeepers such as providers, health systems, policy makers, laboratories, and patients.

Conceptually, personalized medicines should more easily attain market access if they provide improved outcomes to a targeted subset of patients and reduce costs, but this is not always the case. Regulatory, reimbursement, and provider practices and policies are lagging behind this rapidly evolving field. Evidence requirements and decision criteria for market access are heterogeneous across geographies, though several markets, including Australia, Canada, and the EU5, have begun to focus attention on this area [9]. In addition, manufacturers face a steep learning curve in addressing market access challenges for personalized medicines.

All acceptance drivers for conventional pharmaceuticals, such as degree of unmet medical need, balance of benefits and risks versus alternatives, cost versus alternatives, and patient compliance, are also applicable to personalized medicines. Several unique factors also exist for personalized medicines (Table 22.3). Some may be fundamental to economic assessment, while

---

TABLE 22.3 Market Access Factors Unique to Personalized Medicines vs. Conventional Pharmaceuticals

- Expectations for a smaller target population than conventional treatments
- Trial designs that account for test use
- Size of the responder population[a]
- Requirements for clear linkage of test results to treatment selection decisions
- Expectations for greater efficacy and/or safety vs. conventional alternatives[a]
- Some unique economic end points (e.g., avoidance of financial "wastage")[a]
- Implications of test performance on treatment decisions[a]
- Implications of test interpretability (continuous vs. binary)
- Reflexive vs. sequential testing scenarios, where relevant
- Regulatory and reimbursement issues also involve the diagnostic (i.e., FDA approved / CE Marked (indicating compliance with EU legislation) vs laboratory developed tests)
- Cost of the test/testing is considered in value assessment[a]
- HTA channels/funding sources for companion diagnostics can differ from those associated with pharmaceuticals
- Combined budget impact & cost-effectiveness of the test and companion drug considered[a]

[a]Factors commonly considered in current state economic modeling of personalized medicines in some markets.

others address processes such as education and acceptance by health stakeholders, reimbursement systems that did not anticipate treatment use being conditional upon test results, and drivers of appropriate test ordering and use.

Also, HTA pathways for companion diagnostics (i.e., test development in the context of a drug clinical trial) and co-diagnostics (i.e., a test developed separately from the drug but used in treatment selection) are markedly different. For the former, the diagnostic test evaluation can be included with the drug assessment (e.g., as in Germany and the United Kingdom) or follow a device HTA route (e.g., as in France and Australia). Developing evidence of clinical utility is typically much easier for a companion diagnostic because of the ability to directly correlate the evidence linking test use to treatment selection and outcomes. On the other hand, developing similar direct evidence is beyond most diagnostic developers due to health system/market and intellectual property incentive structures. Under this latter scenario, gaps in the evidence base are more frequent, reducing the certainty cost-effectiveness estimation around the test.

Payers also have different perspectives around sufficient clinical evidence to support diagnostic tests, both in terms of acceptable study design and levels of evidence to support reimbursement. Many US (e.g., EGAPP, BCBS TEC, CMTP) and global groups (e.g., the NICE Diagnostic Assessment Committee, EUnetHTA) are considering diagnostic evidence expectations, but as yet, no single accepted perspective has emerged [44].

Economic issues associated with personalized medicine market access can also vary substantially by market [45,46]. For example, in the United States, most commercial payers will look at budget impact versus overall clinical benefits and consider factors such as whether the product combination helps to avoid resource wastage. The HTA body for Great Britain, on the other hand, requires a cost-effectiveness analysis, but many of the unique factors associated with personalized medicines may not be fully considered beyond test performance and cost of the test. The most complex example is evident in Australia, which has explicit personalized medicine assessment criteria [47].

Given this variability, personalized medicine developers must take a broader perspective in value characterization to address this global variability in clinical and health economic evidence requirements. Economic modelers must also consider these additional complexities to develop models that address a range of value metrics.

NEXT-GENERATION SEQUENCING

As we enter the era of whole-genome sequencing and evolution of next-generation sequencing (NGS) testing, it is logical to ask whether we need to question the value of spending $1000 (or less) for an entire genome's worth of information when annual medical expenditures in the United States exceeds $9000 per capita [48]. The inherent difference with NGS tests is that many (e.g., 10−100) genes that a priori are suspected to be causative can be sequenced at the same time using a "gene panel," as opposed to a previous "one gene at a time" sequencing, for about the same cost. Furthermore, NGS can be used to sequence all of the coding regions of the genome (e.g., 10,000 genes) in "whole-exome sequencing," for only a modest increase in effort compared to using directed gene panels.

Three key challenges exist to better understanding the value of NGS technology for clinical care. First, what are the patient and economic impacts of improvements in clinical diagnosis, prognosis, and prediction? If we improve diagnostic yield, what reduction in morbidity and mortality can we expect, and what are the downstream impacts on health care costs? It is likely that for clinical indications where evidence exists that genetic diagnosis and subsequent care alterations provide good patient and economic value, NGS will be cost-effective. For example, Gallego et al. [49] recently evaluated the cost-effectiveness of an NGS colon cancer gene panel for diagnosis of Lynch and related syndromes compared to usual care. The results of the study suggest that NGS in this clinical indication is likely to improve patient outcomes in a cost-effective manner by including highly penetrant genes related to colon cancer and polyposis in addition to those specific to Lynch syndrome.

A second challenge is understanding the value of returning incidental findings to patients that arise as a result of the breadth of genes that can be assessed using NGS. Furthermore, known clinically actionable incidental findings are fairly rare, occurring in approximately 1−3% of patients. Very large (e.g., 100,000 patients) observational, real-world studies will be needed to ascertain impacts on patients and health care systems—and the recently launched Precision Medicine Initiative Cohort promises to provide the platform for conducting such studies. Lastly, creative approaches to modeling long-term impacts will be needed given the diversity of lifetime impacts of various incidental findings.

Bennette et al. [50] recently developed an early-stage Incidental Findings Policy Model. The authors modeled the long-term impacts of returning findings from the American College of Medical Genetics (ACMG) list of 56 genes in which pathogenic or likely pathogenic variants are considered clinically actionable. The preliminary findings suggest that returning incidental findings is likely cost-effective, especially for younger and healthier patient populations, and that additional data on the frequency and penetrance of the incidental finding genes is needed.

Third, there are concerns that the QALY does not adequately capture all potential benefits of personalized medicine, such as the "value of knowing" even when test results do not impact treatment decisions or improved communication between providers and patients [2,51,52]. Individuals may make different personal choices about child-bearing, where to live, how long to work, or how much to save. Quantitative survey techniques such as "discrete choice experiments" can be used to quantify such value. A recent study in Canada by Regier et al. [53] used this technique to assess the value to patients of incidental findings. The survey of 1200 members of the general public revealed that a policy scenario aligned with the ACMG recommendations had a willingness to pay of $446 and a predicted uptake of 66%, although there was significant heterogeneity [53]. Some patients wanted all incidental genomic information, and others wanted none. Future work on the economic value of clinical sequencing using NGS should focus on: (1) creating a framework for measuring and interpreting the economic value of clinical sequencing that is acceptable to technology developers, providers, and health care payers; (2) very large observational studies of the direct and indirect impacts of clinical sequencing; and (3) a better understanding of the value patients place on findings from clinical sequencing.

In summary, the value proposition for personalized medicine offers the same tremendous opportunities and challenges as personalized medicine itself. The key challenges moving forward are: (1) to develop a common understanding among various stakeholders about the general concepts of economic value of personalized medicine; (2) advance the methods of economic evaluation to appropriately consider the "value of knowing" for patients and providers as well as enable the assessment of large gene panels or whole-genome sequencing; and (3) collect sufficient evidence to inform economic evaluations—and develop a broader acceptance of what defines "sufficient" evidence for personalized medicine. As progress is made to address these challenges, a simple answer to addressing the value of personalized medicine will remain elusive as there is not a one-size-fits-all solution. As for specific drugs, biologics, and other medical interventions, scrupulous evaluation of comparative costs and outcomes between specific management strategies will be needed to build an evolving evidence base.

REFERENCES

[1] Lopert R, Elshaug AG. Australia's fourth hurdle' drug review comparing costs and benefits holds lessons for the United States. Health Aff (Millwood) 2013;32:778–87.
[2] Annemans L, Redekop K, Payne K. Current methodological issues in the economic assessment of personalized medicine. Value Health 2013;16:S20–6.
[3] Buchanan J, Wordsworth S, Schuh A. Issues surrounding the health economic evaluation of genomic technologies. Pharmacogenomics 2013;14:1833–47.
[4] Gold MR, Siegel JE, Russell LB, Weinstein M. Cost-effectiveness in health and medicine. New York, NY: Oxford University Press; 1996.
[5] Phillips KA, Sakowski JA, Trosman J, Douglas MP, Liang S, Neumann P. The economic value of personalized medicine tests: what we know and what we need to know. Genet Med 2014;16:251–7.
[6] Hatz MH, Schremser K, Rogowski WH. Is individualized medicine more cost-effective? A systematic review. Pharmacoeconomics 2014;32(5):443–55.
[7] Canestaro WJ, Pritchard DE, Garrison LP, Dubois R, Veenstra DL. Improving the efficiency and quality of the value assessment process for companion diagnostic tests: the Companion test Assessment Tool (CAT). J Manag Care Spec Pharm 2015;21:700–12.

[8] Conti R, Veenstra DL, Armstrong K, Lesko LJ, Grosse SD. Personalized medicine and genomics: challenges and opportunities in assessing effectiveness, cost-effectiveness, and future research priorities. Med Decis Making 2010;30:328−40.

[9] Faulkner E, Annemans L, Garrison L, Helfand M, Holtorf AP, Hornberger J, et al. Challenges in the development and reimbursement of personalized medicine-payer and manufacturer perspectives and implications for health economics and outcomes research: a report of the ISPOR personalized medicine special interest group. Value Health 2012;15:1162−71.

[10] Phillips KA, Van Bebber SL. Measuring the value of pharmacogenomics. Nat Rev Drug Discov 2005;4:500−9.

[11] Guzaukas GF, Garrison LP, Stock J, Au S, Doyle DL, Veenstra DL. Stakeholder perspectives on decision-analytic modeling frameworks to assess genetic services policy. Genet Med 2013;15(1):84−7.

[12] Hornberger J, Cosler LE, Lyman GH. Economic analysis of targeting chemotherapy using a 21-gene RT-PCR assay in lymph-node-negative, estrogen-receptor-positive, early-stage breast cancer. Am J Manag Care 2005;11:313−24.

[13] Lyman GH, Cosler LE, Kuderer NM, Hornberger J. Impact of a 21-gene RT-PCR assay on treatment decisions in early-stage breast cancer: an economic analysis based on prognostic and predictive validation studies. Cancer 2007;109:1011−18.

[14] Oestreicher N, Ramsey SD, Linden HM, McCune JS, van't Veer LJ, Burke W, et al. Gene expression profiling and breast cancer care: what are the potential benefits and policy implications? Genet Med 2005;7:380−9.

[15] Lo SS, Mumby PB, Norton J, Rychlik K, Smerage J, Kash J, et al. Prospective multicenter study of the impact of the 21-gene recurrence score assay on medical oncologist and patient adjuvant breast cancer treatment selection. J Clin Oncol 2010;28:1671−6.

[16] Rayhanabad JA, Difronzo LA, Haigh PI, Romero L. Changing paradigms in breast cancer management: introducing molecular genetics into the treatment algorithm. Am Surg 2008;74:887−90.

[17] Reed SD, Dinan MA, Schulman KA, Lyman GH. Cost-effectiveness of the 21-gene recurrence score assay in the context of multifactorial decision making to guide chemotherapy for early-stage breast cancer. Genet Med 2013;15:203−11.

[18] Paulden M, Franek J, Pham B, Bedard PL, Trudeau M, Krahn M. Cost-effectiveness of the 21-gene assay for guiding adjuvant chemotherapy decisions in early breast cancer. Value Health 2013;16:729−39.

[19] Reed SD, Stewart SB, Scales Jr CD, Moul JW. A framework to evaluate the cost-effectiveness of the NADiA ProsVue slope to guide adjuvant radiotherapy among men with high-risk characteristics following prostatectomy for prostate cancer. Value Health 2014;17:545−54.

[20] Trikalinos TA, Kulasingam S, Lawrence WF. Chapter 10: deciding whether to complement a systematic review of medical tests with decision modeling. J Gen Intern Med 2012;27(Suppl 1):S76−82.

[21] Basu A, Meltzer D. Value of information on preference heterogeneity and individualized care. Med Decis Making 2007;27:112−27.

[22] Claxton K. Bayesian approaches to the value of information: implications for the regulation of new pharmaceuticals. Health Econ 1999;8:269−74.

[23] Woods B, Veenstra D, Hawkins N. Prioritizing pharmacogenetic research: a value of information analysis of CYP2D6 testing to guide breast cancer treatment. Value Health 2011;14:989−1001.

[24] Carlson JJ, Thariani R, Roth J, Gralow J, Henry NL, Esmail L, et al. Value-of-information analysis within a stakeholder-driven research prioritization process in a US setting: an application in cancer genomics. Med Decis Making 2013;33:463−71.

[25] Thariani R, Wong W, Carlson JJ, Garrison L, Ramsey S, Deverka PA, et al. Prioritization in comparative effectiveness research: the CANCERGEN Experience. Med Care 2012;50:388−93.

[26] Wong Y, Meropol NJ, Speier W, Sargent D, Goldberg RM, Beck JR. Cost implications of new treatments for advanced colorectal cancer. Cancer 2009;115:2081−91.

[27] Lange A, Prenzler A, Frank M, Kirstein M, Vogel A, von der Schulenburg JM. A systematic review of cost-effectiveness of monoclonal antibodies for metastatic colorectal cancer. Eur J Cancer 2014;50:40−9.

[28] Health Quality Ontario. KRAS testing for anti-EGFR therapy in advanced colorectal cancer: an evidence-based and economic analysis. Ont Health Technol Assess Ser 2010;10:1−49.

[29] Vijayaraghavan A, Efrusy MB, Göke B, Kirchner T, Santas CC, Goldberg RM. Cost-effectiveness of KRAS testing in metastatic colorectal cancer patients in the United States and Germany. Int J Cancer 2012;131:438−45.

[30] Lawrence D, Maschio M, Leahy KJ, Yunger S, Easaw JC, Weinstein MC. Economic analysis of bevacizumab, cetuximab, and panitumumab with fluoropyrimidine-based chemotherapy in the first-line treatment of KRAS wild-type metastatic colorectal cancer (mCRC). J Med Econ 2013;16:1387−98.

[31] Hoyle M, Crathorne L, Peters J, Jones-Hughes T, Cooper C, Napier M, et al. The clinical effectiveness and cost-effectiveness of cetuximab (mono- or combination chemotherapy), bevacizumab (combination with non-oxaliplatin chemotherapy) and panitumumab (monotherapy) for the treatment of metastatic colorectal cancer after first-line chemotherapy (review of technology appraisal No.150 and part review of technology appraisal No. 118): a systematic review and economic model. Health Technol Assess 2013;17:1−237.

[32] Durkee BY, Qian Y, Pollom EL, King MT, Dudley SA, Shaffer JL, et al. Cost-effectiveness of pertuzumab in human epidermal growth factor receptor 2-positive metastatic breast cancer. J Clin Oncol 2016;34:902−9.

[33] Carlson JJ, Garrison LP, Ramsey SD, Veenstra DL. The potential clinical and economic outcomes of pharmacogenomic approaches to EGFR-tyrosine kinase inhibitor therapy in non-small-cell lung cancer. Value Health 2009;12:20−7.

[34] Bradbury PA, Tu D, Seymour L, Isogai PK, Zhu L, Ng R, et al. Economic analysis: randomized placebo-controlled clinical trial of erlotinib in advanced non-small cell lung cancer. J Natl Cancer Inst 2010;102:298−306.

[35] Ledermann J, Harter P, Gourley C, Friedlander M, Vergote I, Rustin G, et al. Olaparib maintenance therapy in platinum-sensitive relapsed ovarian cancer. N Engl J Med 2012;366:1382−92.

[36] Secord AA, Barnett JC, Ledermann JA, Peterson BL, Myers ER, Havrilesky LJ. Cost-effectiveness of BRCA1 and BRCA2 mutation testing to target PARP inhibitor use in platinum-sensitive recurrent ovarian cancer. Int J Gynecol Cancer 2013;23:846−52.

[37] Veenstra DL, Higashi MK, Phillips KA. Assessing the cost-effectiveness of pharmacogenomics. AAPS PharmSci 2000;2:E29.

[38] Trosman JR, Van Bebber SL, Phillips KA. Health technology assessment and private payers's coverage of personalized medicine. J Oncol Pract 2011;7(3 Suppl):18s−24s.

[39] The AMCP Format for Formulary Submissions Version 3.1. Available from: http://amcp.org/practice-resources/amcp-format-formulary-submisions.pdf [accessed December 2014].

[40] Fox J. Master protocol for squamous cell lung cancer readies for launch. Nat Biotechnol 2014;32:116–18.

[41] Sledge Jr. GW. The challenge and promise of the genomic era. J Clin Oncol 2012;10:203–9.

[42] Schilsky RL. Implementing personalized cancer care. Nat Rev Clin Oncol 2014;11:432–8.

[43] Staren ED, Braun D, Tan B, Gupta D, Kim S, Kramer K, et al. Initial experience with genomic profiling of heavily pretreated breast cancers. Ann Surg Oncol 2014;21:3216–22.

[44] Faulkner E., Tunis S., Branham C., Towse A. Cutting the fog: have we reached clarity on diagnostics and personalized medicine evidence expectations? International Society for Pharmacoeconomics and Outcomes Research, 16th annual international meeting. Washington, DC; 2012.

[45] Fugel HJ, Nuijten M, Faulkner E. The application of economics concepts to stratified medicine-use of health economics data to support market access for stratified medicine interventions. J Med Econ 2014;17:305–11.

[46] Garfield S., Faulkner E., Potsulka A., Berndt K. Navigating a changing landscape: the next generation of diagnostic reimbursement and health care management models. International Society for Pharmacoeconomics and Outcomes Research, 17th annual European Congress. Amsterdam, Netherlands; 2014.

[47] Merlin T, Farah C, Schubert C, Mitchell A, Hiller JE, Ryan P. Assessing personalized medicines in Australia: a national framework for reviewing codependent technologies. Med Decis Making 2013;33:333–42.

[48] Hartman M, Martin AB, Lassman D, Catlin A, National Health Expenditure Accounts Team. National health spending in 2013: growth slows, remains in step with the overall economy. Health Aff (Millwood) 2015;34:150–60.

[49] Gallego CG, Shirts BH, Bennette CS, Guzauskas G, Amendola LM, Horike-Pyne M, et al. Next generation sequencing panels for the diagnosis of colorectal cancer and polyposis syndromes: a cost-effectiveness analysis. San Diego, CA: American Society of Human Genetics; 2014.

[50] Bennette CS, Gallego CJ, Burke W, Jarvik GP, Veenstra DL. The cost-effectiveness of returning incidental findings from next-generation genomic sequencing. Genet Med 2015;17:587–95.

[51] Payne K, McAllister M, Davies LM. Valuing the economic benefits of complex interventions: when maximising health is not sufficient. Health Econ 2013;22:258–71.

[52] Bennette CS, Trinidad SB, Fullerton SM, Patrick D, Amendola L, Burke W, et al. Return of incidental findings in genomic medicine: measuring what patients value—development of an instrument to measure preferences for information from next-generation testing (IMPRINT). Genet Med 2013;15:873–81.

[53] Regier DA, Peacock SJ, Pataky R, van der Hoek K, Jarvik GP, Hoch J, et al. Societal preferences for the return of incidental findings from clinical genomic sequencing. CMAJ 2015;187(6):E190–7.

Chapter 23

Technology Assessment and the Road to Reimbursement of Genomic Based Diagnostics

Robert McDonough and Joanne Armstrong
Aetna, Sugar Land, TX, United States

Chapter Outline

INTRODUCTION

As the science of genomics matures, the number of genetic tests available for use in clinical care has increased greatly. By mid-2016, the National Institutes of Health's Genetic Testing Registry included more than 48,000 tests related to more than 9,700 medical conditions [1]. The number of available tests is expected to increase in the years ahead as technology platforms and bioinformatics capabilities improve.

In the clinical setting, genetic tests are performed to screen for or diagnose disease, provide predictive or prognostic information about a disease, or predict response to drug therapy. The role of some genetic tests in clinical care is well established, such as *BRCA* mutation screening in individuals at high risk for hereditary breast or ovarian cancer [2]. Genetic testing for selected somatic and germline mutations is emerging as a strategy to guide or "personalize" drug therapy in a variety of diseases, most commonly for oncology. In fact, more than 100 therapeutics have biomarker tests included in their Food and Drug Administration (FDA) labels [3].

Notwithstanding the significant advances in genomic medicine, in many cases, the *value* of genetic tests to clinical care is only partially understood.

Genomic and Precision Medicine. DOI: http://dx.doi.org/10.1016/B978-0-12-800681-8.00023-2

Yet, because of the process by which laboratory tests enter the market, virtually all genetic tests—regardless of their level of scientific validation or usefulness in clinical practice—have the potential to be marketed to patients and physicians. The availability and marketing of a genetic test, however, does not ensure adoption in the clinic. A key step influencing the uptake of a genetic test in clinical practice is the *determination of its medical necessity*, which typically determines whether payers will cover the test. Technologies that are not covered under insurance plans are much less likely to be utilized because the cost of testing born by the patient may be prohibitive [4]. This chapter will describe the framework for coverage decision making under commercial insurance plans.

CURRENT REGULATION OF GENETIC TESTING

Genetic tests are regulated by multiple agencies including, the US FDA, the Centers for Medicare and Medicaid Services (CMS), and the Federal Trade Commission (FTC) (Fig. 23.1). The FDA has the authority under the Federal Food, Drug, and Cosmetic Act to regulate medical devices, which include genetic test "kits." A genetic test kit is a group of reagents that are packaged together with instructions on test performance and sold to multiple laboratories. The FDA views genetic test kits as devices and reviews them with the degree of scrutiny varying depending upon the level of risk to the patient. Genetic test kits that are considered lower risk are deemed Class II devices and require clearance based upon 510(k) premarket notification. A 510(k) premarket notification must demonstrate that the device is substantially equivalent to one legally in commercial distribution in the United States: (1) before enactment of medical device amendments to the Food Drug and Cosmetic Act in 1976 or (2) to a device that has been determined by FDA to

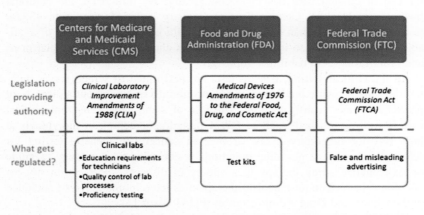

FIGURE 23.1 Regulatory oversight of genetic tests. Source: *From National Human Genome Institute. Regulation of genetic tests, http://www.genome.gov/10002335 [accessed 25.08.15].*

be substantially equivalent [5]. The 510(k) review of low-risk genetic test kits focuses primarily on analytic validity (see below). Examples of Class II genetic tests include prenatal carrier screening tests such as for the causative mutations in cystic fibrosis and Bloom syndrome [6].

Test kits deemed at highest risk (Class III devices) must undergo premarket approval (PMA) [7]. Devices requiring PMAs have been determined by the FDA to pose a significant risk of illness or injury, or devices found not substantially equivalent to a Class I and II predicate through the 510(k) process. The PMA process is more involved than 510(k) review and includes the submission of clinical data to support claims made for the device and, in some cases, may require a clinical trial. PMA of genetic test kits requires evidence of both analytic validity and clinical validity but not clinical utility (see discussion below).

Many genetic tests are novel and may lack a predicate device upon which to be compared. Historically, the FDA classified novel devices without a predicate as Class III risk level and thus subject to the PMA review process. The FDA Modernization Act of 1997 created an alternative pathway to market for these devices called the de novo pathway. This pathway is intended for low-risk devices only. The manufacturer can request a FDA reclassification of the product to Class I or Class II risk. If awarded, the device is reviewed through the 510(k) process and not the more rigorous PMA process. Devices that are determined by the FDA to be low risk (Class I or II) through the de novo process may be marketed and used as predicates for future 510(k) submissions [8].

Most genetic tests are not marketed as test kits and, thus, do not undergo FDA review. Most available genetic tests are categorized as laboratory-developed tests (LDTs). LDTs are tests designed, validated, manufactured, and performed within a single clinical laboratory, and where specimen samples are sent to that laboratory to be tested. LTDs include a wide array of tests performed to detect or measure DNA, RNA, chromosomes, proteins, and certain metabolites using biochemical, cytogenetic, or molecular methods or a combination of these methods in a sample taken from a human body [9]. Historically, LDTs were relatively simple tests, limited in number, and used for the diagnosis of rarely diagnosed disorders, and, thus considered to be low risk to patients. In recent years, the number of LDTs has grown dramatically and increased in complexity. And they are being used in the evaluation of common disorders thus making it more important than ever to ensure that these tests are appropriately validated.

Most LDTs do not have a commercial FDA-approved equivalent. Unlike test kits, LDTs are generally not regulated by the FDA (although some of their components may be) and can find their way to market through a different review process. FDA has the regulatory authority to review LDTs, but it exercises "enforcement discretion" and generally leaves oversight of LDTs to the laboratories performing the tests [10]. The laboratories are regulated under a different review process by authority vested in CMS [11].

CMS, through its authority under the Clinical Laboratory Improvement Act (CLIA), inspects and regulates clinical laboratories where LDTs are developed and performed. CLIA regulations describe quality standards for laboratory testing including certification that the laboratory meets certain standards related to personnel qualifications, quality control procedures, and proficiency testing programs [12]. Under CLIA, the analytic validity of a test as performed in that laboratory is reviewed, but CLIA does not require evaluation of the clinical validity or clinical utility of any particular test. No special requirements exist for laboratories performing DNA-based and other types of genetic tests. No proficiency testing programs in genetics or cytogenetics are required under CLIA [13]. Comparing the quality of tests approved under the FDA process to those reviewed under CLIA programs is not possible. The FDA has indicated that it will increase scrutiny of some LDTs in the future as the number and complexity of LTDs grow. The agency will prioritize review of testing based on the level of risk of the test [14].

Finally, the FTC has broad authority to regulate how genetic tests are marketed to ensure that advertising is not false or misleading. Recently, the FTC has successfully stopped such practices in cases of genetically customized nutritional assessments and genetic assessments of skin health and aging. Still, the overall number of such challenges to false marketing claims of genetic testing companies has been limited [15].

INSURANCE OVERVIEW

Commercial health insurance is a contract between a purchaser (such as an individual or an employer on behalf of its employees) and the insurance provider (such as a commercial health plan). The contract/policy contains a description of the services and benefits covered and excluded. The contract/policy also describes any specific requirements for access to covered services and benefits. The principles that guide the benefits covered under most health plans include the provision that the services covered must relate to the prevention, diagnosis, or treatment of an illness [16]. Consequently, services such as genetic tests that are used for information purposes only such as paternity testing, ancestry testing, or genetic testing used for forensic purposes are usually excluded from coverage. Services covered under the plan must be medically necessary and not experimental and investigational. The information obtained from a covered service must affect the course of treatment of the patient (i.e., health plan member), and that treatment must be likely to improve clinical outcomes. Additionally, the improvements noted must be obtainable outside of investigational settings. Finally, certain medical services may meet the above criteria but may be explicitly excluded from a plan such as tests required because of travel or employment.

Most health plans cover medically necessary services and exclude from coverage services that are considered "experimental," "investigational," or "unproven." Major national health insurance companies and medical professional societies have agreed to use a definition of medical necessity substantially in the following form [17]:

> *"Medically Necessary" or "Medical Necessity" shall mean health care services that a physician, exercising prudent clinical judgment, would provide to a patient for the purpose of preventing, evaluating, diagnosing or treating an illness, injury, disease or its symptoms, and that are: a) in accordance with generally accepted standards of medical practice; b) clinically appropriate, in terms of type, frequency, extent, site and duration, and considered effective for the patient's illness, injury or disease; and c) not primarily for the convenience of the patient, physician or other health care provider, and not more costly than an alternative service or sequence of services at least as likely to produce equivalent therapeutic or diagnostic results as to the diagnosis or treatment of that patient's illness, injury or disease. For these purposes, "generally accepted standards of medical practice" means standards that are based on credible scientific evidence published in peer-reviewed medical literature generally recognized by the relevant medical community or otherwise consistent with the standards set forth in policy issues involving clinical judgment.*

Note that this definition includes standards of evidence and allows an explicit consideration of cost if there is an alternative service that produces equivalent outcomes.

Due to antitrust concerns, each commercial health plan must develop its policies independently. However, the process by which technologies are assessed for effectiveness, and thus covered, under insurance health plans follows common principles of evidence-based review.

TECHNOLOGY ASSESSMENT OF GENETIC TESTS BY PRIVATE PAYERS

The assessment of the effectiveness of health service technologies, including genetic tests, is critical to the determination of medical necessity under the health plans. Health plans conduct health technology assessments (HTAs) to determine whether services are deemed "medically necessary" and eligible for coverage under the insurance benefit plan. Or, the services are deemed "experimental" or "unproven" and ineligible for coverage under the plan.

An HTA is defined a systematic evaluation of properties, effects, and/or impacts of health care technology to inform health care decision makers in health policy or practice [18]. HTAs seek to identify effective technologies and define the most appropriate indications for their use. The HTA process for genetic tests is similar to that for all other technologies.

In evaluating the adequacy of the evidence for medical necessity determination for genetic tests (and other diagnostics), health plans consider the analytic validity, clinical validity, and clinical utility of the test, as well as contextual considerations (ethical, legal, and social)—known as the ACCE model. The ACCE model was formulated by the Center for Disease Control and Prevention's Office of Public Health Genomics as an analytic process to evaluate genetic tests [19] (Fig. 23.2).

The analytic validity of a test describes how accurately and reliably the test measures the genotype of interest: Can the test detect a mutation when it is present (sensitivity) and have a negative test result when the mutation is not present (specificity)? Analytic validity evaluates the consistency and precision of test results within the same or different laboratory using the same or different technology. Clinical validity measures how consistently and accurately the test detects or predicts the intermediate or final outcomes of interest. Clinical validity links genotype to phenotype and seeks to describe the performance of the test as it relates to a clinical condition. It explores how often the test is positive when the disorder is present (sensitivity) and is negative when the disorder is not present (specificity). Clinical utility measures how likely the test is to significantly improve patient outcomes in the "real world." Clinical utility is often referred to the "so what" question.

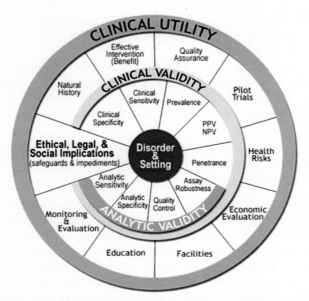

FIGURE 23.2 The ACCE model: an analytical process for evaluating scientific data on emerging genetic tests. Source: *From Haddow JE, Palomaki GE. ACCE: a model process for evaluating data on emerging genetic tests. In: Khoury M, Little J, Burke W, editors. Human genome epidemiology: a scientific foundation for using genetic information to improve health and prevent disease. Oxford University Press; 2003. p. 217–233.*

Is there an intervention that can be applied or withdrawn based on the test result that will result in measurable improvement in patient care and outcomes? There should be either direct evidence or a convincing argument based upon established medical facts that the genetic test alters management such that clinical outcomes are improved. Demonstration of clinical utility is the most frequently cited reason for diagnostic tests failing to obtain coverage. Challenges include the need to link test results to demonstrable improvements in clinical outcomes. Reports of changes in clinician decision making alone are generally considered insufficient evidence of clinical utility. One recent survey found that among the molecular pathology services that failed to obtain coverage by the CMS, 40% were denied due to the lack of adequate clinical utility data [20].

Genetic tests must demonstrate effectiveness in all three domains of test performance to be eligible for coverage by most health plans. Examples of insurance covered genetic tests with proven clinical validity and utility include the Oncotype Dx test for breast cancer and BRCA1 and BRCA2 testing for breast and ovarian cancer [21]. Examples of genetic tests without proven clinical validity or clinical utility that are generally not covered by insurance include the OvaSure test for ovarian cancer and KIF6 genotyping for cardiovascular disease [22,23]. As new evidence emerges, tests can be reevaluated for coverage consideration.

Health plans review multiple sources of evidence to conduct technology assessments including published, peer-reviewed medical literature referenced in National Library of Medicine's PubMed database; evidence-based clinical practice guidelines from leading medical professional associations and governmental health agencies; the regulatory status of technology for those genetic tests that require FDA review; and HTAs such as those indexed in National Library of Medicine, Health Services/Technology Assessment Text database [24]. Of note, a number of organizations perform HTA that might inform payer medical necessity decision related to genetic testing. These include: Blue Cross Blue Shield Technology Evaluation Center, Evaluation of Genomics Application in Practice and Prevention, the Agency for Healthcare Research and Quality, ECRI, Hayes, Inc., and the California Technology Assessment Forum [25−30]. Each of these organizations performs rigorous reviews of the evidence of effectiveness of a technology. However, they may vary with respect to the comprehensiveness of services reviewed, the timeliness of reviews, the populations included for review, and different levels of focus on issues relevant to coverage such as clinical utility [31]. Thus, health plans usually supplement available HTAs with their own evidence review processes. Payers may also consider other payers' coverage positions for benchmarking purposes, but due to antitrust concerns, each payer must develop their coverage positions independently.

Of note, Private payers are required to follow Medicare coverage determinations for their Medicare Advantage members. Private payers are not required to follow Medicare policy for their commercial members. Although Medicare policy may be considered by private payers, the private payers develop their policies independently of Medicare and with each other and the private payer coverage determinations often differ from Medicare policy [32].

Health plans apply standard general methodological principles of study design in evaluating the quality of evidence. The final determination is based on a qualitative review of the quality of the evidence and not a quantitative one based on a scoring system. The following is a list of study designs ranked from most to least methodologically rigorous in their potential ability to minimize systematic bias:

- Randomized controlled trials
- Nonrandomized controlled trials
- Prospective cohort studies
- Retrospective case control studies
- Cross-sectional studies
- Surveillance studies (e.g., using registries or surveys)
- Consecutive case series
- Single case reports.

Consensus statement and expert opinion may be considered in interpreting the data, for contextual considerations, and where data are equivocal or incomplete.

The medical necessity definition provides a limited role for consideration of cost, allowing health plans to limit coverage to the least costly equally effective alternative. However, current standard health benefit plans have no provisions for excluding coverage of interventions because they exceed some predetermined cost-effectiveness threshold.

Health plans follow this process for all medical technologies, both new and established, as required for utilization management accreditation [33]. Draft documents of new clinical policies or updates of existing clinical policies are typically reviewed by the health plans' medical technology committee. The plan's chief medical officer is usually responsible for final approval of clinical policies. Some health plans make their clinical policies publicly available online to provide transparency around the clinical evidence upon which the policy is developed [34]. While it is beyond the scope of health plan function to consult on each manufacturer's product(s) in development, test developers can use published clinical policies to better understand the evidence criteria required by health plans for coverage of genomic tests. All clinical policies are reviewed annually as a health plan credentialing requirement [35,36]. In addition, clinical policies may be updated more frequently based on the availability of new evidence, practice guidelines from relevant professional societies or governmental health agencies, changes in the regulatory status of a medical technology, or inquiry regarding the technology from external or internal groups.

LINKAGE OF CLINICAL POLICY TO PAYMENT

Once a clinical policy is approved, the payment for the technology is considered. Health plans negotiate payment rates with laboratories. Clinical policies may include references to standard Health Insurance Portability and Accountability Act compliant code sets to report services that are provided and to facilitate billing and payment for covered services. Physicians and other providers are required to use the most appropriate code for the service that is provided and the patient's diagnoses. Health plans standardly use ICD-10 codes to report diagnoses on a submission. ICD-10 refers to the 10th revision of the International Statistical Classification of Diseases and Related Health Problems; this code set is created and maintained by the World Health Organization [37]. Current Procedural Terminology (CPT) codes are created and maintained by the American Medical Association, are used to report medical, surgical, and diagnostic services [38]. In addition, Healthcare Common Procedure Coding System (HCPCS) codes, created by CMS, are used to report products, supplies, and services that are not included in the CPT codes [39]. Reimbursement rates are commonly tied to specific CPT and HCPCS codes.

The terms of the member's benefit plan will determine the member's financial responsibility in the form of copayments and deductibles. Further, benefits for laboratory services may vary depending upon whether the laboratory that is performing the test is within the health plan's preferred network. Benefit plans typically require higher copayments and deductibles, or no benefit at all, for services obtained outside of preferred networks.

IMPACT OF NEW TECHNOLOGY PLATFORMS

Advances in technology such as next-generation sequencing (NGS) have made possible the simultaneous and efficient assessment of almost an unlimited number of genes in a single specimen. This has been followed by the rapid introduction of multigene test panels (both somatic and germline) into clinical care. The NIH Genetic Testing Registry lists 933 panel tests available for clinical use today with applications as broad as disease diagnosis or confirmation, prediction, prognosis, and therapeutic management across a broad range of clinical conditions from preimplantation genetic testing of embryos to oncology care [1]. Already, 69 separate panel tests include BRCA mutations. The trend in panel testing raises several concerns. Many panels have little validation data, including clinical validity and clinical utility, that support the effectiveness of the genes included in the panels. The speed with which new panel tests are brought to market is taxing physicians' ability to interpret vast amounts of data and appropriately apply relevant results to clinical care [40,41]. Health plans are also challenged to assess these panels, create clinical policies, and establish reimbursement policies around them. In response to these challenges, health plans (and other policy groups) are deploying expanded resources for support genomic test evaluation. Payers, test developers, and other stakeholders are collaborating to create evidence standards for genomic tests. For example, the Center for Medical Technology Policy (CMTP) is a nonprofit organization with the goal of improving the quality of evidence for coverage decision making. CMTP, through its Green Park Collaborative, has convened multiple stakeholders to create effectiveness guidance documents for the development of new medical technologies. The Green Park Collaborative has published effectiveness guidance documents on NGS and molecular diagnostics in oncology [42].

CONCLUSIONS

Genetic technology is improving almost all aspects of medical care from prevention through treatment of disease. The increasing number, complexity, and speed to market of genetic tests is outpacing the evidence of clinical effectiveness, especially of clinical utility. In this environment, rigorous evidence-based review of new tests is critical to ensure access to tests that are likely to benefit patients and to avoid the medical and economic harms of unproven genetic tests. Evidenced-based clinical policies allow payers to provide access to genetic testing services that have demonstrated value.

REFERENCES

[1] National Institutes of Health Genetic Testing Registry. National Center for Biotechnology Information, US National Library of Medicine. Bethesda, MD. Available from: http://www.ncbi.nlm.nih.gov/gtr/; 2016 [accessed 07.10.16].

[2] Nelson HD, Fu R, Goddard K, et al. Risk assessment, genetic counseling, and genetic testing for BRCA-related cancer: systematic review to update the U.S. preventive services task force recommendation [internet]. Rockville, MD: Agency for Healthcare Research and Quality (US). Available from: http://www.ncbi.nlm.nih.gov/books/NBK179201/; 2013 [accessed 15.10.14]

[3] US Food and Drug Administration. Center for drug evaluation and research. Table of pharmacogenomic biomarkers in drug labeling. Silver Spring, MD. Available from: http://www.fda.gov/drugs/scienceresearch/researchareas/pharmacogenetics/ucm083378.htm; 2015 [accessed 20.05.15].

[4] Deverka PA, Kaufman D, McGuire AL. Overcoming the reimbursement barriers for clinical sequencing. JAMA 2014;312(18):1857–8.

[5] US Food and Drug Administration. Overview of device regulation. Silver Spring, MD: FDA. Available from: http://www.fda.gov/MedicalDevices/DeviceRegulationandGuidance/Overview/#510k; 2015 [accessed 20.05.15]

[6] U.S. Food and Drug Administration (FDA). FDA permits marketing of first direct-to-consumer genetic carrier test for Bloom syndrome. Press release. Rockville, MD: FDA. Available from: http://www.fda.gov/newsevents/newsroom/pressannouncements/ucm435003.htm; 2015 [accessed 02.09.15].

[7]   US Food and Drug Administration. Overview of device regulation. Silver Spring, MD, FDA. Available from: http://www.fda.gov/MedicalDevices/DeviceRegulationandGuidance/ Overview/#510k; 2015 [accessed 30.05.15].

[8]   U.S. Food and Drug Administration (FDA), Center for Devices and Radiologic Health (CDRH). Guidance on pharmacogenetic tests and genetic tests for heritable markers. Silver Spring, MD: FDA. Available from: http://www.fda.gov/MedicalDevices/ DeviceRegulationandGuidance/HowtoMarketYourDevice/PremarketSubmissions/Premarket Notification510k/ucm134578.htm#denovo; 2007 [accessed 03.12.15]

[9]   US Food and Drug Administration (FDA), Center for Devices and Radiologic Health (CDRH). Laboratory developed tests. In vitro diagnostics. Medical devices. Silver Spring, MD: FDA. Available from: http://www.fda.gov/MedicalDevices/ProductsandMedical Procedures/InVitroDiagnostics/ucm407296.htm; 2014 [accessed 10.09.15]

[10] Medical Device Amendment of 1976, Pub. L. No. 94-295, 90 Stat. 539 (1976). Available from: http://www.gpo.gov/fdsys/pkg/STATUTE-90/pdf/STATUTE-90-Pg539.pdf; 2015 [accessed 24.08.15].

[11] U.S. Food and Drug Administration (FDA). In vitro diagnostics: laboratory developed tests. Medical devices. Silver Spring, MD: FDA. Available from: http://www.fda.gov/ medicaldevices/productsandmedicalprocedures/invitrodiagnostics/ucm407296.htm; 2014 [accessed 02.09.15]

[12] Centers for Medicare & Medicaid Services (CMS). Available from: https://www.cms.gov/ Regulations-and-Guidance/Legislation/CLIA/Downloads/LDT-and-CLIA_FAQs.pdf. CLIA overview. What is CMS' authority regarding laboratory developed tests (LDTs) and how does it differ from FDA's authority? Laboratory developed tests: frequently asked questions. v. 2013.10.22. Baltimore, MD: CMS; 2013 [accessed 02.09.15]

[13] Javitt GH, Hudson K. Federal neglect: regulation of genetic testing. Issues Sci Technol 2006;22(3).

[14] U.S. Department of Health and Human Services, Food and Drug Administration, Center for Devices and Radiological Health, Office of In Vitro Diagnostics and Radiological Health, Center for Biologics Evaluation and Research. Draft Guidance for industry, Food and Drug Administration, staff, and clinical laboratories. Framework for regulatory oversight of laboratory developed tests (LDTs). Available from: http://www.fda.gov/downloads/MedicalDevices/DeviceRegulationandGuidance/GuidanceDocuments/UCM416685. pdf; 2014 [accessed 30.08.15].

[15] Federal Trade Commission. FTC approves final consent orders settling charges that companies deceptively claimed their genetically modified nutritional supplements could treat disease. Available from: https://www.ftc.gov/news-events/press-releases/2014/05/ftc-approves-final-consent-orders-settling-charges-companies; 2014 [accessed 30.08.15].

[16] BlueCross BlueShield Association (BCBSA), Technology Evaluation Center (TEC). Technology evaluation center criteria. Chicago, IL: BCBSA. Available from: http://www. bcbs.com/blueresources/tec/; 2015.

[17] Kaminski JL. Defining medical necessity. OLR research report. 2007-R-0055. Revised. Hartford, CT: State of Connecticut. Available from: https://www.cga.ct.gov/2007/rpt/ 2007-r-0055.htm; 2007 [accessed 02.09.15].

[18] International Network of Agencies of Health Technology Assessment: INAHTA glossary. Available from: http://www.inahta.org/HTAGlossary/#_Health_Technolgoy_Assessment; 2015 [accessed 25.08.15].

[19] Haddow JE, Palomaki GE. ACCE: a model process for evaluating data on emerging genetic tests. In: Khoury M, Little J, Burke W, editors. Human genome epidemiology: a scientific foundation for using genetic information to improve health and prevent disease. Oxford University Press; 2003. p. 217–33.

[20] Peabody JW, Shimkhada R, Tong KB, Zubieller M. New Thinking on Clinical Utility: Hard Lessons for Molecular Diagnostics. Am J Manag Care. Available from: http://www.ajmc.com/journals/issue/2014/2014-vol20-n9/new-thinking-on-clinical-utility-hard-lessons-for-molecular-diagnostics/P-2; 2014 [accessed 03.12.15].

[21] BlueCross BlueShield Association (BCBSA), Technology Evaluation Center (TEC). Gene expression profiling in women with lymph node–negative breast cancer to select adjuvant chemotherapy. TEC assessment. Chicago, IL: BCBSA. Available from: http://www.bcbs.com/blueresources/tec/vols/29/29_3.pdf; 2014 [accessed 02.09.15].

[22] U.S. Food and Drug Administration (FDA). OvaSure manufacturer letter. Silver Spring, MD: FDA. Available from: http://www.fda.gov/MedicalDevices/DeviceRegulationand Guidance/IVDRegulatoryAssistance/ucm125130.htm; 2008 [accessed 02.09.15].

[23] Hopewell JC, Parish S, Clarke R, et al. No impact of KIF6 genotype on vascular risk and statin response among 18,348 randomized patients in the Heart Protection Study. J Am Coll Cardiol 2011;57(20):2000–7.

[24] National Institutes of Health, National Library of Medicine. Health services/technology assessment texts (HSTAT). Bethesda, MD: NLM. Available from: http://www.ncbi.nlm.nih.gov/books/NBK16710/; 2015 [accessed 02.09.15].

[25] BlueCross BlueShield Association (BCBSA), Technology Evaluation Center [website]. Chicago, IL: BCBSA. Available from: http://www.bcbs.com/blueresources/tec/; 2015 [accessed 30.11.15].

[26] Evaluation of Genomic Applications in Practice and Prevention (EGAPP) [website]. Atlanta, GA: Office of Public Health Genomics, Centers for Disease Control and Prevention. Available from: http://www.egappreviews.org/; 2014 [accessed 30.11.15].

[27] U.S. Department of Health and Human Services, Public Health Service, Agency for Healthcare Research and Quality (AHRQ) [website]. Rockville, MD: AHRQ. Available from: http://www.ahrq.gov/; 2015 [accessed 30.11.15].

[28] ECRI Institute [website]. Plymouth Meeting, PA: ECRI. Available from: https://www.ecri.org/Pages/default.aspx; 2015 [accessed 30.11.15].

[29] Hayes, Inc. Lansdale, PA. Available from: http://www.hayesinc.com; 2015 [accessed 30.11.15].

[30] California Technology Assessment Forum [website]. San Francisco, CA: CTAF. Available from: http://ctaf.org/; 2015 [accessed 30.11.15].

[31] Trosman JR, Van Bebber SL, Phillips KA. Health technology assessment and private payer's coverage of personalized medicine. Am J Manag Care 2011;17(Suppl. 5): SP53–60.

[32] Chambers JD, Chenoweth M, Thorat T, Neumann PJ. Private payers disagree with Medicare over medical device coverage about half the time. Health Aff (Millwood) 2015;34(8):1376–82.

[33] National Committee for Quality Assurance (NCQA). Certification statuses an organization can achieve. Utilization management. Washington, DC: NCQA. Available from: http://www.ncqa.org/Programs/Certification/UtilizationManagementandCredentialingUMCR/UtilizationManagementUM.aspx; 2015 [accessed 02.09.15].

[34]   Aetna. Medical clinical policy bulletins. Available from: https://www.aetna.com/health-care-professionals/clinical-policy-bulletins/medical-clinical-policy-bulletins.html# [accessed 30.08.15].

[35]   National Committee for Quality Assurance (NCQA). What does NCQA look for when reviewing plans? Washington, DC: NCQA. Available from: http://www.ncqa.org/tabid/1405/Default.aspx; 2015 [accessed 20.10.15].

[36]   URAC. HUM 1: review criteria requirements. Health utilization management 7.2. Standard interpretations. Washington, DC: URAC. Available from: https://www.urac.org/resource-center/standards-interpretations/; 2015 [accessed 20.10.15].

[37]   World Health Organization (WHO). International classification of diseases (ICD). Classifications. Geneva, Switzerland: WHO. Available from: http://www.who.int/classifications/icd/en/; 2015 [accessed 02.09.15].

[38]   American Medical Association (AMA). CPT—current procedural terminology. Chicago, IL: AMA. Available from: http://www.ama-assn.org/ama/pub/physician-resources/solutions-managing-your-practice/coding-billing-insurance/cpt.page; 2015 [accessed 02.09.15].

[39]   Centers for Medicare & Medicaid Services (CMS). HCPCS—general information. Baltimore, MD: CMS. Available from: https://www.cms.gov/Medicare/Coding/MedHCPCSGenInfo/index.html; 2015 [accessed 02.09.15].

[40]   Robson ME, Bradbury AR, Arun B, Domchek SM, Ford JM, Hampel HL, et al. American Society of Clinical Oncology policy statement update: genetic and genomic testing for cancer susceptibility. J Clin Oncol 2015;33(31):3660−7. Available from: http://dx.doi.org/10.1200/JCO.2015.63.0996.

[41]   Domchek SM, Bradbury A, Garber JE, Offit K, Robson ME. Multiplex genetic testing for cancer susceptibility: out on the high wire without a net? J Clin Oncol 2013;31(10):1267−70.

[42]   Center for Medical Technology Policy (CMTP). Effectiveness guidance documents. Resource center. Baltimore, MD: CMTP. Available from: http://www.cmtpnet.org/resource-center/category/effectiveness-guidance-documents/; 2015 [accessed 08.12.15].

Chapter 24

Legal Issues in Genomic and Precision Medicine: Intellectual Property and Beyond

Arti K. Rai

Duke Law Center for Innovation Policy, Durham, NC, United States

Chapter Outline

INTRODUCTION

The legal terrain in genomic and precision medicine encompasses areas as varied as intellectual property, Food and Drug Administration (FDA) regulation, and insurance coverage. This chapter takes as its focus intellectual property. The innovation effects of intellectual property are closely linked, however, to effects generated by regulation and insurance. Thus, after analyzing in depth the most significant recent changes in intellectual property law and policy, the chapter concludes by incorporating these changes into a larger discussion of innovation incentives that encompasses regulation and insurance coverage.

The chapter begins by discussing recent US Supreme Court decisions that shrink the domain of patent-eligible subject matter. As a consequence of these decisions, concerns about numerous patents on upstream discoveries requiring onerous and costly licensing negotiations on the part of downstream genomic innovators have largely dissipated. Instead, commentators are debating whether patents will play a meaningful role going forward, particularly relative to secrecy in sequencing and testing data [1,2].

Perhaps not surprisingly, the prevalence of trade secrecy protection—which simply requires that the owner actually keep data secret and thus may become even more prominent to the extent patents are difficult to secure—has prompted calls for greater data disclosure. Concern about "black box"

Genomic and Precision Medicine. DOI: http://dx.doi.org/10.1016/B978-0-12-800681-8.00024-4

personalized medicine testing has also been a factor in the decision by the US FDA to regulate laboratory-developed tests (LDTs) (including genomic tests) as "medical devices" [3].

Any regulatory moves on the part of the FDA will have a significant impact on the innovation landscape. As matters currently stand, FDA approval will not necessarily improve prospects for insurance coverage. However, a requirement for such approval may solidify the competitive advantage of those companies that have in fact secured such approval.

Approval requirements may also make having a significant intellectual property stake more important for new entrants. Given challenges in securing patents, this intellectual property is likely to include trade secrecy in data generated to meet FDA requirements.

THE SUPREME COURT DECISIONS AND THEIR IMPACT

Although the US patent statute contains many patentability requirements, the Supreme Court has recently been particularly interested in the issue of whether the claimed subject matter is eligible for patenting in the first instance. Patent lawyers have typically seen patent eligibility as a cursory inquiry, particularly relative to such fact-intensive and complex analyses as whether the patent covers territory that is nonobvious. The Supreme Court has not shared this view. To the contrary, the Court has, since 2010, decided four cases on patentable subject matter. As discussed below, only two of these cases involved biomedical research (and only one genomic patenting). However, because the subject matter test adopted by the Court is very similar across all four cases, the two cases involving business method and software patents have had an influence on genomics.

In fact, the Court's view that poor patent quality is a problem across all technologies appears to have been influenced by the vast literature decrying persistent problems of vagueness and overbreadth in business method and software patents. Moreover, while patent lawyers typically see quality as an issue to be addressed through patentability requirements such as nonobviousness and adequate disclosure, the Court has adopted subject matter as its cudgel.

In its subject matter cases, the Supreme Court has emphasized that it intends three exceptions to patent-eligible subject matter to be a robust screen. These exceptions, enunciated in the Court's case law as early as the 19th century, are abstract ideas, laws of nature, and products of nature. The two unanimous decisions involving biomedical research are *Mayo Collaborative Services v. Prometheus Laboratories, Inc.* (132 S. Ct. 1289, 2012), handed down by the Court in March 2012 and *Association for Molecular Pathology v. Myriad Genetics, Inc.* (133 S. Ct. 2017, 2013), handed down by the Court in 2013.

In both *Mayo* and *Myriad*, the Court struck down patent claims associated with diagnostic medical practice, including genome-related diagnostic practice. *Mayo* rejected method claims on measuring a thiopurine drug metabolite to adjust doses of a thiopurine drug. *Myriad* rejected claims to DNA associated with breast cancer (BRCA1 and BRCA2) that had merely been isolated (genomic DNA or gDNA). Such claims are typically associated with diagnostic genetic medicine (Table 24.1).

The Court's interest in diagnostic genetic patents came after many years of heated public controversy over whether such patents pose an undue impediment to patient access as well as patient, physician, and scientist autonomy (132 S. Ct. 1289, 2012). For most of the public, access and autonomy questions were more compelling than the relatively technical economic question typically asked by patent lawyers and legal scholars: On balance, does the shelter against competition provided by the patent in question do more to promote innovation than to hinder it.

The Supreme Court's policy analysis in *Mayo* and *Myriad* addressed this economic question, at least implicitly. The *Myriad* decision drew a line that preserved economically important claims to DNA that codes for full proteins (cDNA) while striking down other, more broadly written claims to gDNA. Critics had charged that Myriad's particular gDNA claims inhibited incremental innovation in BRCA1/2 testing, and that gDNA claims as a category created upstream patent licensing barriers for downstream developers of whole-genome sequencing. Dissipating the shadow of infringement liability over whole-genome sequencing was extremely important to officials at the National Institutes of Health and the US Office of Science and Technology Policy. They successfully convinced the Solicitor General to reject the US Patent and Trademark Office's (PTO) long-held, but only lightly theorized, position of allowing claims on all "isolated" DNA molecules [4].

TABLE 24.1 Highlights of *Mayo* and *Myriad*

	Mayo	Myriad
Subject matter of patent claims in dispute	Method of measuring thiopurine drug metabolite to adjust dosage of thiopurine drug	BRCA1 and BRCA2 gDNA and cDNA
Test applied	"Two step": disallows claims that cover: (1) law/product of nature; *and* have (2) no additional "inventive step"	Claims must cover subject matter "markedly different" from product of nature
Result	Method claims not allowed	gDNA claims not allowed; cDNA claims allowed
Consequences for patent claims in precision medicine	Potentially very significant	Relatively modest

In *Myriad*, the Court made the acute observation that the claims it struck down where deliberately written in a manner that made them very difficult to invent around. As the Court noted, Myriad claimed not "the specific chemical composition of a particular molecule" but, instead, the information "encoded in the BRCA1 and BRCA2 genes." To be sure, the Court did uphold Myriad's cDNA claims, even though it viewed those as drawn to information as well. Moreover, the Court failed to enunciate explicitly *why* claims to information in the form of cDNA are less problematic than claims to information in the form of gDNA. As discussed below, this failure renders the opinion's own language less useful to lower tribunals struggling to distinguish between different types of claims to information.

Meanwhile the *Mayo* case has prompted justifiable concern that its resuscitation of old, and long-criticized, approaches to subject matter eligibility will undermine not just overly broad diagnostic patents but virtually all such patents. *Mayo* employed a two-step analysis that looks first at whether a patent-ineligible category is claimed. If so, the court must determine whether a sufficient "inventive step" has been added to the ineligible category to make the claim as a whole patent-eligible. Applying this two-step analysis, the *Mayo* Court held that the claims at issue were invalid because they added only conventional activity (e.g., administering a thiopurine drug and measuring metabolite levels) to the "natural law" that individuals metabolize thiopurine drugs differently.

Mayo's two-step procedure resuscitates a much-criticized 1978 software patent case, *Parker v. Flook* (437 U.S. 584, 1978), in which the Court had, for purposes of determining patent eligibility, dissected out the patent-ineligible "abstract idea" and then analyzed whether what remained was novel. The *Mayo* decision suggests the continuing viability of this "point of novelty" approach even though (as numerous commentators as pointed out) *Flook* failed to advance a robust innovation policy justification for why abstract ideas, products of nature, and laws of nature should be dissected out.

To its credit, the *Mayo* opinion did more than *Flook* to address policy considerations. Throughout the opinion, the Court alluded to the possibility that claims on abstract ideas, products of nature, and laws of nature could unduly "preempt" future research. In this context, *Mayo* even recognized arguments about the importance of distinguishing broad laws that interfere with large areas of future innovation from narrower laws. After recognizing these arguments, the Court further acknowledged that the law of nature it was addressing—that individuals metabolize thiopurine-containing drugs differently—was in fact quite narrow.

Unfortunately, the Court did not follow through on the promise of its reasoning. Instead, the Court insisted that it needed to enunciate a "bright-line prohibition" striking down all patents covering laws of nature, no matter how narrow.

Notably, the result in *Myriad* did not adhere to the Mayo two-step/"point of novelty" rationale. Although the *Myriad* Court rejected gDNA claims, it affirmed claims to DNA with regions that do not code for a protein excised. The Court accepted claims to complementary or cDNA even though they represent nothing more than the conventional application of routine laboratory techniques to a product of nature—chromosomal DNA.

On the other hand, the Court's most recent decision on patent-eligible subject matter, its 2014 opinion in *CLS Bank v. Alice* (134 S. Ct. 2347, 2014), squarely relied upon the *Mayo* two-step framework. In that case, the Court held that if a patent claims a patent-ineligible abstract idea, then it must add an "inventive concept" in order to take the claim into the realm of eligibility. *CLS Bank*'s reliance on *Mayo* suggests that it, not *Myriad*, may be the more important decision going forward, at least for method or process claims.

The full effects of *Mayo* and *Myriad* remain to be seen. Only a small number of cases directly implicating the decisions have thus far made it up to the Court of Appeals for the Federal Circuit, the court immediately below the Supreme Court that hears all appeals involving patent law. The Federal Circuit cases highlight both the potential breadth of the *Mayo* ruling as well as the challenge lower courts face in interpreting both *Mayo* and *Myriad*.

One case already decided by the Federal Circuit is a sequel to the *Myriad* litigation that reached the Supreme Court [5]. After the Supreme Court struck down Myriad's gDNA claims, a number of competitors entered the markets offering diagnostic testing services for BRCA1 and BRCA2 mutations. Myriad asserted a variety of claims that had not been at issue in the litigation that reached the Supreme Court. The Federal Circuit opinion considered claims directed to synthetically created DNA primers used for DNA polymerization and method claims involving comparisons between wild-type BRCA sequences with a patient's BRCA sequences. Invoking the Supreme Court's discussion of structural similarity between isolated DNA and naturally occurring DNA, the court found the primer claims sufficiently structurally and functionally similar to naturally occurring DNA that they were patent-ineligible. The Federal Circuit did not, however, suggest what factors would create dissimilarity sufficient to reach the threshold of patent eligibility.

Faced with this confusing case law, the PTO has attempted to provide guidance to its examiners [6]. The PTO guidance analyzes Supreme Court and lower court cases dating from the 1930s to the present in order to isolate the factors examiners should consider in determining the patent eligibility of a claim that involves a natural product. The PTO guidance emphasizes not only *Myriad* but also a landmark 1980 Supreme Court case, *Diamond v. Chakrabarty* [7]. The latter case emphasized that claims involving natural products could be patent-eligible so long as the covered product was "markedly different" from the product found in nature. According to the PTO guidance, marked differences include biological, pharmacological, chemical, or physical properties as well as functional and structural characteristics. Notably, the PTO states explicitly an economic distinction only implied by the Supreme Court's *Myriad* decision: Because the structural difference between a natural gene and cDNA ensures that cDNA claims do not "improperly [tie] up future use of the ... gene," cDNA is "markedly different" from the product of nature and falls within the realm of patent eligibility [10, p. 30]. The guidelines also emphasize that purified (and thus structurally identical) versions of products that occur in nature in their impure form may be patent-eligible if the purification results in a functionally different product.

If interpreted through the lens put forward in the PTO guidelines, the impact of *Myriad* may be relatively modest. The impact of *Mayo* is likely to be more dramatic. *Ariosa v. Sequenom* [7] a case involving a patented method for detecting paternal cell-free fetal-derived DNA (cffDNA) in maternal blood by amplifying the paternally inherited nucleic acid from a maternal serum or plasma sample, suggests the case's potential impact. At the Federal Circuit, a three-judge panel of the court stated that the Supreme Court's ruling in *Mayo* required that the panel find the method patent-ineligible. As the majority noted, the method in question failed the broad two-step test set forth in *Mayo*: cffDNA is a natural phenomenon, and Sequenom's standard methodology for amplifying the cffDNA did not add an inventive step. A concurrence by Judge Linn did observe, however, that, for purposes of striking down the metabolite-measurement patent at issue in the *Mayo* case, the Court did not have to define inventive step as broadly as it did. Because physicians had been measuring metabolite levels well before the patent in *Mayo*, and the inventor in that case had simply specified a particular metabolite range, the Court could have struck down the patent using a narrower definition of inventive step. Under that narrower definition, Sequenom's patent would have survived, as amplifying *cffDNA*, whose existence was previously unknown, was in fact inventive. The arguments in Judge Linn's concurrence have been reiterated and amplified by other Federal Circuit judges. However, as these judges all note, only the Supreme Court has the power to reinterpret *Mayo* in this fashion.

Despite *Mayo*, the PTO continues to issue at least some patents on what appear to be correlations. A potentially instructive case can be found in a recent patent issued to VM Institute of Research [8]. The patent claims a method of treating breast cancer in a patient diagnosed with triple-negative (TN) breast cancer—that is, cancer that does not generally express estrogen receptor, progesterone receptor, and expresses only normal levels of human epidermal growth factor receptor. Specifically, the inventors were able to determine that expression levels of certain proteins were indicative of a recurrent TN tumor. They claimed "a method of predicting the overall survival potential of a TN breast cancer in a subject" that relied on comparing the expression level of relevant proteins in a test sample from a subject to that of the relevant proteins in a "standard sample indicative of a recurrent TN breast cancer."

The patent application was initially rejected as claiming a patent-ineligible natural correlation. After the initial rejection, the applicant amended the claim, adding the step of "modifying a treatment regimen of the subject based on the results" of the biomarker comparison. The Examiner determined, however, that "modifying a treatment regimen" could be "accomplished mentally by thinking or designing a new treatment plan, and does not set forth any physical steps." The Examiner suggested "amending the claims to recite an active step of administering a specified treatment regimen based on specific results." After the applicant added the term "administering an aggressive cancer treatment regimen" based on test results, the Examiner allowed the claim [9].

Arguably, the amendments required by the Examiner were relatively cosmetic. Whether such amended claims would survive court challenge is open to question. Moreover, adding steps to correlation claims, particularly steps that are not likely to be performed by the same entity, that does the correlation could raise challenges for patent enforcement. Under current infringement law, a claim can be infringed only if a single entity performs all of its steps [10].

Some have argued that the next step for at least certain diagnostic testing firms is likely to be trade secrecy with respect to data that emerges from their testing [1,2]. Indeed, in the case of Myriad itself, the firm stopped contributing data to public databases in November 2004, well before the challenge to its patents. Because of its comprehensive database, which includes data from over 1 million patients, Myriad can uniquely claim a variant of unknown significant rate of about 3%, substantially lower than that of its competitors.

Notably, the challenge of trade secrecy arises to some extent independent of whether patents exist. To be sure, conventional legal doctrine holds that patents must disclose enough information to allow those in the field to "make and use the invention" [11]. Thus, on the conventional view, inventors must choose between having a patent and keeping information as a trade secret. However, the available evidence indicates that disclosure is often insufficient to permit replication [12]. Although failure to disclose may lead to invalidation of the patent, invalidating a patent is expensive and thus the threat of such invalidation may fail to incentivize complete disclosure.

Meanwhile, trade secrecy surrounding complex diagnostic tests that rely on nontransparent algorithms has engendered concern at the FDA. In the context of its general determination that it should now regulate LDTs, the FDA has emphasized that many modern LDTs are "highly complex" and use nontransparent algorithms and/or complex software to generate device results [3]. The FDA's work in this area follows on a prominent Institute of Medicine ("IOM") report, issued in 2012, that recommended FDA "develop and finalize a risk-based guidance or regulation on bringing omics-based tests to FDA for review" [13, p. 136]. More generally, the IOM report called for greater transparency in the data, metadata, analysis plans, computer code, and computational procedures at issue in complex genomics tests [13].

Of course, FDA regulation of LDTs does not mean that the underlying data and algorithms would be transparent. However, the FDA does provide a summary review of its approval determinations [14]. Moreover, the FDA would of course have access to this information and would thereby presumably serve as a check on quality, particularly clinical quality. The FDA has emphasized that the current regulatory scheme, under which the Center for Medicare and Medicaid Services (CMS) assesses a test's analytical validity under the Clinical Laboratory Improvement Amendments (CLIA), does not assess the accuracy with which a test measures the presence or absence of a particular clinical condition.

THE FDA'S RISK-BASED APPROACH

The FDA has proposed a "risk-based" approach to the regulation of LDTs [3]. Under this approach, "low-risk" LDTs, LDTs for rare diseases and unmet needs, and "traditional LDTs" used and interpreted directly by physicians and pathologists working within a single institution responsible for patient care would be subject only to notification and/or registration and listing, and adverse event reporting. Premarket review and quality systems requirements would not apply. For tests that were found moderate or high risk, applicable requirements would depend on whether the tests in question were considered to be the highest-risk (Class III) devices or more moderate risk (Class II). According to the FDA, it is particularly concerned about LDTs in three categories: those that use complex, nontransparent algorithms; those that act like companion diagnostics—that is, tests that claim to guide selection of therapy; and those that have the same intended use as an FDA-approved medical device.

If and when FDA premarket review requirements ultimately apply, meeting these requirements will likely require substantial capital investment. Moreover, this investment may come with no particular guarantee of insurance coverage. While drugs that garner FDA approval generally get covered by the CMS, no such presumption currently exists for molecular diagnostics [14]. Conversely, insurers have covered certain tests that did not go through FDA approval. One prominent example is Genomic Health's Oncotype Dx breast cancer assay, developed as an LDT under the provisions of CLIA. The test is reimbursed in the United States by CMS and by all major payers. It has also been incorporated in published treatment guidelines [14].

CONCLUSIONS

Going forward, to the extent FDA premarket review applies to certain tests, the requisite investment may not be forthcoming without some credible intellectual property stake on the part of the test developer. As a historical matter, medical device manufacturers regulated by the FDA have typically had substantial patent portfolios. Additionally, at least one survey has found that venture capital investors in the medical device arena ranked "intellectual property risk" as the third most important risk, immediately after regulatory and reimbursement risk [15].

In the area of genomic diagnostics, intellectual property protection is likely to take the form of *both* patents and trade secrecy, with a particular emphasis on trade secrecy to the extent that the *Mayo* decision calls into question future patents on diagnostic correlations and also threatens patents that already exist.

For those tests that do pass premarket review, insurance coverage may or may not be forthcoming. Coverage may turn on the extent to which FDA and CMS can coordinate their requirements. However, for those who can overcome the hurdle, the FDA requirement may erect a powerful barrier against potential future competitors.

In sum, the historical situation was one of reasonably high innovation and competition. Of course, serious gaps remained, particularly with respect to previously approved drugs that worked for only a small subset of the population for which they were prescribed [16]. And challenges in obtaining insurance coverage created challenges for innovators. Even so, low barriers to entry and some ability to secure patents made entry attractive.

Going forward, FDA regulation and further dilution of patent protection are highly salient concerns. Significant FDA regulation might lead to higher test quality. But combined with further dilution of patent protection, it could also lead to fewer competitors and potentially lower overall rates of innovation.

REFERENCES

[1]   Cook-Deegan R, Conley JM, Evans JP, Vorhaus D. The next controversy in genetic testing: clinical data as trade secrets. Eur J Hum Genet 2013;21:585—8.

[2]   McElligott S, Field RI, Bristol-Demeter M, Domchek SM, Asch DA. How genetic variant libraries effectively extend gene testing patents: implications for intellectual property and good clinical care. J Clin Oncol 2012;30:2943—5.

[3]   U. S. Department of Health and Human Services, FDA. Framework for regulatory oversight of laboratory developed tests; draft guidance for industry, Food and Drug Administration staff, and clinical laboratories. 79 Federal Register 59776-59779. October 3, 2014.

[4]   Rai, A.K. Diagnostic patents at the Supreme Court. Intellectual Property L.Rev. 18: 1—9.

[5]   In re BRCA1- and BRCA2-based hereditary cancer test patent litigation (Fed. Cir. 2014).

[6]   Department of Commerce, U.S. Patent and Trademark Office. Interim guidelines on patent subject matter eligibility. 79 Federal Register 74618-74633. December 16, 2014.

[7]   Ariosa Genetics v. Sequenom (Fed. Cir. 2015).

[8]   U.S. Patent No. 8,642,270.

[9]   Bosman J. Patent profile: VM Institute of Research receives patent for method of treating triple negative breast cancer. March 2, 2014.

[10]   Muniauction Inc. v. Thomson Corp., 532 F.3d 1318 (Fed. Cir. 2008).

[11] 35 U.S.C. Section 112.

[12] Lemley M. Ignoring patents. Michigan State Law Review 2008;19−55.

[13] Institute of Medicine. Evolution of translational omics: lessons learned and the path forward. 2012.

[14] Institute of Medicine. Genome-based diagnostics: clarifying pathways to clinical use-workshop summary. 2012.

[15] Ackerly DC, Valverde AM, Diener LW, Dossary KL, Schulman KA. Fueling innovation in medical devices (and beyond): venture capital in health care. Health Aff 2009;28(1): w68−75.

[16] Schork NT. Personalized medicine: time for one-person trials. Nature 2015;520:609−11.

Index

374 Index

Printed in the United States
By Bookmasters